Distribution-Free Methods for Statistical Process Monitoring and Control

Markos V. Koutras · Ioannis S. Triantafyllou
Editors

Distribution-Free Methods for Statistical Process Monitoring and Control

 Springer

Editors
Markos V. Koutras
Department of Statistics
and Insurance Science
University of Piraeus
Piraeus, Greece

Ioannis S. Triantafyllou
Department of Computer Science
and Biomedical Informatics
University of Thessaly
Volos, Greece

ISBN 978-3-030-25083-6   ISBN 978-3-030-25081-2   (eBook)
https://doi.org/10.1007/978-3-030-25081-2

This Springer imprint is published by the registered company Springer Nature Switzerland AG
The registered company address is: Gewerbestrasse 11, 6330 Cham, Switzerland

Preface

Statistical process control is widely used to monitor the quality of the final product of a process. In any production process, no matter how carefully it is maintained, a natural variability is always present. Control charts facilitate the practitioners to identify assignable causes so that corrective actions are carried out and the process is restored to the desirable in-control state.

In most control charts, the process output is assumed to follow a specified probability distribution (usually normal); therefore the techniques applied for them are parametric ones and are affected by the distributional assumption used each time. However, this assumption may not be fulfilled in practice and therefore the resulting control charts cannot be applied, or, if applied, may not be accurate enough. Therefore, the development of nonparametric methods which can be efficiently used for hypothesis-testing problems without making any specific assumptions about the distribution of the underlying process is crucial. These techniques can also be exploited to develop control charts which can be used under a nonparametric framework. Throughout the present manuscript, some distribution-free monitoring schemes will be introduced and studied in detail. In addition, a review of the majority of recent advances in the field of Nonparametric Statistical Process Control will be presented.

In Chap. 1, an up-to-date overview of nonparametric Shewhart-type univariate control charts is provided. In this chapter, the most recent developments on the topic are reviewed, and more specifically only the advances that appeared during the last decade are discussed in detail. For each distribution-free control chart, the general setup and several performance characteristics are presented.

Chapter 2 reviews the recent literature on nonparametric control charts, giving emphasis on multivariate schemes, which make use of order statistics, signs, or ranks for the computation of the test statistic that is exploited for the decision making. In addition, a simulation study is carried out in order to evaluate the performance of these charts when compared to each other, as well as to their parametric counterparts. Finally, some concluding remarks are given, as well as some ideas and directions for future work.

In Chap. 3, the problem of having observations tied to the monitored population quantile (e.g., the median) is considered, and it is indicated that, when ties occur, the Shewhart Sign (SN) control chart is no longer distribution-free. Some procedures to handle the occurrence of ties are proposed. The authors demonstrate that the best strategy is to implement a Bernoulli trial approach and point out that this approach allows the distribution-free properties of the Shewhart SN to be generally preserved.

Chapter 4 deals with the issue of assuming in practice the normality and independence when a process is statistically monitored. Since these assumptions play an important role in the construction of the monitoring scheme exploited for the problem at hand, it is crucial that they are thoroughly investigated and properly validated, so that the results can be depended on. Some examples with real data are provided, and taking into account the corresponding consequences when the assumptions are not fulfilled, the authors conclude that a nonparametric approach consists of a safer option in practice.

In Chap. 5, change-point analysis-based distribution-free control charts, designed for Phase I applications especially for individual observations, are constructed for retrospectively detecting single or multiple changes in location and dispersion of univariate variables. A real example is included to unfold the capabilities of the developed methodologies.

In Chap. 6, six distribution-free exponentially weighted moving average (EWMA) schemes for simultaneously monitoring the location and scale parameters of a univariate continuous process are investigated. More precisely, a well-known distribution-free EWMA scheme based on the Lepage statistic is considered, and five new EWMA schemes are introduced. Finally, a Monte-Carlo simulation study is carried out for the comparison of the suggested schemes.

In Chap. 7, two nonparametric Shewhart-type control charts based on order statistics with signaling runs-type rules are introduced. Exact formulae for the alarm rate, the variance of the run length distribution, and the average run length (ARL) for both charts are derived, along with extensive tables that may facilitate the practitioner for the implementation of the proposed schemes. In addition, several numerical comparisons against competitive nonparametric control charts reveal that the new monitoring schemes are quite efficient in detecting the shift of the underlying distribution.

In Chap. 8, a novel and effective new method for disease early detection is proposed. To use this method, a patient's risk to the disease is first quantified at each time point using survival data modeling and variable selection, and then the longitudinal pattern of the risk is monitored sequentially over time by a control chart. A signal is produced once the cumulative difference between the risk pattern of the patient under monitoring and the risk pattern of a typical person without the disease in concern exceeds a control limit.

The chapters included in this book were all refereed; we would like to express our sincere gratitude to all reviewers for their diligent work and commitment. Generally speaking, it has been a very pleasant experience corresponding with all the authors that contributed to this volume. Our sincere thanks go to all of them for their great support and cooperation throughout the course of this project. Finally, we would like to thank the Springer production team for their help and patience during the preparation of this book.

Piraeus, Greece   Markos V. Koutras
Volos, Greece   Ioannis S. Triantafyllou
March 2020

Contents

Recent Advances on Univariate Distribution-Free Shewhart-Type Control Charts

Markos V. Koutras and Ioannis S. Triantafyllou

Abstract In this chapter, we provide an up-to-date overview of nonparametric Shewhart-type univariate control charts. The monitoring schemes incorporated in the present literature review depict the most recent developments on the topic, since it has been chosen to discuss only the advances appeared during the last decade. For each distribution-free control chart, the general setup and several performance characteristics are presented in some detail.

Keywords Nonparametric statistical process monitoring · Shewhart-type control charts · Run length distribution · Order statistics · Sign statistic · Wilcoxon-type statistics · Rank-based statistics

1 Introduction

Statistical process control is widely used to monitor the quality or the final product of a process. In any production process, no matter how carefully it is designed, a natural variability always occurs. Control charts help the practitioners to identify assignable causes so that the state of statistical control is achieved. Intuitively, in the event of having an undesirable shift in the process, a control chart should detect it as quickly as possible and give an out-of-control signal.

A great amount of control charts, already introduced in the literature, is based on the assumption that the process follows a specified probability distribution. However, this argument is not always true in practice and therefore the resulting control charts may not be reliable. To overcome this problem and simultaneously keep the traditional structure of a monitoring scheme, several nonparametric control charts

M. V. Koutras
Department of Statistics and Insurance Science, University of Piraeus, Piraeus, Greece
e-mail: mkoutras@unipi.gr

I. S. Triantafyllou (✉)
Department of Computer Science and Biomedical Informatics, University of Thessaly, Volos, Greece
e-mail: itriantafyllou@uth.gr

have been proposed in the literature. Celano et al. (2016a, b, c) provided some performance comparisons between several parametric and nonparametric control charts for jointly monitoring location and scale. Chakraborti et al. (2001) provided a thorough overview of the univariate nonparametric control charting literature up to 2000, while Chakraborti (2011) updated that review covering much of the literature up to 2010. McCracken and Chakraborti (2013) presented a recent literature overview for joint monitoring of control schemes, including the schemes with known and unknown parameters. Chakraborti and Graham (2019b) brought the aforementioned reviews forward to 2017, discussing the most recent developments on the topic. For a detailed and thorough study on nonparametric statistical process control, the interested reader is referred to Qiu (2018, 2019) or the excellent textbooks of Qiu (2014) and Chakraborti and Graham (2019a).

In the control charts literature, there have been introduced several traditional control charts which focus on improving the sensitivity of the traditional Shewhart-type monitoring schemes; one of them is the class of the so-called synthetic control charts. Khilare and Shirke (2010) proposed a synthetic control chart based on the sign statistic for detection of possible shift in the process median, while Khilare and Shirke (2012) introduced nonparametric synthetic and side-sensitive synthetic monitoring schemes for controlling fraction non-conforming due to increase in the process variation by combining the traditional Shewhart-type sign and the confirming run length control charts. For an up-to-date overview of the class of synthetic control charts, the interested reader is referred to Rakitzis et al. (2019).

In this chapter, we provide an overview of the most recent advances on Shewhart-type univariate nonparametric control charts. More specifically, we focus mainly on charts that have appeared in the literature during the last decade (2009–2019). Over 100 publications on univariate Shewhart-type nonparametric control charts and related topics are reviewed here. All monitoring schemes incorporated in the present literature review are classified in three distinct groups in terms of their plotted statistic: charts based on order statistics, sign charts and charts based on ranks. For each control scheme, its performance and several characteristics are discussed in some detail.

2 Distribution-Free Control Charts Based on Order Statistics

In this section, we focus on the distribution-free control charts, which utilize as charting statistics, one or more order statistics from random samples drawn from the underlying process. One of the early published works on this topic, authored by Janacek and Meikle (1997), introduced a two-sided Shewhart–type control chart based on the median with control limits determined from an in-control random sample (usually referred to as reference sample). Chakraborti et al. (2004) discussed in detail the aforementioned scheme, studied several characteristics of it and in addition, suggested further generalizations of it. In what follows, the order of appearance of the reviewed publications is chronological in an effort of setting up an easily accessible and well-structured flow.

2.1 A Median Control Chart Based on Bootstrap Methods

Park (2009) considered the construction of a nonparametric control chart which utilizes the sample median as monitoring statistic. The control limits of the proposed distribution-free control chart are determined by exploiting an estimate of the sample median variance. More precisely, the center line of the control chart established by Park (2009) coincides to the median value of all medians of the available samples of size n. The upper and lower control limits is obtained by applying appropriate bootstrap methods for the estimation of the variance of the sample median. For illustration purposes, the monitoring of real data obtained from the Ford Motor Company (see also Alloway and Raghavachari (1991)).

Let us assume that a sample of size n, say $X_1, X_2, \ldots, X_n$, is drawn from the underlying process with an unknown continuous distribution F. If θ and $\widetilde{X}$ denote the population and sample median respectively, then the limiting distribution of $\sqrt{n}(\widetilde{X} - \theta)$ as $n \to \infty$ is shown to be normal with mean 0 and variance $(2f(\theta))^{-2}$, where f is the corresponding probability density function of F. The main idea relies on obtaining bootstrap control limits by approximating of the distribution for the sample median through a Monte-Carlo approach. More precisely the Monte-Carlo bootstrap procedure can be described briefly as follows:

- A bootstrap sample with replacement is obtained from the original test sample $X_1, X_2, \ldots, X_n$.
- Using the bootstrap sample, the bootstrap sample median is determined
- The aforementioned two steps are repeated at least 1000 times.

Park (2009) considered the construction of median control charts by applying several bootstrap methods for obtaining the corresponding control limits, namely the Standard Bootstrap, the Bootstrap Percentile, the Bootstrap Bias-Corrected Percentile, the Bootstrap-t and the Bootstrap Hybrid methods. For more details, the interested reader is referred to Sect. 3 therein. The proposed control chart is constructed as follows:

- The center line is determined as the median value of all medians of the available reference samples of size n drawn from the process.
- The Lower and Upper Control Limit is chosen as the median of $\{L_1^*, L_2^*, \ldots, L_r^*\}$ and $\{U_1^*, U_2^*, \ldots, U_r^*\}$ respectively, where $(L_i^*, U_i^*), i = 1, 2, \ldots, r$ corresponds to a bootstrap confidence interval obtained by the implementing one of the abovementioned bootstrap methods.

2.2   Precedence Control Charts with Signaling Rules

Chakraborti et al. (2009) proposed two Phase II nonparametric control charts based on precedence statistics. The plotted statistic can be any order statistic in a Phase II test sample, while two different runs-rules are applied, namely the signaling rules proposed by Derman and Ross (1997) and Klein (2000). More precisely, the proposed nonparametric control charts called the 2-of-2 *DR* chart and the 2-of-2 *KL* chart respectively, produce a signal whenever the following occurs:

- two consecutive plotted points fall on or outside the control limits (for the 2-of-2 *DR* chart)
- two consecutive plotted points both fall on or above the upper control limit or both fall on or below the lower control limit (for the 2-of-2 *KL* chart).

In other words, Chakraborti et al. (2009) considered generalizing the standard 1-of-1 precedence chart, introduced earlier by Chakraborti et al. (2004) by incorporating the aforementioned signaling rules involving runs of the plotted statistic above and/or below the control limits. The plotted statistic for the hth test sample of size n is the jth order statistic $Y_{j:n}^h$ of the corresponding test sample. The control limits of both monitoring schemes are two specific order statistics $X_{a:m}, X_{b:m}$ of a reference sample of size m drawn while the process is in-control. The waiting time until the first signal (or alternatively the run length) of the proposed monitoring schemes is studied in detail. If the random variables Z_h and Z_h' are defined as

$$Z_h = \begin{cases} 1, & \text{if } Y_{j:n}^h \notin (LCL, UCL) \\ 0, & \text{if } Y_{j:n}^h \in (LCL, UCL) \end{cases}, \quad h = 1, 2, 3, \dots$$

$$Z_h' = \begin{cases} 0, & \text{if } Y_{j:n}^h \in (LCL, UCL) \\ 1, & \text{if } Y_{j:n}^h \geq UCL \\ 2, & \text{if } Y_{j:n}^h \leq LCL \end{cases}, \quad h = 1, 2, 3, \dots$$

then the corresponding waiting time of the proposed 2-of-2 *DR* and 2-of-2 *KL* chart can be expressed as

$$T_2 = \min\{t : Z_{t-1} = 1, Z_t = 1\}$$

and

$$T_2' = \min\{T_2^{(1)}, T_2^{(2)}\},$$

respectively, where

$$T_2^{(1)} = \min\{t : Z_{t-1}' = Z_t' = 1\}, \quad T_2^{(2)} = \min\{t : Z_{t-1}' = Z_t' = 2\}.$$

Denoting by $\lambda_t = P(Z_1 = 1, \ldots, Z_t = 1), t = 1, 2, \ldots$, the unconditional distribution of T_2 is given by

$$P(T_2 = x) = \sum_{y=1}^{x-2} \sum_{j=0}^{\min(y,[(x-y-2)/2])} \sum_{i=0}^{y} (-1)^j (-1)^i \binom{y}{j}\binom{y}{i}\binom{x-2(j+1)-1}{y-1}\lambda_{x-y+i}, \quad x \geq 3$$

and $P(T_2 = 0) = P(T_2 = 1) = 0$, $P(T_2 = 2) = \lambda_2$.

In addition to the above result, Chakraborti et al. (2009) obtained the distribution of the waiting time of T_2'. Given $X_{a:m}, X_{b:m}$, let us denote by T_2^* the waiting time for two consecutive 1's or two consecutive 2's in the sequence of independent and identically distributed trials $Z_1', Z_2', \ldots$. Applying some well-known results appeared in Fu and Lou (2003), the distribution of T_2^* can be expressed as

$$P(T_2^* = x | X_{a:m}, X_{b:m}) = \xi N^{x-1}(I - N)\mathbf{1}', \quad x \geq 2,$$

where

$$N = \begin{bmatrix} 0 & 1 - p_L - p_U & p_U & p_L \\ 0 & 1 - p_L - p_U & p_U & p_L \\ 0 & 1 - p_L - p_U & 0 & p_L \\ 0 & 1 - p_L - p_U & p_U & 0 \end{bmatrix}, \quad \xi = \begin{bmatrix} 1 & 0 & 0 & 0 \end{bmatrix}, \quad \mathbf{1} = \begin{bmatrix} 1 & 1 & 1 & 1 \end{bmatrix},$$

with the quantities p_L, p_U denoting the following conditional probabilities:

$$p_L = P(Y_{j:n} \leq X_{a:m} | X_{a:m} = x_1), \quad p_U = P(Y_{j:n} > X_{b:m} | X_{b:m} = x_2).$$

The unconditional distribution of the waiting time T_2' can be readily deduced by averaging over the joint distribution $h_{a,b}(x_1, x_2)$ of the order statistics $X_{a:m}, X_{b:m}$ as

$$P(T_2' = x) = \int_{-\infty}^{\infty} \int_{-\infty}^{x_2} P(T_2^* = x | X_{a:m} = x_1, X_{b:m} = x_2) h_{a,b}(x_1, x_2) dx_1 dx_2.$$

In the publication of Chakraborti et al. (2009), explicit expressions for the mean and standard deviation of the run length of the new control charts have been derived, and a discussion on the computation of the corresponding false alarm rate is included. Several numerical comparisons of the proposed charts versus the basic 1-of-1 precedence chart and the classical parametric Shewhart X-chart have been carried out revealing that the runs-type signaling rules improve the chart's sensitivity to a location shift.

2.3   A Two-Chart Nonparametric Monitoring Scheme Based on Order Statistics

Balakrishnan et al. (2010) introduced a distribution–free Shewhart–type control chart that takes into account the location of a single order statistic of the test sample (such as the median) as well as the number of observations that lie between the control limits. The proposed monitoring scheme involves the construction of two separate control charts. The plotted statistic of the first chart is the jth order statistic of each test sample drawn from the underlying process, while the charting statistic of the second one is related to the number of observations that lie between the control limits.

It goes without saying that, before starting the monitoring of the process, a reference sample $X_1, X_2, \ldots, X_m$ should be drawn from the in–control distribution F in order to establish the control limits and consequently test samples of size n, say $Y_1, Y_2, \ldots, Y_n$ will be obtained from the underlying process. The decision whether the process is in-control or has shifted to an out–of–control distribution with cumulative distribution function G is based on two specific order statistics from the reference sample which are used as control limits, namely

$$LCL = X_{a:m}, \quad UCL = X_{b:m}$$

with $1 \leq a < b \leq m$. It is worth mentioning that the abovementioned general framework of constructing control charts has been adopted by many authors. Since their nonparametric monitoring schemes are reviewed later on, whenever we describe control charts which follow the aforementioned general setup, a relative quotation to the present subsection will be mentioned.

For constructing their control chart, Balakrishnan et al. (2010) suggested that after the test sample is collected, its jth order statistic $Y_{j:n}$ should be computed along with the statistic

$$R = R(Y_1, Y_2, \ldots, Y_n; X_{a:m}, X_{b:m}) = |\{i \in \{1, 2, \ldots, n\} : X_{a:m} \leq Y_i \leq X_{b:m}\}|$$

which enumerates the observations in the test sample that lie between the control limits. Then, the process is declared in-control, if the following two conditions hold true:

$$LCL \leq Y_{j:n} \leq UCL \quad \text{and} \quad R \geq r.$$

The false alarm rate of the proposed monitoring scheme is given by

$$FAR = 1 - \sum_{r-1 \leq c+d \leq n-1} \frac{\binom{j-c+a-2}{a-1}\binom{m+n-b-d-j}{n-j-d}\binom{b+c+d-a}{c+d+1}}{\binom{m+n}{n}}.$$

If $f(s, t)$ denotes the joint density function of two order statistics from a random sample of size n from the Uniform distribution, the in-control average run length of the proposed monitoring scheme is given by

$$ARL_{in} = \int_0^1 \int_0^t \frac{1}{1 - q(s, t; r)} f(s, t) ds dt,$$

where

$$q(v, w; r) = \sum_{r-1 \leq c+d \leq n-1} \binom{n}{j-c-1, c+d+1, n-j-d} v^{j-c-1} (w - v)^{c+d+1} (1 - w)^{n-j-d},$$

$$0 \leq v < w \leq 1$$

and

$$\binom{n}{n_1, n_2, n_3} = \frac{n!}{n_1! n_2! n_3!}, \quad n = n_1 + n_2 + n_3.$$

In Balakrishnan et al. (2010), the out-of-control performance of the aforementioned control chart has been studied in some detail, while explicit expressions for the computation of its alarm rate and average run length under the well-known Lehmann alternatives $G = F^\gamma, \gamma > 0$ (see, e.g., Lehmann (1953)) have been derived. In addition, several numerical results carried over by the authors, shed light on the robustness and the efficacy of the new monitoring scheme.

2.4 A Nonparametric Control Chart Based on the Pooled Median

Graham et al. (2010) proposed a distribution-free Shewhart-type scheme for monitoring the location parameter of a continuous distribution in a Phase I process control setting. The monitoring scheme utilizes the pooled median of the available Phase I samples and the charting statistic is the number of observations in each sample that are less than the pooled median. More precisely, the m available random samples of size n are first pooled together and the pooled median M of the combined sample of $N = mn$ observations is computed. Then, in each sample of size n the number $U_i, i = 1, 2, \ldots, m$ of observations which are less than M is computed and plotted on the proposed control chart. Graham et al. (2010) proved that the in-control joint distribution of the random variables $U_1, U_2, \ldots, U_m$ is the multivariate hypergeometric distribution with probability density function $f_{U_1, U_2, \ldots U_m}(u_1, u_2, \ldots, u_m)$.

For the proposed chart, the probability of at least one false alarm out of the m samples when the process is in-control, known as *False Alarm Probability* (*FAP*), is investigated and the following expression is derived:

$$FAP = 1 - \sum_{u_1=LCL+1}^{UCL-1} \sum_{u_2=LCL+1}^{UCL-1} \cdots \sum_{u_m=LCL+1}^{UCL-1} f_{U_1,U_2,\ldots U_m}(u_1, u_2, \ldots, u_m),$$

where *LCL* and *UCL* is the corresponding lower and upper control limit of the proposed monitoring scheme. The general setup of a Phase I control chart entails that for the fixed *m* and *n*, one may determine appropriate control limits acquiring a prespecified FAP value. A detailed numerical study is carried out, along with several simulation-based comparisons which reveal that the proposed monitoring scheme performs well and, in some cases, better than other existing competitive control charts. For illustration purposes, the authors indicate how the proposed median control chart can be exploited for monitoring the inside diameters of the well-known piston-rings data given in Montgomery (2009).

2.5 A Nonparametric Control Chart Based on Grouped Observations

Bakir (2012) developed a distribution-free Shewhart-type control chart for detecting a change in more than one parameters in the probability distribution of the underlying process. The proposed monitoring scheme is designed for grouped observations and calls for reference (or training) data, namely a group of *m* observations $X_{01}, X_{02}, \ldots, X_{0m}$ drawn from the process when it is in-control with cumulative distribution function F_0. The new chart uses as charting statistic a modified version of the well-known two-sample Kolmogorov–Smirnov test statistic and allows the exact determination of the conditional average run length of the proposed scheme over the family of all symmetric and non-symmetric continuous distributions.

Let us denote by $S_0(z)$ the empirical distribution function of the reference random sample, i.e.,

$$S_0(z) = \begin{cases} 0, & \text{if } z < X_{0(1)}, \\ j/m, & \text{if } X_{0(j)} \leq z < X_{0(j+1)}, \quad j = 1, 2, \ldots, m-1, \\ 1, & \text{if } z \geq X_{0(m)} \end{cases}$$

where $X_{0(j)}$ is the *j*th order statistic of the reference sample. At each sampling instance $t, t = 1, 2, \ldots$, a test sample $Y_{t1}, Y_{t2}, \ldots, Y_{tn}$ with cumulative distribution function F_y is drawn from the process. If $Y_{t(i)}$ denotes the order statistics of the *t*th test sample, the empirical distribution function of each test sample is given as

$$S_t(z) = \begin{cases} 0, & \text{if } z < Y_{t(1)}, \\ i/n, & \text{if } Y_{t(i)} \le z < Y_{t(i+1)}, \quad i = 1, 2, \ldots, n-1. \\ 1, & \text{if } z \ge Y_{t(n)} \end{cases}$$

Bakir (2012) considered three different scenarios for the possible shift that may occur in the underlying process. According to the first scenario (*Case* 1, hereafter), the process may tend to produce stochastically smaller observations than the observations of the in-control state. In other words, the attention heads for testing the null hypothesis $H_0^- : F_y(z) \le F_0(z)$ (for all $-\infty < z < +\infty$) versus the alternative $H_a^- : F_y(z) \ge F_0(z)$ (for all z) and $F_y(z) > F_0(z)$ for at least one z. Under the second scenario (*Case* 2, hereafter), the proposed control chart needs to detect whether or not the process tends to produce stochastically larger observations than the observations of the in-control phase. In statistical terms, one has to test the null hypothesis $H_0^+ : F_y(z) \ge F_0(z)$ (for all $-\infty < z < +\infty$) versus the alternative hypothesis $H_a^+ : F_y(z) \ge F_0(z)$ (for all z) and $F_y(z) < F_0(z)$ for at least one z. The last possible shift studied by Bakir (2012) (*Case* 3, hereafter), refers to a process tending to produce smaller and/or larger observations than the in-control state, the interest focuses in testing the null hypothesis $H_0 : F_y(z) = F_0(z)$ (for all $-\infty < z < +\infty$) versus the alternative $H_a : F_y(z) \ne F_0(z)$ for at least one z. The charting statistics of the proposed Shewhart-type control chart for each case takes on the following forms:

$$\psi_t^- = \min_{z=x_{0j}} [S_0(z) - S_t(z)] \quad (Case\ 1)$$

$$\psi_t^+ = \max_{z=x_{0j}} [S_0(z) - S_t(z)] \quad (Case\ 2)$$

$$\psi_t = \max_{z=x_{0j}} |S_0(z) - S_t(z)| \quad (Case\ 3).$$

Note that the abovementioned statistics are modified versions of the traditional two-sample Kolmogorov-Smirnov statistic, where the maximization is taken only over the reference sample observations (see, also Conover (1999)). Under *Cases* 1, 2 and 3, the process is declared out-of-control if

$$S_0(z_{\max}) - S_t(z_{\max}) \le -L$$

or

$$S_0(z_{\max}) - S_t(z_{\max}) \ge L$$

or

$$|S_0(z_{\max}) - S_t(z_{\max})| \ge L$$

respectively. Note that $z = z_{\max}$ corresponds to the value at which the quantity $|S_0(z) - S_t(z)|$ takes on its maximum. Bakir (2012) noted that, given the reference sample, $n S_t(z_{\max})$ becomes a binomial random variable $B_{n,\pi}$ with n trials and success probability $\pi = F_y(z_{\max})$. The exact conditional probability of a signal of all three proposed control charts are given below

$$P^- = P(B_{n,\pi} \geq n[S_0(z_{\max}) + L] | X_{01}, X_{02}, \ldots, X_{0m}) \qquad \text{(Case 1)}$$

$$P^+ = P(B_{n,\pi} \leq n[S_0(z_{\max}) - L] | X_{01}, X_{02}, \ldots, X_{0m}) \qquad \text{(Case 2)}$$

$$P = P^- + P^+, \qquad \text{(Case 3)}$$

while the corresponding exact conditional average run length is equal to the reciprocal of the respective abovementioned signaling probability. Bakir (2012) stated that the required unconditional expectations over the reference sample cannot be expressed via a closed form. Therefore, a simulation-based study was carried out in order to estimate the unconditional probability of producing a signal and the unconditional average run length for the proposed monitoring schemes. At each simulation run, a reference sample and a test random sample were generated and both conditional signaling probability and conditional average run length were computed via the corresponding explicit expressions. The above-mentioned conditional values were averaged out over the number of simulations in order to reach the desired unconditional expectations.

Moreover, a simulation-based study was carried out for investigating the sensitivity of the proposed nonparametric control chart against skewness and possible presence of outliers.

2.6 One-Sided Control Charts Based on Precedence Statistics

Balakrishnan et al. (2015) considered one-sided control charts based on precedence statistics. Since the aim was to study early failures, the choice of one-sided charts seems to be more appropriate. The in- and out-of-control alarm rate as well as the average run length of the proposed monitoring schemes were studied in some detail, while suitable randomized procedures for selecting the best precedence control chart have also been proposed.

Let us denote by $X_1, X_2, \ldots, X_m$ a reference sample drawn from the in−control distribution F and assume that the test samples of size n, say $Y_1, Y_2, \ldots, Y_n$, are coming from a process with common cumulative distribution function G. Let us next consider the precedence statistic $P_{(b)}$ of order b that counts the number of Y-observations that are smaller than the bth ordered X-observation. The probability mass function of the random variable $P_{(b)}$ is given as

$$P(P_{(b)} = c) = \frac{\binom{n}{c}}{B(b, m-b+1)} \int_0^1 (GF^{-1}(u))^c (1 - GF^{-1}(u))^{n-c} u^{b-1} (1-u)^{m-b} du,$$

where $B(\cdot, \cdot)$ is the complete beta function. The random variable $P_{(b)}$ plays the role of the monitoring statistic with an upper control limit equal to the bth order statistic of the reference sample, namely $UCL = X_{b:m}$. Moreover, Balakrishnan et al. (2015) implemented the above formula for obtaining the survival function of $P_{(b)}$ by summation, while under the assumption $F = G$ the following ensues:

$$P(P_{(b)} = c) = \frac{\binom{b+c-1}{c} \binom{m+n-b-c}{n-c}}{\binom{m+n}{n}}.$$

Balakrishnan et al. (2015) pointed out that large values of $P_{(b)}$ provide evidence that the process is out-of-control. Hence, the critical region for the proposed control chart takes on the form $W_b = \{P_{(b)} \geq c_{b,a}\} \Leftrightarrow \{Y_{c_{b,a}:n} < X_{b:m}\}$, where a is the pre-specified nominal level of *False Alarm Rate*. Based on the aforementioned result, one may readily determine the desired value of $c_{b,a}$. More specifically, let c be such that

$$\sum_{i=c+1}^{n} \frac{\binom{b+i-1}{i} \binom{m+n-b-i}{n-i}}{\binom{m+n}{n}} \leq a < \sum_{i=c}^{n} \frac{\binom{b+i-1}{i} \binom{m+n-b-i}{n-i}}{\binom{m+n}{n}}.$$

Therefore, the critical threshold related to the random variable $P_{(b)}$ can be expressed as

$$c_{b,a} = \begin{cases} c, & \text{if } a - \sum_{i=c+1}^{n} P(P_{(b)} = i) \geq \sum_{i=c}^{n} P(P_{(b)} = i) - a \\ c+1, & \text{otherwise.} \end{cases}$$

It is straightforward that the distribution of the monitoring statistic $P_{(b)}$ depends on F and G. Therefore, the out-of-control performance of the proposed control chart can be evaluated only for given choices of G (in terms of F). Balakrishnan et al. (2015) considered the well-known Lehmann-type alternatives $G = F^{\gamma}, \gamma \in (0, 1)$ and $G = 1 - (1 - F)^{\gamma}, \gamma \in (1, +\infty)$ and proved that the probability mass function of the variable $P_{(b)}$ can be expressed as

$$P(P_{(b)} = c) = \frac{\binom{n}{c}}{B(b, m - b + 1)} \sum_{h=0}^{n-c} \binom{n-c}{h}(-1)^h B(b + (c + h)\gamma, m - b + 1)$$

and

$$P(P_{(b)} = c) = \frac{\binom{n}{c}}{B(b, m - b + 1)} \sum_{h=0}^{c} \binom{c}{h}(-1)^h B(b, m - b + \gamma(n + h - c) + 1)$$

respectively.

In addition, with regard to the run length distribution of the proposed one-sided nonparametric control chart they proved that the average run length of the new monitoring scheme takes on the form

$$ARL(F, G) = \frac{1}{B(b, m - b + 1)} \int_0^1 \frac{u^{b-1}(1 - u)^{m-b}}{1 - p_{F,G}(u)} du,$$

where

$$p_{F,G}(u) = \frac{1}{B(c_{b,a}, n - c_{b,a} + 1)} \sum_{h=0}^{c_{b,a}-1} \binom{c_{b,a} - 1}{h}(-1)^h \frac{(1 - GF^{-1}(u))^{n+h-c_{b,a}+1}}{n + h - c_{b,a} + 1}.$$

For illustration purposes, the proposed scheme was implemented for monitoring coal mining disaster data (see also Jarett (1979)).

The aforementioned precedence statistic $P_{(b)}$ can be viewed as the sum of the first b placement statistics $N_1, N_2, \ldots, N_b$, namely

$$P_{(b)} = \sum_{j=1}^{b} N_j,$$

where N_j corresponds to the number of Y-observations between the order statistics $X_{j-1:m}$ and $X_{j:m}$ (for $j = 1, 2, \ldots, m$). Balakrishnan et al. (2015) proposed also the following weighted version of the precedence statistic:

$$P_{(b)}^* = \sum_{j=1}^{b} (m - j + 1)N_j.$$

The interested reader may refer to the full publication for a detailed performance study of the weighted precedence control chart, where the alarm rate and the average run length are studied in both in- and out-of-control situations.

2.7 The Minimum and the Median Precedence Control Chart with Improved Runs-Rules

Malela-Majika et al. (2016) studied two members of the class of precedence control charts which has been first introduced by Chakraborti et al. (2004). According to the general setup, a reference Phase I sample of size m is available from the in-control process, while a test sample $Y_1, Y_2, \ldots, Y_n$ is also drawn. The monitoring statistic of the precedence control charts coincide to the jth order statistic $Y_{j:n}$ of the test sample. The performance of the minimum and the median precedence control chart is examined in terms of their in- and out-of-control run length properties in a detailed simulation study. The aforementioned one-sided control schemes utilize the minimum $(Y_{1:n})$ and the median $(Y_{(n+1)/2:n}$ or $(Y_{n/2:n} + Y_{(n+2)/2:n})/2)$ of each test sample as the monitoring statistic respectively, while their upper control limit is properly determined as the bth order statistic of the reference sample. The theoretical results which have been established by Chakraborti et al. (2004) are used in order to evaluate the performance and robustness of the monitoring schemes. For the numerical study which has been carried out, several distributions have been considered, such as the standard normal, the student's t or the gamma distribution.

Moreover, Malela-Majika et al. (2016) suggested adding supplementary runs-rules in order to improve the performance of the distribution-free precedence control charts. More precisely, the following three runs-rules were practiced (see, also Khoo and Ariffin (2006) and Antzoulakos and Rakitzis (2008)):

- The 2-of-2 Runs-Rule, where a signal is produced when two charting statistics, say $Y_{j:n}^h$, $Y_{j:n}^{h+1}$ from two consecutive test samples, $h = 1, 2, \ldots$, both plot on or above the upper control limit or both plot on or below the lower control limit.
- The 2-of-2 Improved Runs-Rule, where some warning limits are introduced in addition to the traditional control limits. The improved runs-rule signals, when one charting statistic $Y_{j:n}^h$ plots on or above the upper control limit or on or below the lower control limit or when two consecutive charting statistics, say $Y_{j:n}^h$, $Y_{j:n}^{h+1}$, $h = 1, 2, \ldots$, plot between the upper warning and the upper traditional control limit or between the lower traditional and the lower warning control limit.
- The 2-of-2 Improved Modified Runs-Rule, where apart from the traditional control limits, two additional warning limits and a center line are used. A signal is produced when one charting statistic $Y_{j:n}^h$ plots on or above the upper control limit or plots on or below the lower control limit or when considering two consecutive charting statistics, say $Y_{j:n}^h$, $Y_{j:n}^{h+1}$, $h = 1, 2, \ldots$, one plots between the center line and the upper control limit and another one plots between the upper warning control limit and the upper control limit, or one plots between the lower control limit and the center line and another one plots between the lower control limit and the lower warning control limit.

A detailed discussion on the determination of the design parameters was provided, while the impact of the size of the reference sample on the performance of the resulting chart was also investigated. In addition, an extensive numerical experimentation that was carried out provided several interesting concluding remarks about the in-control robustness and out-of-control performance of the proposed monitoring schemes under several distributions, such as the standard normal distribution, the Student's distribution with 4 degrees of freedom or the Gamma distribution. Here is a brief synopsis of the conclusions stated by Malela-Majika et al. (2016). Under the standard normal distribution, the precedence control chart which utilizes as monitoring statistic the minimum of each test sample, namely $Y_{1:n}^h$, performs best for small shifts when the 2-of-2 Runs-Rule and the 2-of-2 Improved Modified Runs-Rule is applied. On the other hand, for moderate to large shifts of the underlying distribution, the 2-of-2 Improved Modified Runs-Rule and the 2-of-2 Improved Runs-Rule outperform. In addition, if the median of each test sample is used as a plotted statistic, then the corresponding precedence control chart enhanced with the 2-of-2 Improved Modified Runs-Rule and the 2-of-2 Improved Runs-Rule seems to be preferable. However, it should be stressed that under different distributional assumptions for the monitored characteristic, the abovementioned remarks are slightly divergent.

2.8 Control Charts Based on the Total Median Statistic

Figueiredo and Gomes (2016) utilized the total median statistic in order to construct a control chart for monitoring symmetric contaminated normal distributions with heavier-than-normal tails. The proposed method is based on the so-called bootstrap sample, which is simply obtained by randomly sampling with replacement from the observed reference sample. More precisely, let $X_1, X_2, \ldots, X_m$ be a random sample of size m from the in-control distribution F and denote by $X_1^*, X_2^*, \ldots, X_m^*$ the corresponding bootstrap sample with cumulative distribution function estimated by the classical empirical distribution function of the observed sample, i.e., by $F_m^*(x) = (1/n) \sum_{i=1}^m I_{\{x_i \leq x\}}$, where I_A is the indicator function of the set A.

The plotted statistic of the new chart is the total median statistic, which is calculated as a linear combination of all possible values of the median of the bootstrap sample or equivalently by

$$
TMd = \sum_{i=1}^m \sum_{j=i}^m a_{i,j,m} \frac{X_{i:m} + X_{j:m}}{2} = \sum_{i=1}^m a_{i,m} X_{i:m},
$$

where $a_{i,m} = (1/2)\left(\sum_{j=i}^m a_{i,j,m} + \sum_{j=1}^i a_{j,i,m}\right)$, $1 \leq i \leq n$. The coefficients $a_{i,j,m}$ appearing in the last expression are given by

$$a_{i,j,m} = \begin{cases} \frac{1}{m^m} \sum\limits_{k=0}^{(m-1)/2} \frac{m!(i-1)^k}{k!(m-k)!} \sum\limits_{r=[m/2]-k+1}^{m-k} \frac{(m-k)!(m-i)^{m-k-r}}{r!(m-k-r)!}, & \text{if } 1 \leq i = j \leq m \\ \frac{m!\{i^{m/2}-(i-1)^{m/2}\}\{(m-j+1)^{m/2}-(m-j)^{m/2}\}}{m^m((m/2)!)^2}, & \text{if } m \text{ even and } 1 \leq i < j \leq m \\ 0, \text{ if } m \text{ odd and } 1 \leq i < j \leq m. \end{cases}$$

As Figueiredo and Gomes (2016) mentioned, it is not feasible to obtain the exact distribution of the monitoring statistic TMd in the general case, however one may determine accurate quantiles of it by simulation and use them as lower and upper control limits of the proposed control chart.

The performance of the proposed control was studied through an extensive numerical experimentation and compared to competitive schemes under contaminated normal data and in particular scaled- and student contaminated normal distributions.

2.9 Multiple-Chart Nonparametric Monitoring Schemes Based on Order Statistics

Triantafyllou (2018a) introduced a nonparametric Shewhart-type monitoring scheme that takes into account the location of two order statistics of the test sample as well as the number of observations from the test sample that lie between the control limits. In fact, the proposed scheme is a generalization of the control chart established by Balakrishnan et al. (2010) which has been reviewed in Sect. 2.3 of the present manuscript. The monitoring scheme follows the general setup of the nonparametric control charts proposed by Balakrishnan et al. (2010) and an additional monitoring statistic is used.

The construction of the proposed nonparametric control scheme calls for two order statistics of the test sample drawn from the process. More specifically, after the test sample is collected, the jth and the kth order statistic $Y_{j:n}, Y_{k:n}$ are chosen and made use of along with the statistic R mentioned before. Then, the process is declared in-control, if the following conditions hold true:

$$LCL \leq Y_{j:n} \leq Y_{k:n} \leq UCL \text{ and } R \geq r,$$

where r is a positive integer. The false alarm rate of the monitoring scheme is given by

$$FAR = 1 - \sum_{c_1=0}^{n-2} \sum_{c_3=\max(0,r-c_1-c_2-2)}^{n-c_1-c_2-2}$$

$$\frac{\binom{a+j-c_1-2}{a-1}\binom{m+n-b-k-c_3}{m-b}\binom{b+c_1+c_2+c_3-a+1}{b-a-1}\binom{m+n}{b+k+c_3-1}}{\binom{m+n}{n}\binom{j+c_2+m+n-k-1}{j+b+c_2+c_3-2}}$$

where $c_2 = k - j - 1$.

Triantafyllou (2018a) provided the next explicit expression for the operating characteristic function p (the probability that the proposed scheme does not signal) in the case of Lehmann alternatives $G = F^\gamma$

$$p = P(X_{a:m} \leq Y_{j:n} \leq Y_{k:n} \leq X_{b:m} \text{ and } R(Y_1, Y_2, \ldots, Y_n; X_{a:m}, X_{b:m}) \geq r)$$

$$= \sum_{c_1=0}^{n-2} \sum_{c_3=\max(0,r-c_1-c_2-2)}^{n-c_1-c_2-2} \frac{m!n!}{(a-1)!(b-a-1)!(m-b)!(j-c_1-1)!c_1!c_2!c_3!(n-k-c_3)!}$$

$$\times B(c_1 + c_2 + 2, c_3 + 1) B_\gamma(\gamma(j - c_1 - 1) + a, b - a; c_1 + c_2 + c_3 + 2)$$

$$\times B(c_1 + 1, c_2 + 1) B_\gamma(\gamma(c_2 + c_3 + j + 1) + b, m - b + 1; n - k - c_3).$$

where

$$B_\gamma(a, b; l) = \int_0^1 x^{a-1}(1-x)^{b-1}(1-x^\gamma)^l dx = \sum_{k=0}^l (-1)^k \binom{l}{k} B(a + k\gamma, b).$$

Explicit expressions for the in- and out-of-control average run length are also derived by applying similar arguments as the ones exploited in Chakraborti et al. (2004) or Balakrishnan et al. (2010). An extensive numerical experimentation illustrated its efficacy for detecting possible shifts in the distribution of the underlying process. A comparative numerical study carried out for comparing it to the control charts established by Balakrishnan et al. (2009), Mukherjee and Chakraborti (2012) and Chowdhury et al. (2014) revealed that the proposed distribution-free control scheme performs better in all cases considered.

In addition, a class of nonparametric Shewhart–type control charts based on the location of three order statistics of the test sample as well as the number of observations in that sample that lie between the control limits was established by Triantafyllou (2018b). Within the context described above, Triantafyllou (2018b) proposed the construction of nonparametric control schemes that exploit the location of three test sample observations drawn from the process. More precisely, the ith, jth and the kth order statistic $Y_{i:n}, Y_{j:n}, Y_{k:n}$ are selected and made use of along with the test statistic R defined previously in Sect. 2.3 of the present manuscript. The class of distribution-free control charts, introduced by Triantafyllou (2018b), makes use of an in-control rule, which embraces the following conditions:

Condition 1. The observations $Y_{i:n}, Y_{j:n}, Y_{k:n}$ of the test sample should lie between the order statistics $X_{a:m}$ and $X_{b:m}$ of the reference sample, namely $X_{a:m} \leq Y_{i:n} \leq Y_{j:n} \leq Y_{k:n} \leq X_{b:m}$.

Condition 2. The number of observations of the Y-sample that lie between the order statistics $X_{a:m}$ and $X_{b:m}$ should be equal to or more than r, namely $R \geq r$

The conditions stated above, define four separate plotted statistics and the process is declared in-control, if the following conditions hold true:

$$X_{a:m} \leq Y_{i:n}, Y_{j:n}, Y_{k:n} \leq X_{b:m}, \quad R \geq r, \text{ for } i < j < k.$$

The False Alarm Rate of the new monitoring scheme can be computed as

$$FAR = 1 - \sum_{c_1=0}^{n-3} \sum_{c_4=\max(0,r-c_1-c_2-c_3-3)}^{n-c_1-c_2-c3-3}$$

$$\frac{\binom{a+i-c_1-2}{a-1}\binom{m-b+n-k+c_4}{m-b}\binom{b+c_1+c_2+c_3+c_4-a+2}{b-a-1}}{\binom{m+n}{n}}.$$

where

$$c_2 = j - i - 1, c_3 = k - j - 1.$$

By exploiting the condition–uncondition technique (see, e.g., Balakrishnan et al. (2010)) Triantafyllou (2018b) deduced an expression for the exact run length distribution. More specifically, the average run length (*ARL*) of the proposed nonparametric control chart, can be written as

$$ARL = \int_0^1 \int_0^t \frac{1}{1 - q(G \circ F^{-1}(s), G \circ F^{-1}(t); r)} f(s, t) ds dt,$$

where $f(s, t)$ is the joint density function of two order statistics of a random sample from the Uniform distribution in the interval $(0, 1)$ (see, e.g., Balakrishnan and Ng (2006)). Note that the quantity $q(v, w; r)$ is defined as

$$q(v, w; r) = \sum_{c_1=0}^{n-3} \sum_{c_4=\max(0,r-c_1-c_2-c_3-3)}^{n-c_1-c_2-c_3-3} q_{c_1,c_2,c_3,c_4}(v, w), \quad 0 \leq v < w \leq 1,$$

where

$$q_{c_1,c_2,c_3,c_4}(v, w) =$$

$$\frac{n!}{(i-c_1-1)!(n-k-c_4)!(c_1+c_2+c_3+c_4+3)!} v^{i-c_1-1}(w-v)^{c_1+c_2+c_3+c_4+3}(1-w)^{n-k-c_4}.$$

Several numerical results, displayed for the proposed family of nonparametric control charts, depict that the proposed monitoring scheme attains competitive performance in comparison with well-known distribution-free control charts. More precisely, the proposed distribution-free control scheme performs better than the other three competitive nonparametric charts established by Mukherjee and Chakraborti

(2012), Chowdhury et al. (2014) and Triantafyllou (2018a) for all the cases considered. Note that the abovementioned comparative results were produced under the Normal distribution and the Laplace distribution. For illustration purposes, the proposed monitoring scheme was implemented for reliability monitoring, where a data-driven application by Xie et al. (2002) was discussed in some detail.

2.10   Distribution-Free Precedence Control Charts with the 2-of-(h+1) Supplementary Runs-Rule

Malela-Majika et al. (2019) improved the precedence monitoring schemes which were introduced by Chakraborti et al. (2004), by applying 2-out-of-($h + 1$) supplementary runs-rules. The authors deliberated both non-side and side-sensitive runs-rules for their proposed schemes. More specifically, the following runs-rules were considered:

- the non-side-sensitive w-of-($w + v$) rule suggested by Derman and Ross (1997). Based on this particular rule (DR rule, hereafter), a signal is produced whenever w out of $w + v$ successive samples fall on or outside the control limits, which are separated by at most v samples that fall between the control limits. The rationale of this runs-rule does not focus on which control limit is violated, namely there is no need to care about whether the test samples fall above the upper control limit or below the lower control limit of the chart.
- the side-sensitive w-of-($w + v$) rule suggested by Klein (2000). Based on this particular rule (KL rule, hereafter), a signal is produced whenever w out of $w + v$ successive samples fall below (above) the lower (upper) control limit, which are separated by at most v samples that fall above (or below) the lower (upper) control limit.

Malela-Majika et al. (2019) investigated the ability of the proposed control charts for detecting possible shifts, in terms of several performance measures. Both in- and out-of-control behavior of the resulting two-sided control chart was studied in detail, while its zero- and steady-state performance was investigated via a Markov chain approach. Among their main theoretical results, one may highlight the construction of the corresponding transition probability matrix and an explicit expression for the computation of the average extra quadratic loss function of the new monitoring scheme given below

$$AEQL = \frac{1}{\delta_{\max} - \delta_{\min}} \int_{\delta_{\min}}^{\delta_{\max}} \left(\delta^2 \times ARL(\delta) \right) \times f(\delta) d\delta$$

where $\delta \neq 0$ and $ARL(\delta)$ correspond to the specific shift in the location parameter and the average run length of the proposed chart respectively. Note that $f(\delta)$ was written as the probability density function of a Uniform distribution with parameters 0 and 1. For both zero-state and steady-state modes, the unconditional average run length of the precedence 2-of-$(h + 1)$ schemes is investigated and closed formulae are delivered. For example, the unconditional zero-state average run length of the proposed precedence control chart with 2-of-$(h + 1)$ DR runs-rules can be expressed as

$$UARL_{ZS}(\delta) = \int_0^1 \int_0^t \left[\frac{2 - p_2^h}{1 - p_2 - p_2^h - p_2^{h+1}} \right] f_{a,b}(s, t) ds dt,$$

where $f_{a,b}(s, t)$ denotes the joint probability mass function of the ath and the bth order statistic of a random sample of size m from the Uniform distribution $(0, 1)$ and

$$p_2 = P(X_{a:m} \le Y_{j:n} \le X_{b:m} | X_{a:m} = x_{a:m}, X_{b:m} = x_{b:m}).$$

Employing analogous arguments Malela-Majika et al. (2019) obtained respective formulae for the evaluation of the steady-state and zero-state unconditional average run length of both proposed schemes with DR or KL runs-rules. For more details, the interested reader is referred to Sects. 3.1 and 3.2 therein.

2.11 A Four-Chart Monitoring Scheme Based on Order Statistics

Triantafyllou (2019a) introduced a nonparametric Shewhart–type control scheme considering not only the position of specific ordered observations from both reference and test data drawn from the process, but also the test data that are placed between the control limits. The proposed monitoring scheme adopts the idea of Balakrishnan et al. (2010) and appends an additional condition, which seems to make the decision rule more accurate. In words, the proposed distribution-free chart, enables the double checking for detecting possible shifts of the underlying distribution. The testing procedure that was implemented in order to decide whether the in-control distribution of the process has been shifted or not, looks for some evidence for the equality between the in-control distribution and the distribution of test samples through three different ways. The decision rule of the control chart requires the verification of three conditions.

Let $X_1, X_2, \ldots, X_m$ denote a reference sample drawn from a process with in-control distribution F and three particular ordered observations, say $X_{a:m}, X_{i:m}, X_{b:m}, 1 \le a < i < b \le m$, are picked out. The integers a, i, b are design parameters and should be appropriately selected so that a specific level of performance is reached. Suppose next that random samples $Y_1, Y_2, \ldots, Y_n$ with distribution G, are picked independently in order to detect a plausible shift in the underlying distribution.

The proposed monitoring scheme takes advantage of the position of single ordered observations from both test and reference sample. More precisely, the new monitoring scheme asks for an order statistic of each test sample to be enveloped by two pre-specified observations of reference sample, while at the same time an ordered observation of the reference sample be enclosed by two pre-determined values of the test sample.

Firstly, the test sample should be drawn. Then, the cth, jth and the dth order statistic $Y_{c:n}, Y_{j:n}, Y_{d:n}$ are selected and implemented together with test statistic R defined earlier.

The class of distribution-free monitoring schemes, introduced in this paper, makes use of an in-control rule, which embraces the following three conditions:

Condition 1. The statistic $Y_{j:n}$ of the test sample should lie between the observations $X_{a:m}$ and $X_{b:m}$ of the reference sample, namely $X_{a:m} \leq Y_{j:n} \leq X_{b:m}$.

Condition 2. The interval $(Y_{c:n}, Y_{d:n})$ formulated by two appropriately chosen order statistics of the test sample should enclose the value $X_{i:m}$ of the reference sample, namely $Y_{c:n} \leq X_{i:m} \leq Y_{d:n}$.

Condition 3. The number of observations of the Y-sample that are placed enclosed by the observations $X_{a:m}$ and $X_{b:m}$ should be equal to or more than r, namely $R \geq r$.

Condition 1 asks for a single observation of each test sample to be enclosed between two specific order statistics of the reference sample. A popular choice would be to call for the median of each test sample to lie between two pre-determined ordered observations of the reference sample. The implementation of Condition 1 offers the first test report concerning the equality of distributions F, G. In addition, Condition 2 makes an attempt to compare the distributions of reference and test samples by a different point of view. More precisely, Condition 2 checks out whether a single observation of reference sample is sealed up by two pre-specified order statistics of the test sample. Finally, Condition 3 seems to strengthen the testing procedure that has already been established by Condition 1. The main motivation behind the aforementioned conditions is to produce enough evidence about the equality (or not) of the underlying in-control distribution of the process and the distribution of test samples drawn independently from each other. The conditions stated above, define four separate plotted statistics. More precisely, the proposed distribution-free scheme requires the construction of four different control charts.

The process is claimed to be in-control, if the next restrictions are satisfied

$$X_{a:m} \leq Y_{j:n} \leq X_{b:m}, \quad Y_{c:n} \leq X_{i:m}, \quad Y_{d:n} \geq X_{i:m}, \quad R \geq r.$$

The large amount of design parameters of the proposed monitoring scheme, e.g., $m, n, a, i, b, c, j, d, r$, gives the practitioner a notable flexibility for achieving a pre-specified level of in-control or out-of-control performance of the resulted chart.

The operating characteristic function of the proposed control scheme can be expressed as

$$p = p(m, n, a, i, b, c, j, d, r; F, G) = \int_0^1 \int_0^z \int_0^t q(GF^{-1}(s), GF^{-1}(t), GF^{-1}(z); r, d, j) f(s, t, z) ds dt dz$$

where

$$q(v_1, v_2, v_3; r, d, j) = \sum_{k_3=0}^{n-3} \sum_{k_4=0}^{n-3} \sum_{k_6=\max(r-d+j-1,0)}^{n-k_3-k_4-3} q_{k_3,k_4,k_6}(v_1, v_2, v_3), 0 \le v_1 < v_2 < v_3 \le 1,$$

with

$$q_{k_3,k_4,k_6}(v_1, v_2, v_3) = \frac{n!}{(j - k_3 - k_4 - 1)!k_3!(k_4 + k_6 + d - j + 1)!(n - d - k_6)!}$$
$$\times v_1^{j-k_3-k_4-1}(v_2 - v_1)^{k_3}(v_3 - v_2)^{k_4+k_6+d-j+1}(1 - v_3)^{n-d-k_6},$$

while

$$f(s, t, z) = \frac{m!}{(a - 1)!(i - a - 1)!(b - i - 1)!(m - b)!} s^{a-1}(t - s)^{i-a-1}(z - t)^{b-i-1}(1 - z)^{m-b}.$$

Based on the above expressions, Triantafyllou (2019a) proved that the *False Alarm Rate* of the proposed four-chart monitoring scheme is given as

$$FAR = 1 - \sum_{k_3=0}^{n-3} \sum_{k_4=0}^{n-3} \sum_{k_6=\max(r-d+j-1,0)}^{n-k_3-k_4-3} \frac{\binom{i+k_3-a-1}{k_3}\binom{a+j-k_3-k_4-2}{a-1}\binom{m+n-b-d-k_6}{m-b}}{\binom{m+n}{n}},$$

The conditioning argument mentioned before (see also Chakraborti et al. (2004)) has been used for delivering suitable expressions for in- and out- of control *Average Run Length* of the proposed monitoring scheme and the numerical computations of *Average Run Length* have been accomplished through appropriate numerical approximations for the corresponding integral by using suitable adaptive algorithms, which recursively subdivide the integration region as needed.

Moreover, the operating characteristic function of the monitoring scheme, under the Lehmann alternative $G = F^\gamma$, can be expressed as

$$p_{\text{Lehmann}} = p(m, n, a, i, b, c, j, d, r; F, G)$$

$$= \sum_{k_3=0}^{n-3} \sum_{k_4=0}^{n-3} \sum_{k_6=\max(r-d+j-1,0)}^{n-k_3-k_4-3} Q \cdot B_\gamma(\gamma(j - k_3 - k_4) + a, i - a; k_3)$$

$$B_\gamma(\gamma(k_6 + d + 2) + b - 1, m - b + 1; n - d - k_6)$$
$$\times B_\gamma(\gamma(j - k_4) + i, b - i; k_4 + k_6 + d - j + 2),$$

where Q is defined as

$$Q = \frac{m!n!B(j - c - k_3 - k_4, c)B(k_4 + 1, d - j)B(k_6 + 1, k_4 + d - j + 1)}{(a - 1)!(i - a - 1)!(b - i - 1)!(m - b)!(c - 1)!(j - c - k_3 - k_4 - 1)!k_3!k_4!(d - j - 1)!k_6!(n - d - k_6)!}$$

For the proof of the abovementioned theoretical results, the interested reader is referred to the Appendix section in Triantafyllou (2019a).

The proposed chart was compared to the so-called *W-CUSUM* and *W-EWMA* control charts established by Li et al. (2010), to *W-EWMA-BRSS* proposed by Malela-Majika and Rapoo (2016), as well as to the Mann-Whitney-based chart (*MW chart* hereafter) instituted by Chakraborti and van de Wiel (2008). From the numerical comparisons carried out by Triantafyllou (2019a), the proposed distribution-free monitoring scheme becomes substantially efficient. More specifically, the new chart seems to outperform the *MW chart*, the *W-CUSUM chart* and *W-EWMA chart* in all cases considered, while against *W-EWMA-BRSS chart* the proposed chart gives better results for almost all cases considered. The only exception (among the cases considered) appears when, under the *Gamma* (3, 1) distribution, the shift is assumed to be equal to 0.50. In addition, numerical comparisons between the proposed monitoring scheme and the control charts introduced by Balakrishnan et al. (2015) are also considered. The comparisons between the aforementioned control chart and the monitoring scheme established by Triantafyllou (2019a) were based on the alarm rates that the both control charts manage to attain for several Lehmann-type alternatives. The nonparametric control scheme introduced by Triantafyllou (2019a) meets the same or even better performance, compared to the control chart established by Balakrishnan et al. (2015), in detecting the possible shift of the underlying distribution process.

3   Distribution-Free Control Charts Based on Sign Statistics

In this section, we focus on the distribution-free control charts, which are based on sign statistics. For each control scheme reviewed, the main results and characteristics are discussed in some detail. One of the pioneer works on this topic, was accomplished by Amin et al. (1995), where a 1-of-1 Shewhart-type sign chart for monitoring the median of the underlying process was established. However, the first work on the topic after 2010, was the publication of Human et al. (2010). The order of appearance of the publications discussed in this section, is chronological for making the reading more coherent.

3.1   Nonparametric Sign Control Charts Based on Runs

Human et al. (2010) considered a class of nonparametric Shewhart-type control charts based on the sign statistic. The proposed charts utilize a single charting statistic and exploit at the same time runs-rules. The monitoring schemes use the information from multiple samples including the most recent one to reach a decision whether a signal will be triggered. This is achieved by taking into account a run of values of the charting statistic that fall outside the control limits. The resulting charts, which are similar in spirit to the distribution-free control charts established by Chakraborti and Eryilmaz (2007), seem to be user-friendly, while a high level of out-of-control performance can be achieved.

Let $X_{i1}, X_{i2}, \ldots, X_{in}$ denote a random sample of size $n > 1$ drawn at sampling stage $i = 1, 2, \ldots$ with cumulative distribution function F. The monitoring statistic of the proposed control chart is given as

$$T_i = \sum_{j=1}^{n} I_{\{X_{ij} > \theta_0\}},$$

where θ_0 is a specified percentile. Therefore, T_i corresponds to the number of observations of the ith sample which are larger than θ_0. Apparently the variable T_i follows the binomial distribution with parameters n and $p = P(X_{ij} > \theta_0)$. When the percentile of interest equals to its target value, namely $\theta = \theta_0$, the process is in-control and the probability of reaching a control decision equals

$$p_0 = P(X_{ij} > \theta_0 | \text{the process is in-control}).$$

The lower and the upper control limit of the proposed monitoring scheme established by Human et al. (2010) are determined as

$$LCL = a, UCL = n - b,$$

where the charting constants a and b are integers in the interval $(0, n)$ under the restriction that the LCL is smaller than UCL, i.e., $a + b < n$.

The proposed control charts consider a run of charting statistics that fall outside the control limits in order to decide whether an out-of-control signal should be produced or not. More precisely, Human et al. (2010) introduced the so-called k-of-k ($k \geq 2$) and the k-of-w ($w \geq k \geq 1, w \geq 2$) control charts based on the sign statistic. According to the k-of-k chart, an out-of-control signal is produced when k consecutive plotted points fall outside the control limits. The k-of-w control chart seems to be a natural generalization of the aforementioned scheme, since it produces a signal whenever k out of w charting statistics plot outside the control limits.

Human et al. (2010) studied the distribution of the run length of the proposed nonparametric control charts by applying an appropriate Markov chain approach. In order to derive explicit expressions for the computation of the *Average Run Length* or the *Standard Deviation of the Run Length*, each charting statistic T_i should be classified into one of two (three) categories for the one-sided (two-sided) control chart depending on whether T_i plots on or above the *UCL*, on or below the *LCL*, and/or between the *LCL* and *UCL*. In order to keep track of the aforementioned classification, a new sequence of random variables is defined and its distribution is used to assess the desired performance characteristics of the proposed monitoring schemes.

The authors compared their sign control charts based on runs, namely the 2-of-2 and 2-of-3 runs-rules charts to the competing nonparametric monitoring schemes introduced by Chakraborti and Eryilmaz (2007) and Amin et al. (1995) under the normal, the double exponential and the Cauchy distribution. According to their numerical results, the proposed sign control charts have substantially better out-of-control performance when compared to Amin et al. (1995), while it seems to compete well with the scheme established by Chakraborti and Eryilmaz (2007).

3.2   Improved Shewhart-Type Runs-Rules Nonparametric Sign Charts

Kritzinger et al. (2014) incorporated runs-ruless to the well-known nonparametric sign control chart introduced by Amin et al. (1995) in order to increase its capability to detect large shifts in the process. The in- and out-of-control run length distribution of the proposed control charts was studied by the aid of a Markov chain approach and several important performance characteristics were investigated in some detail.

The monitoring statistic used in the control chart is the classical sign statistic $T_i, i = 1, 2, 3, \ldots$ defined earlier. According to the proposed setup of the two-sided monitoring schemes by Kritzinger et al. (2014), four control limits and a center line are appropriately determined, namely $LCL_B, LCL_A, UCL_A, UCL_B$ and CL respectively which partition the control region into nine zones. A Markov chain approach is applied in order to study the distribution of the run length of the improved runs-rules sign control charts (for a thorough and well-documented relative text, see Fu and Lou (2003)). Appropriate transition probability matrices useful for obtaining the waiting time distributions associated with the proposed monitoring schemes are given, while explicit formulae for their several performance characteristics of them, such as the mean and the variance of the run length variable (for more details, the interested reader is referred to Sect. 5 in Kritzinger et al. (2014)). Needless to say that the two-sided control chart can be routinely modified to an upper (lower) one-sided improved sign control chart by simply dropping the control limits $LCL_A, LCL_B(UCL_A, UCL_B)$.

A numerical performance analysis carried out by the authors reveals that their monitoring schemes seem to be quite competitive under four different distributions (Normal, *Student t*, Exponential and Uniform) for the underlying process. In fact, the improved runs-rules sign charts seem to be more capable in detecting large process shifts, while maintaining the same sensitivity as the one acquired by the runs-rules sign control charts in detecting smaller shifts.

3.3 Shewhart-Type Sign Control Charts for Monitoring a Finite Horizon Process

Celano et al. (2016a, b, c) investigated the statistical performance of the nonparametric Shewhart-type sign control chart introduced by Amin et al. (1995) for monitoring the position of a continuous quality characteristic in a process with a finite production horizon.

According to the set up examined by them, we assume that a manufacturing process is scheduled to produce a finite number of N parts during a production horizon of finite length, say H hours. If I denotes the number of scheduled inspections within the production horizon H, then the sampling interval between two consecutive inspections equals $h = H/(I + 1)$ hours, since it is assumed that no inspection is scheduled at the end of the run. If we denote by $X_{i1}, X_{i2}, \ldots, X_{in}$ the subgroup of observations collected at inspection $i = 1, 2, \ldots, I$, the monitoring statistic of the Shewhart-type sign control chart is defined as

$$SN_i = \sum_{j=1}^{n} \text{sign}(X_{ij} - T_M),$$

where T_M is the target value, while

$$\text{sign}(x) = \begin{cases} 1, & \text{if } x > 0 \\ 0, & \text{if } x = 0 \\ -1, & \text{if } x < 0. \end{cases}$$

The in-control distribution of the variable SN_i can be readily deduced on observing that

$$SN_i = 2D_i - n,$$

where D_i enumerates the positive signs in the sequence $X_{ij} - T_M$, $j = 1, 2, \ldots, n$ and follows a binomial distribution with parameters n and $p_0 = P(X_{ij} > T_M | T_M = \theta_0) = 0.5$ for all $i = 1, 2, \ldots, I$; the parameter θ_0 appearing in the last expression stands for the in-control value of the process median.

The design of the nonparametric sign control chart calls for the determination of two parameters n and d; the second one denoting the minimum number of positive signs within a sample such that the control chart should trigger a signal. The *False Alarm Rate* of the sign control chart turns out to be

$$FAR = 1 - F_B\left(\frac{n+c}{2} - 1\middle|n, p_0\right) + F_B\left(\frac{n-c}{2} - 1\middle|n, p_0\right),$$

where $c = 2d - n$ and F_B denotes the cumulative distribution function of the binomial random variable with parameters n and $p_0 = 0.5$.

On the other hand, let us assume that the process shifts away from the target value and its out-of-control median becomes $\theta_1 = T_M + \delta\sigma_0$, where σ_0 is the in-control process standard deviation. Then, the out-of-control probability for an observation to take on value larger than the target equals $T_M\, p_\delta = P(X_{ij} > T_M|\delta) = 1 - F_X(T_M|\delta)$ and the alarm of the sign control chart will be given by the expression

$$AR = 1 - F_B\left(\frac{n+c}{2} - 1\middle|n, p_\delta\right) - F_B\left(\frac{n-c}{2} - 1\middle|n, p_\delta\right).$$

The truncated average run length and several other statistical performance metrics of the control charts have been investigated by Celano et al. (2016a, b, c) (see also Nenes and Tagaras (2010)), while both in- and out-of-control performance studies have been carried out for evaluating the sensitivity for detecting a location shift of the underlying process. The performance comparison with the Shewhart t control chart proposed by Celano et al. (2011) provided strong evidence that the out-of-control behavior of the Shewhart-type sign control chart is acceptable for a finite horizon production.

3.4 Shewhart-Type Sign Control Charts for Low-Volume Production

Celano et al. (2016a, b, c) considered the well-known Shewhart sign test statistic under the assumption that the population size is small. A new approach was proposed for extending its implementation to finite batch sizes of work to be produced. A pioneer work that deals with a control chart for a process with a finite horizon can be found in Ladany (1973), where an economic optimization problem for a Shewhart-type p-chart is discussed. Shewhart-type control charts for short runs have been introduced by Del Castillo and Montgomery (1993, 1996) and Tagaras (1996), who considered adaptive design parameters varying according to a Bayesian rule.

Celano et al. (2016a, b, c) implemented the Shewhart-type sign control charts on the online monitoring of finite populations by treating the sign statistic as a hypergeometric distributed random variable, when the population size is finite. The monitoring statistic remains the traditional sign statistic $SN_i = \sum_{j=1}^{n} \text{sign}(X_{ij} - T_M)$ defined earlier. Employing analogous arguments with those implemented by Celano et al. (2016a, b, c), similar outcomes for the computation of Type I and II probability error of the proposed statistical procedure were also derived by Celano et al. (2016a, b, c). A simple rule for selecting the design parameters and the number of inspections to be scheduled during a production run is suggested.

The numerical experimentation carried out by Celano et al. (2016a, b, c) depict the statistical properties of the Shewhart sign control chart for low-volume production. More precisely, the authors pointed out that their approach can be implemented for any size of the finite population such that a specific level of the *False Alarm Rate* during a production run is reached. For illustration purposes, the proposed monitoring scheme was applied for the production of mechanical parts undergoing a drilling operation which generates a hole with specific parameters to be positioned at a target location point.

3.5 Nonparametric Signed-Rank Control Charts with Variable Sampling Intervals

Coehlo et al. (2017) considered a nonparametric Shewhart-type control chart based on the variable sampling interval technique. The new chart utilized the well-known signed-rank statistic and was compared to fixed sampling interval signed-rank charts which have been already introduced in the literature. Since the signed-rank test has been proved to be more powerful than the sign test (see, e.g., Gibbons and Chakraborti (2010)), Coehlo et al. (2017) introduced a new nonparametric Variable Sampling Interval control chart on the basis of the Wilcoxon signed-rank statistic. The proposed monitoring scheme seems to be quite capable in detecting changes in a specified location parameter of the distribution of the underlying process.

The key idea for the so-called Variable Sampling Interval charts is to monitor the process by drawing samples at different time intervals, which are based upon where the most recent charting statistic plots on the control chart. For example, if it plots closer to the center line of the chart, then the time until the next test sample to be collected will be larger. Let us denote by $\mathbf{X}_i = (X_{i1}, X_{i2}, \ldots, X_{in})$ the random sample of size n which is drawn from the process at the ith time point. For the Variable Sampling Interval chart, test samples are taken at a finite number of intervals of length $d_1, d_2, \ldots, d_\eta$. In the special case $\eta = 2$, the lengths d_1, d_2 are chosen such that $l_1 < d_i < l_2$, where $l_1(l_2)$ is the minimum (maximum) possible interval length. Generally speaking, the region between the lower and the upper control limit is partitioned into η regions, say $I_1, I_2, \ldots I_\eta$ where d_j, $j = 1, 2, \ldots, \eta$ denotes the time unit drawing the next sample when the charting statistic plots in I_j. The time to signal on a Variable Sampling Interval chart depends on both the time to signal and the number of samples to signal. Thus, two popular performance measures, namely the average number of samples (ANOS) to signal and the average time to signal (ATS), are most commonly used.

Following the framework established by Coehlo et al. (2017), when the process is in-control, the median θ takes on a specified value θ_0. In addition, the observations drawn from the process are independent and follow an unknown but symmetric continuous distribution. Under these assumptions, the process median coincides to the corresponding mean and the Wilcoxon signed-rank statistic is distribution-free.

At each sampling instance i, the aforementioned monitoring statistic is defined as

$$W_i^+ = \sum_{j=1}^{n} I_{\{X_{ij}-\theta_0>0\}} R_{ij}^+, \quad i = 1, 2, \ldots,$$

where

$$R_{ij}^+ = 1 + \sum_{k=1}^{n} I_{\{|X_{ik}-\theta_0|-|X_{ij}-\theta_0|>0\}}.$$

Note that R_{ij}^+ corresponds to the so-called within-group absolute rank of the deviations $|X_{ik} - \theta_0|$. The lower and upper control limit of the resulting monitoring scheme are given as $LCL = \frac{n(n+1)}{2} - c, UCL = c$ respectively, where c is a positive integer appropriately chosen so that a pre-specified in-control performance is achieved. Coehlo et al. (2017) followed an analogous procedure as the one exploited by Amin and Widmaier (1999) and partitioned the set of distinct value by the monitoring statistic into the following two time intervals:

$$I_1 = \left\{ \frac{n(n+1)}{2} - c, \ldots, \frac{n(n+1)}{2} - k - 1 \right\} \cup \{k+1, \ldots, c\}, I_2 = \left\{ \frac{n(n+1)}{2} - k, \ldots, k \right\}.$$

If the value of the monitoring statistic $W_i^+ \in I_1 (W_i^+ \in I_2)$, then we should wait $d_1 (d_2 > d_1)$ time units before collecting the next test sample. The proposed control chart will produce a signal whenever a plotted point is located inside the region $\{0, \ldots, n - c - 1\} \cup \{c + 1, \ldots, n\}$. The probability for the monitoring scheme to produce a signal while the process is in-control and out-of-control is given by

$$a_0 = P\left[\left(W_i^+ < \frac{n(n+1)}{2} - c \right) \cup (W_i^+ > c) | \theta = \theta_0 \right]$$

and

$$a_1 = P[(W_i^+ < n - c) \cup (W_i^+ > c) | \theta = \theta_1]$$

respectively.

In order to compare the performance of the proposed signed-rank control to other competitive schemes, Coehlo et al. (2017) considered four different symmetric distributions (standard normal, Uniform (0, 1), Laplace (0, 1) and t distribution). The numerical results spoke in favor of implementing the Variable Sampling Intervals signed-rank control chart when data follow an unknown symmetric distribution.

3.6 Sign Control Charts Based on Ranked Set Sampling

McIntyre (1952) introduced the so-called ranked set sampling (*RSS*) scheme for collecting data when observing the actual measurements is difficult while ranking a set of sample units is relatively easier and reliable. According to the *RSS* procedure, at step 1 we collect k samples of size k, say $X_{1,i}, X_{2,i}, \ldots, X_{k,i}, i = 1, 2, \ldots, k$, while at step 2 each sample is arranged in ascending order. Finally, the rth smallest observation from the rth sample is selected for all $r = 1, 2, \ldots, k$. The aforementioned steps can be repeated m times (cycles) in order to produce a balanced *RSS* of size $N = mk$. Asghari et al. (2018) introduced a nonparametric control chart based on *RSS* and the sign statistic which is advantageous for the monitoring of a process whose units are difficult and/or costly to measure.

Let us denote by $X_{(r:k)i}$ the rth order statistic from the rth sample in the ith cycle, so that $\{X_{(r:k)i}^t, r = 1, 2, \ldots, k; i = 1, 2, \ldots, m\}$ corresponds to a *RSS* scheme of size $N = mk$ from the process with in-control median θ_0 at time t. The charting statistic of the proposed monitoring scheme is given by

$$SN_{RSS}^t = \sum_{r=1}^{k} \sum_{i=1}^{m} \text{sign}(X_{(r:k)i}^t - \theta_0), \quad t = 1, 2, \ldots$$

or equivalently

$$SN_{RSS}^t = 2Y_{RSS}^t - mk,$$

where

$$Y_{RSS}^t = \sum_{r=1}^{k} \sum_{i=1}^{m} I_{\{X_{(r:k)i}^t - \theta_0 > 0\}} = \sum_{r=1}^{k} Y_{m(r)}.$$

Hettmansperger (1995) provided the mean and variance of the variable Y_{RSS}^t, by exploiting the fact that the random variable $Y_{m(r)}$ follows the binomial distribution with parameters m and p_r where

$$p_r = \sum_{l=0}^{r-1} \binom{k}{l} [F(\theta - \theta_0)]^l [1 - F(\theta - \theta_0)]^{k-l}, \quad r = 1, 2, \ldots, k,$$

while Koti and Jogeph Babu (1996) proved that its probability mass function can be expressed as

$$P(Y_{RSS}^t = y) = \sum \prod_{r=1}^{k} P(Y_{m(r)} = i_r),$$

where the summation is carried over all $(i_1, i_2, \ldots, i_r)$ satisfying the conditions $\sum_{r=1}^{k} i_r = y; 0 \leq i_r \leq m, r = 1, 2, \ldots, k$. In addition, Barabesi (1998) derived the exact distribution of the random variable Y_{RSS}^t by a generating function approach. Based on the fact that the in-control distribution of the statistic Y_{RSS}^t is symmetric about zero, the control limits of the two-sided and one-sided RSS monitoring scheme can be determined by using the following equation:

$$P_{\theta_0}(SN_{RSS}^t \geq UCL) = P_{\theta_0}\left(Y_{RSS}^t \geq \frac{N + UCL}{2}\right).$$

The performance of the above scheme was investigated, under three different distributions, by Asghari et al. (2018). Their numerical results depict the superiority of the proposed method versus the simple random sampling technique.

4 Distribution-Free Control Charts Based on Ranks

In this section, we focus on distribution-free monitoring schemes, which use rank-based monitoring statistics. One of the early works on the topic is attributed to Altukife (2003) who proposed nonparametric control charts based on the so-called sum-of-rank test for detecting effectively small shifts from the process mean especially when a heavy tailed distribution is considered.

4.1 Nonparametric Control Charts Based on the Wilcoxon-Type Rank-Sum Statistic

Balakrishnan et al. (2009) introduced three distribution-free Shewhart-type control charts that exploit run and Wilcoxon-type rank-sum statistics to detect possible shifts of the underlying continuous process. In this subsection, we shall discuss only the one which utilizes ranks for constructing its monitoring statistic.

Let us then denote by $X_1, X_2, \ldots, X_m$ a random sample of size m from the in-control distribution F and assume that two specific order statistics are used as control limits, namely $LCL = X_{a:m}, UCL = X_{b:m} (1 \leq a < b \leq m)$. Suppose next that test samples $Y_1, Y_2, \ldots, Y_n$ are drawn independently of each other (and also independently of the reference sample) and that we are interested in checking whether the process is still in-control or not. In statistical terms, if G denotes the cumulative distribution function of the Y_i's, the aim is to test the hypothesis $F = G$. Apparently, under the hypothesis $F = G$, the number of test sample observations Y_j that fall between successive X-observations should not attain "extreme" values, with extremes being determined based on the proportion n/m. With this in mind, Balakrishnan et al.

(2009) proposed to use as a monitoring statistic for deciding whether the process is in-control the sum of ranks (from the joint sample of X and Y-*observations*) of the Y-observations that lie between the control limits. The aforementioned statistic may be expressed through the so-called "exceedance statistics" whose distributional properties have been discussed by a number of authors including Fligner and Wolfe (1976) and Randles and Wolfe (1979); an elaborate discussion on this topic has been provided in the monograph by Balakrishnan and Ng (2006). More specifically, let us denote by M_i, $i = 1, 2,..., m$, the number of test sample observations Y_j that fall between the $(i$-$1)$th and ith order statistics of the X-sample (with the convention that $X_{(0)} = -\infty$). The Wilcoxon-type rank-sum statistic mentioned above can be expressed as

$$W = \sum_{i=a+1}^{b} W_i,$$

where W_i denotes the sum of the ranks of the Y-observations falling between $X_{(i-1)}$ and $X_{(i)}$. (see, e.g., Wilcoxon (1945)). The monitoring statistic W can be expressed in terms of M_i, as follows

$$W = \frac{1}{2}\left(\sum_{i=a+1}^{b} M_i\right)^2 + \sum_{i=a+1}^{b} i M_i + \left(M_0 + a - \frac{3}{2}\right) \sum_{i=a+1}^{b} M_i,$$

where $M_0 = \sum_{i=1}^{a} M_i$ denotes the number of observations of the Y-sample before the LCL. The process is declared to be in-control if the following two conditions hold true:

$$W \leq w \quad \text{and} \quad M_0 \leq r_0,$$

with w, r_0 being the design parameters of this chart.

Balakrishnan et al. (2009) indicated that the support of W is

$$R_W = \left\{0, 1, \ldots, \frac{n(n + 2b - 1)}{2}\right\}$$

and derived the distribution of it by making use of the joint distribution of $(M_0, M_{a+1}, M_{a+2}, \ldots, M_b)$. More precisely, the distribution of the rank-sum statistic W can be expressed as

$$P(W = w) = \sum_{m_0, m_{a+1}, m_b} p_{a,b}(m_0, m_{a+1}, \ldots, m_b), \quad w \in R_W,$$

where

$$p_{a,b}(m_0, m_{a+1}, m_{a+2}, \ldots, m_b) = \frac{\binom{m_0 + a - 1}{a - 1} \binom{m + n - m_0 - \sum\limits_{i=a+1}^{b} m_i - b}{m - r}}{\binom{m + n}{n}},$$

while the summation is carried over all nonnegative integers $m_0, m_{a+1}, \ldots, m_b$ satisfying the conditions

$$m_0 + \sum_{i=a+1}^{b} m_i \leq n \text{ and } \frac{1}{2}\left(\sum_{i=a+1}^{b} m_i\right)^2 + \sum_{i=a+1}^{b} im_i + (m_0 + a - \frac{3}{2}) \sum_{i=a+1}^{b} m_i = w.$$

It is straightforward that the *False Alarm Rate* of the proposed monitoring scheme is given by

$$FAR = 1 - P(W \leq w \text{ and } M_0 \leq r_0).$$

Since the signaling events are not independent, the average run length (*ARL*) of the chart cannot be calculated as the reciprocal of the signaling probability. However, Balakrishnan et al. (2009) made use of an appropriate conditioning argument (see also Chakraborti (2000)) in order to establish a formula for the exact run length distribution and its mean. More specifically, if we denote by $f_{a:b}$ the joint density function of the uniform order statistics $(U_a, U_{a+1}, \ldots, U_b)$ from a random sample of size m (see, e.g., David and Nagaraja (2003)), the *Average Run Length* of the proposed Wilcoxon-type monitoring scheme can be expressed as

$$ARL = \iint \cdots \int_{0 \leq u_a \leq u_{a+1} \leq \ldots \leq u_b \leq 1} \frac{1}{1 - Q(GF^{-1}(u_a), GF^{-1}(u_{a+1}), \ldots, GF^{-1}(u_b))}$$
$$\times f_{a:b}(u_a, u_{a+1}, \ldots, u_b) du_a du_{a+1} \cdots du_b,$$

where

$$Q(v_a, v_{a+1}, \ldots, v_b) = \sum_{(m_0, m_{a+1}, \ldots, m_b) \in A} q(v_a, v_{a+1}, \ldots, v_b), \quad A \subseteq \{0, 1, 2, \ldots\}^{b-a},$$

and

$$q(v_a, v_{a+1}, \ldots, v_b) =$$

$$\frac{n!}{m_0! \left(\prod\limits_{j=a+1}^{b} m_j!\right)\left(n - m_0 - \sum\limits_{j=a+1}^{b} m_j\right)!} v_a^{m_0} \prod_{j=a+1}^{b} (v_j - v_{j-1})^{m_j} (1 - v_b)^{n - m_0 - \sum\limits_{j=a+1}^{b} m_j}$$

is the multinomial probability, with $0 \leq v_a \leq v_{a+1} \leq \ldots \leq v_b \leq 1$.

The *Average Run Length* can be evaluated numerically for any choice of the design parameters under both in-control and specific out-of-control situations. For more details, the interested reader is referred to Sects. 5 and 6 in Balakrishnan et al. (2009).

4.2 A Nonparametric Control Chart Based on the Mood Statistic

Murakami and Matsuki (2010) proposed charting statistics based on the two-sample Mood statistic for detecting scale shifts of a distribution. They studied in detail the new monitoring scheme, while in the case of small sample sizes, they derived exact control limits for the suggested scheme.

Suppose that a reference sample of size m, say $X_1, X_2, \ldots, X_m$ is drawn from the in-control process, while an arbitrary test sample $Y_1, Y_2, \ldots, Y_n$ is also available. Let $R_1 < R_2 < \cdots < R_m$ denote the combined-samples ranks of the X −observations in ascending order. The monitoring statistic of the proposed control chart can be expressed as

$$M_{m,n} = \sum_{i=1}^{m} \left(R_i - \frac{N+1}{2} \right)^2,$$

where $N = m + n$ and the Mood chart produces a signal whenever the value of the statistic $M_{m,n}$ plots outside the control limits (L_{mn}, U_{mn}). For large sample sizes, Murakami and Matsuki (2010) stated that the lower and upper control limit of the proposed chart can be evaluated through the following formulae:

$$L_{mn} = E(M_{m,n}) - c\sqrt{\text{Var}(M_{m,n})}, \quad U_{mn} = E(M_{m,n}) + c\sqrt{\text{Var}(M_{m,n})},$$

where the parameter c is appropriately selected in order to achieve a specified level of performance, while

$$E(M_{m,n}) = \frac{m(N^2 - 1)}{12}, \ \text{Var}(M_{m,n}) = \frac{mn(N+1)(N^2 - 4)}{180}.$$

In addition, Murakami and Matsuki (2010) applied the well-known Lugannani-Rice formula (see also Jensen (1995) or Wood et al. (1993)) for the upper-tail probability for intermediate sample sizes in order to derive a saddlepoint approximation of the cumulative distribution function of the random variable $M_{m,n}$. Based on the aforementioned approximation, Murakami and Matsuki (2010) provided a table of the estimated control limits of the Mood charting statistic for a large range of the design parameters. They also carried out a simulation study, using several distributions such as the lognormal, the logistic or the Cauchy distribution and concluded that the normal approximation is quite accurate only for the case $m, n \geq 150$.

4.3 A Shewhart-Lepage Control Chart for the Joint Monitoring of Location and Scale

Mukherjee and Chakraborti (2012) established a single distribution-free control chart for monitoring simultaneously the unknown location and scale parameters of a continuous distribution of the underlying process. The proposed plotted statistic SL combines two well-known nonparametric test statistics: the Wilcoxon rank-sum test for location and the Ansari–Bradley test for scale. The test statistic SL has been introduced by Lepage (1971) in order to deal with a nonparametric two-sample test for both location and dispersion.

Let $X_1, X_2, \ldots, X_m$ be a random sample of size m from the in-control distribution F and assume that test samples $Y_{i1}, Y_{i2}, \ldots, Y_{in}, i = 1, 2, \ldots$ are drawn independently of each other (and also independently of the reference sample) from a continuous distribution G. For the comparison between the ith test sample and the reference sample, two different nonparametric testing procedures are followed. According to the first one, the well-known Wilcoxon rank-sum statistic, say T_1, should be computed for testing the equality of the corresponding location parameters. If we denote by Z_k an indicator variable which takes on the value 1(0) when the kth order statistic of the combined N ($N = m + n$) observations belongs to the test (reference) sample, then the test statistic T_1 is defined as

$$T_1 = \sum_{k=1}^{N} k Z_k.$$

The second statistic, the Ansari-Bradley test statistic, quantifies the differences in scale between the $X-$ and $Y-$ observations and is defined as follows:

$$T_2 = \sum_{k=1}^{N} |k - 0.5(N + 1)| Z_k.$$

Note that formulae for the computation of the mean and the variance of both test statistics T_1, T_2 have already appeared in the literature (see, e.g., Gibbons and Chakraborti (2010)). More precisely, we have

$$E(T_1 | IC) = \mu_1 = 0.5n(N + 1), \quad \text{Var}(T_1 | IC) = \sigma_1^2 = \frac{1}{12} mn(N + 1),$$

$$E(T_2|IC) = \mu_2 = \begin{cases} \frac{nN}{4}, & \text{if } N \text{ is even} \\ \frac{n(N^2-1)}{4N}, & \text{if } N \text{ is odd} \end{cases},$$

$$\text{Var} E(T_2|IC) = \sigma_2^2 = \begin{cases} \frac{mn(N^2-4)}{48(N-1)}, & \text{if } N \text{ is even} \\ \frac{mn(N+1)(N^2+3)}{48N^2}, & \text{if } N \text{ is odd}. \end{cases}$$

The single (combined) monitoring statistic of the proposed Shewhart-Lepage control chart for the ith test sample is given by

$$SL_i = SW_i^2 + SAB_i^2, i = 1, 2, \ldots,$$

where SW_i, SAB_i are the standardized Wilcoxon rank-sum and Ansari-Bradley statistics, i.e.,

$$SW_i = \frac{T_{1i} - \mu_1}{\sigma_1} \quad \text{and} \quad SAB_i = \frac{T_{2i} - \mu_2}{\sigma_2}.$$

The proposed chart produces an out-of-control signal whenever the statistic SL_i exceeds an upper control limit H. Note that since $SL_i \geq 0$, the lower control limit equals to zero. When the process is declared to be out-of-control at the ith test sample, the values of both statistics SW_i^2, SAB_i^2 are compared to specified constants $H_1, H_2 < H$ respectively in order to identify the type of violation. In other words, if only $SW_i^2(SAB_i^2)$ exceeds $H_1(H_2)$, then a shift in location (scale) is implied. On the other hand, if SW_i^2 exceeds H_1 and SAB_i^2 exceeds H_2, a shift in both location and scale is indicated.

Mukherjee and Chakraborti (2012) provided a thorough investigation of the in- and out-of-control performance of the control chart in terms of the mean, standard deviation, median and further percentiles of the run length distribution. Some interesting remarks referring to the influence of the reference sample size have also been included in their study.

4.4 A Shewhart–Cucconi Control Chart for the Joint Monitoring of Location and Scale

Chowdhury et al. (2014) proposed a distribution-free Shewhart-type control chart based on the Cucconi statistic which achieves joint monitoring of both the location and scale parameters of the underlying process. Several nonparametric tests for jointly monitoring location and scale parameters are based on the combination of two separate statistical tests, one for location and one for scale (see Sect. 4.3). The approach suggested by Cucconi (1968) addresses the location-scale problem by considering the squares of ranks and contrary ranks (see also Marozzi (2009)).

If we denote by Z_k an indicator variable which takes on the value 1 (0) whenever the kth order statistic of the combined sample of $N = m + n$ observations comes from the reference (test) sample, then Cucconi (1968) suggested using again the statistic T_1 exploited in Sect. 4.3 and an additional statistic (Wilcoxon-type rank-sum statistic) defined as

$$S_1 = \sum_{k=1}^{N} k^2 Z_k.$$

The statistic S_1 represents the sum of the squares of the ranks of test sample observations in the combined sample. Finally, the next variable is defined (see also Mood (1954))

$$S_2 = \sum_{k=1}^{N} (N + 1 - k)^2 I_k = n(N + 1)^2 - 2(N + 1)T_1 + S_1.$$

Under the assumption $F = G$, one may prove that

$$E(S_1|IC) = E(S_2|IC) = \frac{n(N + 1)(2N + 1)}{6}$$

and

$$\mathrm{Var}(S_1|IC) = \mathrm{Var}(S_2|IC) = \frac{mn(N + 1)(2N + 1)(8N + 11)}{180}.$$

Therefore, the corresponding standardized statistics will read

$$U = \frac{S_1 - E(S_1|IC)}{\sqrt{\mathrm{Var}(S_1|IC)}} = \frac{6S_1 - n(N + 1)(2N + 1)}{\sqrt{mn(N + 1)(2N + 1)(8N + 11)/5}}$$
$$V = \frac{S_2 - E(S_2|IC)}{\sqrt{\mathrm{Var}(S_2|IC)}} = \frac{6S_2 - n(N + 1)(2N + 1)}{\sqrt{mn(N + 1)(2N + 1)(8N + 11)/5}}.$$

Marozzi (2009) argued that when $F = G$, the correlation coefficient between the variables U and V is given as

$$p = \mathrm{Corr}(U, V|IC) = \frac{2(N^2 - 4)}{(2N + 1)(8N + 11)} - 1.$$

The monitoring statistic of the proposed Shewhart–Cucconi control chart for simultaneous testing of shift in location and scale parameters is defined as

$$C(U, V) = \frac{U^2 + V^2 - 2pUV}{2(1 - p^2)}.$$

Manifestly $\min C(u, v) = 0$ while the equation $C(u, v) = \text{constant}$ represents an ellipse. The proposed chart produces an out-of-control signal whenever the statistic C exceeds an upper control limit $H > 0$. When the process is declared to be out-of-control, the p-values p_1, p_2 for both testing procedures (Wilcoxon test for location and Mood test for scale) are computed. If p_1 is very low but not p_2, there exists evidence for a shift in the location only. On the other hand, if p_1 is relatively high but p_2 is low, only a shift in scale is suspected. Finally, if both p-values are very low, then an evidence for shift in both location and scale is provided.

Marozzi (2009) supplied control limits for the Shewhart–Cucconi control scheme for some typical nominal in-control average run length values. He also proceeded to several numerical comparisons with existing nonparametric control charts, which revealed that the proposed monitoring scheme performs competitively.

4.5 Nonparametric Control Charts Based on the Hogg-Fisher-Randle and the Savage Rank Statistic

Mukherjee and Sen (2015) proposed two nonparametric control charts based on the Hogg-Fisher-Randle statistic and the Savage rank statistic. Suppose that a reference sample $X_1, X_2, \ldots, X_m$ from the in-control distribution F is available a priori and a test sample (or Phase II observations) of size n is sequentially observed from the underlying process with cumulative distribution function G. In the classical nonparametric inference, there have been considered linear test statistics of the form

$$T = \sum_{j=1}^{n} a_N(R_j),$$

where $a_N(R_j)$ denotes the so-called score function of ranks (see Hogg et al. (1975) or Kössler (2010)). Gastwirth (1965) floated the idea of sacrificing the ranks of some test sample observations; more specifically he suggested using a sum of partial ranks starting from or ending at a certain pre-specified percentile. Hogg et al. (1975) proposed the following score function for detecting possible shift in location parameter when the underlying density is right-skewed:

$$a_N(R_j) = \begin{cases} R_j, & \text{if } j = 1, 2, \ldots, n \text{ and } R_j \leq [(N+1)/2] \\ 0, & \text{otherwise} \end{cases},$$

where $[z]$ denotes the greatest integer less than or equal to z. Then he introduced a test statistic T_H obtained by summing up the ranks of the test sample observations, which lie below the combined sample median. Generally speaking, the Gastwirth-type score function for right-skewed density is defined as

$$a_N(R_j) = \begin{cases} R_j, & \text{if } j = 1, 2, \ldots, n \text{ and } R_j \leq K, [(N+1)/2] \leq K < N \\ 0, & \text{otherwise} \end{cases}$$

An empirical study showed that K closer to $[(3m+1)/4]$ gives satisfactory in-control performance

Mukherjee and Sen (2015) used Gastwirth-type score functions to establish a monitoring statistic of the form

$$T_G = \sum_{j:R_j \leq K}^{n} R_j,$$

thereof obtaining a Hogg-Fisher-Randle control chart. On the other hand, a Savage test statistic is introduced as

$$T_S = \sum_{j=1}^{n} \left(1 - \sum_{l=R_j}^{N} \frac{1}{l} \right).$$

Then, the corresponding standardized statistics denoted by T_{HFR}, T_{SAV} can be defined via the following equations:

$$T_{HFR} = \frac{T_H - \frac{nK(K+1)}{2N}}{\sqrt{\frac{mn(K+1)\{-3K^2+(4N-3)K+2N\}}{12(N-1)N^2}}}$$

$$T_{SAV} = \frac{T_S}{\sqrt{\frac{mn}{N-1}\left(1 - \frac{\sum_{l=1}^{N} \frac{1}{l}}{N} \right)}},$$

respectively.

The proposed control charts produce an out-of-control signal whenever the corresponding monitoring statistic T_{HFR} and T_{SAV} exceed an upper control limit H_{HFR} and H_{SAV} respectively. Mukherjee and Sen (2015) compared the aforementioned schemes to well-known rank-based control charts, which have already been established in the literature. A simulation-based analysis revealed that the new charts are quite capable in detecting location shift especially when the underlying process distribution is non-normal, while for small reference sample, the corresponding false alarm rate seems to be quite competitive. In addition, its in-control robustness was investigated, while detailed comparisons in terms of out-of-control run length under various amounts of location shift were provided. The Hogg-Fisher-Randle chart appears to be an excellent alternative to the traditional Wilcoxon chart, as it is usually less influenced by bias in the Phase I sample and also has lower *False Alarm Rate* than its competitors in most cases.

4.6 Distribution-Free Phase II Mann–Whitney Control Charts with Runs-Rules

Malela-Majika et al. (2016) considered adding runs-rules to enhance the performance of the distribution-free Phase II Shewhart-type chart based on the well-known Mann–Whitney statistic which was introduced by Chakraborti and van de Wiel (2008). Assume that a reference sample $X_1, X_2, \ldots, X_m$ is drawn from the in-control distribution F and test samples $Y_1, Y_2, \ldots, Y_n$ are sequentially observed from the underlying process with cumulative distribution function G. The monitoring statistic of the proposed Mann-Whitney control chart can be expressed as

$$M_{XY} = \sum_{i=1}^{m} \sum_{j=1}^{n} I_{\{Y_j - X_i > 0\}}.$$

When implementing the classical 1-*of*-1 Mann-Whitney control chart, the monitoring statistic M_{XY} is plotted for each test sample and an out-of-control signal is produced whenever the plotted point falls outside the control limits (LCL, UCL). Since the in-control distribution of the charting statistic is symmetric about $mn/2$, it is reasonable to set $LCL = mn - UCL$.

The unconditional *Average Run Length* of the Mann–Whitney control chart can be expressed as an m-dimensional integral. More specifically, conditioning on the observed values $\mathbf{x} = (x_1, x_2, \ldots, x_m)$ of the reference sample, the probability that a test sample leads to an out-of-control signal is given as

$$p_G(\mathbf{x}) = P_G(M_{XY} \leq mn - UCL) + P_G(M_{XY} \geq UCL).$$

Consequently, the unconditional *Average Run Length* of the aforementioned monitoring scheme may be expressed as

$$ARL = \int_{-\infty}^{+\infty} \int_{-\infty}^{+\infty} \ldots \int_{-\infty}^{+\infty} \frac{1}{p_G(\mathbf{x})} dF(x_1) dF(x_2) \ldots dF(x_m).$$

Malela-Majika et al. (2016) stated that it is difficult and time-consuming to compute the *ARL* based on the abovementioned formula. Instead, a Monte-Carlo simulation study with 10000 simulations was carried out for evaluating the in- and out-of-control performance of the Mann–Whitney control chart. Malela-Majika et al. (2016) considered the 2-of-2 KL runs-rule (see Klein (2000)) and the improved runs-rule established by Khoo and Ariffin (2006) which is a combination of the classical 1-*of*-1 and the 2-of-2 runs-rules. The simulation procedure for determining the chart constants and obtaining the characteristics of the run length distribution for the control chart with runs-rules is described in detail by Malela-Majika et al. (2016).

The numerical comparisons reveal that the enhanced Mann-Whitney control charts outperform other competing schemes in detecting possible shifts of the underlying distribution.

4.7 Economically Designed Nonparametric Control Chart Based on the Wilcoxon Rank-sum Statistic

Li et al. (2016) dealt with the problem of setting up an economical design for monitoring an unknown location parameter via a nonparametric control chart that makes use of the Wilcoxon rank-sum statistic. Duncan (1956) seems to be the first to coin the term economic design for a control chart in a seminal paper. The major objective of an economic design is the maximization of the profit or the minimization of the cost incurred in the course of process monitoring. Since then, the area has attracted a lot of research interest. For example, Celano et al. (2012) or Zhang et al. (2014) considered economically designed schemes for short production runs, while Yeh et al. (2011) and Su et al. (2014) developed economic models under non-normality. The technique practiced by Li et al. (2016) aims at blending the advantage of economic design with the benefit of the nonparametric approach for process monitoring.

The rationale behind the monitoring scheme has been already presented in detail in Sect. 4.3 of the present chapter. In the economic model proposed by Li et al. (2016), it is assumed that the process starts in the in-control state and the quality characteristic of interest follows an unknown univariate continuous distribution. The aim is to detect the occurrence of a single assignable cause that leads to a fixed shift in the (unknown) location parameter. The time until the assignable cause occurs is assumed to be exponentially distributed with rate λ. The process is not self-correcting, that is, once the process shifts out-of-control, it remains at this state until the assignable cause is detected and its effect is removed. Following standard practice as described in Duncan (1956) or Montgomery (2009), the expected loss per unit time $E(L)$ is given by the expression

$$E(L) = \frac{(a_1 + a_2 n)}{h} + a_4 - \frac{\frac{a_4}{\lambda} - a_3 - \frac{a'_3 ae^{-\lambda h}}{1 - e^{-\lambda h}}}{\frac{1}{\lambda} + \frac{h}{1 - \beta} - \xi + gn + D},$$

where h denotes the sampling interval, a (β) represents the probability of Type I (II) error at each sample, a_1 (a_2) corresponds the fixed (variable) cost of sampling, a_3 and a'_3 account for the cost of finding the assignable cause and investigating a false alarm respectively, while a_4 denotes the penalty cost per unit time associated with the out-of-control production. The parameter ξ stands for the expected time between the last sampling and the shift while in-control and the symbols g, D denote respectively the time to sample and chart one item and the time to spot out the assignable cause after an action signal is triggered. For a modified version of this loss function taking into consideration different process features, the interested reader is referred to Lorenzen and Vance (1986).

The optimal control chart design refers to the problem of minimizing $E(L)$. To achieve that, an additional cycle of observing m reference observations from the in-control distribution is required. The *False Alarm Rate* of the two-sided Wilcoxon rank-sum control chart can be obtained via enumeration or using normal approximation. For large sample sizes, one may apply the following approximation:

$$FAR = \Phi\left(\frac{LCL + 0.5 - \frac{n(m+n+1)}{2}}{\sqrt{\frac{mn(m+n+1)}{12}}}\right) + 1 - \Phi\left(\frac{UCL + 0.5 - \frac{n(m+n+1)}{2}}{\sqrt{\frac{mn(m+n+1)}{12}}}\right),$$

where $\Phi(\cdot)$ denotes the cumulative distribution function of the standard normal distribution. On the other hand, the exact distribution of the monitoring statistic when a shift has occurred, depends on the underlying unknown process distribution G. Li et al. (2016) proposed two different approaches for calculating the probability β of the monitoring scheme. According to the first one, a bootstrapped sample of size n without replacement should be drawn from the m reference observations, while a target shift δ to each of the n observations is added (see, also Chatterjee and Qiu (2009), Capizzi and Masarotto (2009) or Abbasi and Guillen (2013)). Afterwards, the Wilcoxon rank-sum statistic is computed based on the m reference observations, while n new observations obtained by adding target shift δ. The last step is repeated 10000 times and therefore 10000 replicates of Wilcoxon rank-sum statistic are computed for the m reference observations and n shifted data. The desired probability β can now be empirically estimated based on the 10000 bootstrap from the sampling distribution of the Wilcoxon rank-sum statistic for the underlying shift. The second approach presented by Li et al. (2016) for evaluating the probability β is based on the asymptotic normality of the aforementioned test statistic when the sample sizes m and n are large.

The monitoring scheme with control limits $LCL = -l$, $UCL = l$ suggested by Li et al. (2016) deploys in two stages. The first one, namely the Design stage, calls for the collection of the reference sample of size m and the computation of probability a from the exact in-control distribution of the Wilcoxon rank-sum statistic as a function of n and l, for $n = 1, 2, ...$ Afterwards, the probability β can be estimated by applying one of the two proposed approaches mentioned earlier. For the two-sided Wilcoxon rank-sum control chart the valid range of control limits requires a minimum (or maximum) value equal to $n(n + 1)/2$ (or $n(2m + n + 1)/2$). Consequently, the optimization calls for a grid search for values of n and l and a golden section search for h in order to minimize the objective function $E(L)$. Finally, a global minimum of $E(L)$ is achieved and once an optimal charting scheme (n, h, l) is determined, the procedure moves to the so-called charting stage. Please mention that the charting stage includes the classical steps of process monitoring, namely the computation of the proposed statistic for each test sample drawn from the process and the characterization of the process as in- or out-of-control, depending on the location of the observed points in comparison with the corresponding control limits. A detailed simulation study carried out by Li et al. (2016) revealed that the performance of the procedure based on the bootstrapping is strongly encouraging and robust for several continuous distributions.

4.8 Distribution-Free Lepage-Type Circular-Grid Charts for Joint Monitoring of Location and Scale Parameters

Mukherjee and Marozzi (2017) introduced a graphical tool, named circular-grid charts, for simultaneously monitoring the location and scale of a process via the Lepage-type statistics. The well-known Lepage test for jointly testing the location and scale differences between two samples has already been presented in Sect. 4.3. Tamura (1963) proposed a generalization of the classical Wilcoxon rank-sum statistic $T_1 = \sum_{k=1}^{N} k Z_k$ by considering the quantity

$$S_p = \sum_{k=1}^{N} k^p Z_k$$

for $p > 0$.

In addition, aiming at the comparison of the scale parameters of the samples, Tamura (1963) introduced the following class of statistics:

$$M_q = \sum_{k=1}^{N} \left| k - \frac{N+1}{2} \right|^q Z_k, \quad q > 0.$$

Note that S_1 reduces to the classical Wilcoxon rank-sum statistic, while M_1 and M_2 are the statistics employed in the Ansari-Bradley and Mood nonparametric scale tests respectively. Following the technique described in Sect. 4.3 it seems sensible to construct a Lepage-type control chart which makes use of the sum of squares

$$L_{p,q} = \left(\frac{S_p - E_{H_0}(S_p)}{\sqrt{\mathrm{Var}_{H_0}(S_p)}} \right)^2 + \left(\frac{M_q - E_{H_0}(M_q)}{\sqrt{\mathrm{Var}_{H_0}(M_q)}} \right)^2.$$

Manifestly the special case $p = q = 1$, yields the traditional Lepage test. It is worth mentioning that several modifications of the Lepage test have been proposed in the literature (see, e.g., Büning and Thadewald (2000), Murakami (2007) or Neuhäuser (2011)).

After collecting the reference sample of size m, namely $X_1, X_2, \ldots, X_m$, several test samples of size n, say $Y_{i1}, Y_{i2}, \ldots, Y_{in}, i = 1, 2, \ldots$ are drawn from the process. For each test sample the standardized statistics are computed as follows:

$$S_{pi}^* = \frac{S_{pi} - E_{H_0}(S_p)}{\sqrt{\mathrm{Var}_{H_0}(S_p)}} \quad \text{and} \quad M_{qi}^* = \frac{M_{qi} - E_{H_0}(M_q)}{\sqrt{\mathrm{Var}_{H_0}(S_q)}},$$

where the quantities $E_{H_0}(S_1), \mathrm{Var}_{H_0}(S_1), E_{H_0}(M_1), \mathrm{Var}_{H_0}(M_1)$ can be readily obtained by recalling the equations appeared in Sect. 4.3 for the special case $p = q = 1$. In addition, the following ensue:

$$E_{H_0}(M_2) = \frac{n(N^2 - 1)}{12}, \quad \mathrm{Var}_{H_0}(M_2) = \frac{mn(N + 1)(N^2 - 4)}{180}.$$

The testing procedure for the control chart suggested by Tamura (1963) is as follows. We set up a two-dimensional chart where the $X-$ axis stands for the S_p^* values while the $Y-$ axis stands for the M_q^*. We draw a circle with radius r, centered at the origin and a square with its sides being placed at distance $\pm r$ away from the two axes. The area inside the circle is considered to be the in-control region of the proposed control chart. Note that the parameter r is a design parameter and should be appropriately determined so as to achieve a specified in-control performance.

According to the previous description, the proposed Shewhart-type Lepage monitoring scheme is based on plotted points (S_{pi}^*, M_{qi}^*), while the monitoring statistic of the new chart is simply the square of the Euclidean distance of each point (S_{pi}^*, M_{qi}^*) from the origin $(0, 0)$. The process is declared to be in-control (out-of-control) whenever the plotted point (S_{pi}^*, M_{qi}^*) falls inside (outside) the circle. If the control chart signals at the ith step, an interesting question is whether the shift has occurred in the location and/or the scale parameter of the process. To this end, the circular area and the square described earlier can be proved to be very useful. Mukherjee and Marozzi (2017) considered two different out-of-control situations depending on the location of the plotted point. Under the first case, the point lies outside both the circle and the square, while the second scenario is activated whenever a point lies outside both the circle but inside the square. In all cases considered, a numerical comparison between the values of the statistics (S_{pi}^*, M_{qi}^*) and the parameter r, leads to a conclusion about the kind of shift which has occurred in the distribution process.

A detailed numerical study based on Monte-Carlo simulations for the performance of the new distribution-free Shewhart-type chart has been carried out by Mukherjee and Marozzi (2017), while a guideline for practitioners is also offered. An implementation strategy based on both average and median run length has been considered, while a search algorithm for determining the chart constants of the proposed Phase II charts is discussed as well.

4.9 Distribution-Free Fuzzy Shewhart–Lepage Control Schemes

Chong et al. (2017) introduced a distribution-free Shewhart-Lepage-Type control chart and further proposed a fuzzy control scheme for joint monitoring of the location and scale parameters of the underlying process. Along with the stupendous research growth in the area of nonparametric process control, the fuzzy logic control schemes have also drawn significant attention among researchers. These schemes are very useful in practice when the information is not well organized or vague. Fuzzy control schemes are more sensitive compared to traditional Shewhart control schemes; see Sabegh et al. (2014) for a thorough literature review on fuzzy control schemes.

Chong et al. (2017) discussed a Shewhart-Lepage monitoring scheme based on the max-type statistic. The general framework does not differ significantly from the one followed by Mukherjee and Chakraborti (2012) (see Sect. 4.3), where the standardized Wilcoxon rank-sum and Ansari-Bradley statistics S_{1i} and S_{2i} were used and the squared distance of the point (S_{1i}, S_{2i}) from the origin was exploited for the simultaneous monitoring of the location and the scale parameters. As Lepage (1971) mentioned a max-type statistic, namely $\Psi_j = \max\{|S_{1j}|, |S_{2j}|\}$ can also be practiced for simultaneous monitoring. McCracken et al. (2013) mentioned that the max-type scheme seems to be suitable to detect large shifts in mean accompanied by no other small shifts in the variance of a normally distributed process. Motivated by this, in the nonparametric joint monitoring context, Chong et al. (2017) established a one-sided monitoring scheme based on the statistic Ψ_j. According to his approach, if $|S_{2j}| < H_M < |S_{1j}|(|S_{1j}| < H_M < |S_{2j}|)$, a shift in location (scale) is declared, while if both $|S_{1j}|, |S_{2j}|$ exceed H_M, shifts in both location and scale are declared.

Chong et al. (2017) introduced also a fuzzy Shewhart-Lepage scheme, using as plotted points the couples $(|S_{1j}|, |S_{2j}|)$. They argue that a strong evidence for an in-control process is achieved when both the max and distance type statistics satisfy the in-control conditions. On the other hand, when the process is declared in-control by either the max or distance type statistics (but not by both of them) they argue that a fuzzy in-control state is evidenced. Otherwise, the process is declared out-of-control. For more details about the schematic form and the practical implementation of the proposed fuzzy control chart, the interested reader is referred to Chong et al. (2017). A detailed numerical simulation-based study carried out by them shed light on the performance of their monitoring scheme in terms of several characteristics associated with its run length distribution.

4.10 Generalized Shewhart-Lepage-Type Control Schemes

Mukherjee and Sen (2018) generalized the distribution-free Shewhart-type Lepage scheme described in Sect. 4.3 applying an adaptive approach. This approach is known in the statistical literature as percentile modifications of ranks (or adaptive Gastwirth Score) and offers a powerful tool to improve the classical rank tests.

Let $X_1, X_2, \ldots, X_m$ be a random sample of size m from the in-control distribution F and assume that test samples $Y_{i1}, Y_{i2}, \ldots, Y_{in}, i = 1, 2, \ldots$ are drawn independently from a continuous distribution Gastwirth (1965) and Kossler (2006) introduced the following statistics for $p, r \geq 1/N, p + r \leq 1$

$$T_p = \begin{cases} \sum_{i=N-P+1}^{N} (i - (N - P))Z_i, & \text{if } N \text{ is odd} \\ \sum_{i=N-P+1}^{N} [i - (N - P) - 1/2]Z_i, & \text{if } N \text{ is even} \end{cases}$$

$$B_r = \begin{cases} \sum_{i=1}^{R} ((R - i + 1))Z_i, & \text{if } N \text{ is odd} \\ \sum_{i=1}^{R} [(R - i + 1/2)]Z_i, & \text{if } N \text{ is even} \end{cases}$$

where Z_k denotes the indicator variable defined earlier in Sect. 4.3, while $P = [Np]$, $R = [Nr]$. Based on T_p, B_r two adaptive statistics are defined as

$$W_g = T_p - B_r, \quad S_g = T_p + B_r.$$

It is evident that test statistics W_g, S_g are similar to the Wilcoxon and Ansari–Bradley statistic respectively accounting for appropriate percentile modification. Consequently, an adaptive Lepage-type Gastwirth statistic based on the Mahalanobis distance between W_g, S_g can be expressed as

$$L_g = \frac{(W_g - \mu_{W_g})^2 \sigma_{W_g}^2 + 2\sigma_{W_g S_g}(W_g - \mu_{W_g})(S_g - \mu_{S_g}) + (S_g - \mu_{S_g})^2 \sigma_{S_g}^2}{\sigma_{W_g}^2 \sigma_{S_g}^2 - (\sigma_{W_g S_g})^2},$$

where $\mu_{W_g}(\mu_{S_g})$ and $\sigma_{W_g}^2(\sigma_{S_g}^2)$ are the mean and variance of the random variable $W_g(S_g)$ respectively, while $\sigma_{W_g S_g}^2$ is the corresponding covariance.

Note that since L_g is a positive semi-definite quadratic form, therefore it seems plausible to consider the lower control limit of the chart as zero. The proposed control chart produces an out-of-control signal whenever the plotted statistic L_g exceeds an upper control limit H. If the process is declared out-of-control at the ith test sample, the next step is to investigate whether the shift occurs in the location parameter or in the scale parameter on in both. To this end, Mukherjee and Sen (2018) suggested to compute the $p-$ values of the Wilcoxon test for location and the Ansari-Bradley test for scale using both the reference and the ith test sample and use them to reach a decision (a similar procedure has been described at the end of Sect. 4.4). The optimal implementation strategies of the proposed class of Shewhart-type Lepage-Gastwirth schemes to achieve lower out-of-control *ARL* and *FAR* have been well studied by Mukherjee and Sen (2018).

4.11 Nonparametric Control Charts Based on Modified Wilcoxon-Type Statistics

Koutras and Triantafyllou (2018) introduced a general class of nonparametric Shewhart-type control charts based on modified Wilcoxon-type rank-sum statistics. The proposed nonparametric control charts make use of the length M_i of runs of test sample observations $Y_1, Y_2, \ldots, Y_n$ (from a continuous distribution G) that lie between successive observations of the reference sample $X_1, X_2, \ldots, X_m$ with cumulative distribution function F (see Sect. 4.1). The in-control rules of these charts embrace the following two conditions:

Condition 1. The number of observations of the Y-sample before the observation $X_{a:m}$ is less or equal to c_1, namely $M_0 = \sum_{i=1}^{a} M_i \leq c_1$, where c_1 is a positive-valued parameter.

Condition 2. A properly defined continuous function g of M_0 and $M_i, i = a + 1, a + 2, \ldots, b$ does not exceed a threshold, namely $g(M_0, M_{a+1}, M_{a+2}, \ldots, M_b) \leq c_2$, where c_2 is a positive-valued parameter

Each one of the conditions stated above, defines a separate plotted statistic. As a consequence, the proposed distribution-free control scheme requires the construction of two one-sided charts that are used simultaneously to reach a decision whether the process is in-control or has shifted out-of-control. In the first one, the integer test statistic M_0 will be plotted versus an appropriate upper control limit (design parameter c_1), while in the second chart the function $g(M_0, M_{a+1}, M_{a+2}, \ldots, M_b)$ shall be plotted versus a second upper control limit (design parameter c_2). The *False Alarm Rate* may is given by

$$FAR = 1 - P(M_0 \leq c_1 \text{ and } g(M_0, M_{a+1}, M_{a+2}, \ldots, M_b) \leq c_2)$$

$$= 1 - \sum_{(m_0, m_{a+1}, \ldots, m_b) \in A} P_{in}(M_o = m_0, M_{a+1} = m_{a+1}, M_{a+2} = m_{a+2}, \ldots, M_b = m_b),$$

where A is the space containing the values of the random vector $(M_0, M_{a+1}, M_{a+2}, \ldots, M_b)$ that do not issue an alarm for the control chart in use, while

$$P_{in}(M_0 = m_0 \text{ and } M_j = m_j \text{ for } a + 1 \leq j \leq b) = \frac{\binom{m_0 + a - 1}{a - 1}\binom{m + n - m_0 - \sum_{i=a+1}^{b} m_i - b}{m - b}}{\binom{m + n}{n}}.$$

Under the Lehmann alternatives $G(x) = (F(x))^{\gamma}$, the corresponding *Alarm Rate* of the nonparametric control charts of the proposed class is given by

$$AR = 1 - P(M_0 \le c_1 \text{ and } g(M_0, M_{a+1}, M_{a+2}, \ldots, M_b) \le c_2)$$

$$= 1 - \sum_{(m_0, m_{a+1}, \ldots, m_b) \in A} P_L(M_o = m_0, M_{a+1} = m_{a+1}, M_{a+2} = m_{a+2}, \ldots, M_b = m_b),$$

where

$$P_L(M_0 = m_0 \text{ and } M_j = m_j \text{ for } a+1 \le j \le b)$$

$$= d_1 d_2 \gamma^{-(b-a+1)} \prod_{j=a}^{b-1} B\left(\frac{j}{\gamma} + m_0 + \sum_{i=a+1}^{j} m_i, m_{j+1} + 1\right)$$

$$\times \sum_{\ell=0}^{m-b} (-1)^\ell \binom{m-b}{j} B\left(\frac{b+\ell}{\gamma} + m_0 + \sum_{i=a+1}^{b} m_i, n - m_0 - \sum_{i=a+1}^{b} m_i + 1\right),$$

while

$$d_1 = \frac{m!}{(a-1)!(m-b)!}, \quad d_2 = \frac{n!}{m_0! \left(\prod_{j=a+1}^{b} m_j!\right)\left(n - m_0 - \sum_{i=a+1}^{b} m_j\right)!},$$

and $B(a, b)$ denotes the well-known beta function.

It is worth mentioning that exploiting the condition-uncondition technique (see, e.g., Chakraborti (2000)) a formula for the exact run length distribution and its mean can be established.

Following the lines of the general setup, described in Koutras and Triantafyllou (2018), for constructing nonparametric control charts, they introduced three distribution-free control schemes by plotting into two separate one-sided charts the quantity M_0 and one of the statistics W_{min}, W_{max} or W_E. For instance, the minimum value of the Wilcoxon's test statistic (W_{min}, hereafter) is achieved when all the remaining $\left(n - M_0 - \sum_{i=a+1}^{b} M_i\right)$ observations of the test sample occur between the bth and $(b+1)$-th observations of the reference sample. The monitoring statistic W_{min} can be expressed as

$$W_{min} = (n - m_o)\frac{n + m_o + 2b + 1}{2} + \sum_{i=a+1}^{b} (a - b - 2 + i)m_i.$$

The distribution-free control scheme described above is referred as W_{min}-chart.

Moreover, the Wilcoxon's test statistic receives its maximum value when all the remaining $n - M_0 - \sum_{i=a+1}^{b} M_i$ observations of the test sample occur after the mth observation of the reference sample. The maximal rank-sum statistic, as it is usually referred, can be readily expressed as

$$W_{\max} = (n - m_o)\frac{n + m_o + 2m + 1}{2} + \sum_{i=a+1}^{b} (a - m - 2 + i)m_i.$$

Employing analogous arguments, a second nonparametric control scheme could be constructed by using as plotted statistics the random variables M_0, $W_{\max}$. The resulting distribution-free control scheme is referred as $W_{\max}$-chart.

Finally, Koutras and Triantafyllou (2018) proposed a rank-sum statistic, say W_E, using the expected rank-sums of observations from the test sample that fall between the bth and the $(b + 1)$th observation, the $(b + 1)$th and the $(b + 2)$th observation,..., and finally those that are placed after the mth observation of the reference sample. It is straightforward that W_E is simply the average of $W_{\min}$ and $W_{\max}$, i.e.,

$$W_E = \frac{W_{\min} + W_{\max}}{2} = (n - m_0)(n + m_0 + b + m + 1) + \sum_{i=a+1}^{b} (2a - b - m - 4 + 2i)m_i.$$

In this scenario, W_E plays now the role of the function g and along with M_0 consist the plotted statistics of a third distribution-free control scheme, referred as W_E-chart. A numerical study carried out showed that the proposed control charts attain competitive in- and out-of-control performance.

4.12 Distribution-Free Control Charts for Subgroup Location and Scale Based on the Multi-sample Lepage Statistic

Li et al. (2019) introduced a distribution-free procedure for Phase I analysis which is capable of assessing stability of both location and scale parameters of a process using a single plotted statistic. The suggested procedure is based on the multi-sample Lepage statistic, which was developed by Rublik (2005) by combining the rank-based Kruskal Wallis statistic (KW, hereafter) and the multi-sample Ansari-Bradley statistic.

Suppose that $X_{i1}, X_{i2}, \ldots, X_{in_i}$ denote independent random samples from the distribution $F_i(x)$, $i = 1, 2, \ldots, k$, while $R_{i1}, R_{i2}, \ldots, R_{in_i}$ correspond to the ranks of the observations from the ith subgroup in the pooled sample of size $N = \sum_{i=1}^{k} n_i$. The KW statistic is given by

$$T_{KW} = \frac{1}{w_N^2} \sum_{i=1}^{k} n_i \left(\frac{S_i^{(KW)}}{n_i} - \frac{N+1}{2} \right)^2,$$

where $w_N^2 = \frac{N(N+1)}{2}$ and $S_i^{(KW)}$ are the partial sums

$$S_i^{(KW)} = \sum_{j=1}^{n_i} R_{ij}, i = 1, 2, \ldots, k.$$

The multi-sample Ansari–Bradley statistic is defined as

$$T_{AB} = \frac{1}{v_N^2} \sum_{i=1}^{k} n_i \left(\frac{S_i^{(AB)}}{n_i} - \mu_N \right)^2,$$

where

$$S_i^{(KW)} = \sum_{j=1}^{n_i} \left(\frac{N+1}{2} - \left| R_{ij} - \frac{N+1}{2} \right| \right), i = 1, 2, \ldots, k$$

and

$$\mu_N = \begin{cases} \frac{N+2}{4}, & \text{if } N \text{ is even} \\ \frac{(N+1)^2}{4N}, & \text{if } N \text{ is odd} \end{cases}, \quad v_N^2 = \begin{cases} \frac{N(N^2-4)}{48(N-1)}, & \text{if } N \text{ is even} \\ \frac{(N+1)(N^2+3)}{48N}, & \text{if } N \text{ is odd} \end{cases}.$$

The multi-sample version of the Lepage statistic equals $T = T_{KW} + T_{AB}$. Assuming that all subgroup observations are independent and identically distributed, it is quite straightforward that $E(T) = 2(k - 1)$. Rublík (2005) showed that if k and $\min(n_1, \ldots, n_k)$ are sufficiently large, then T has asymptotically a χ^2 distribution with $2(k - 1)$ degrees of freedom. By denoting

$$LC_i = \frac{n_i}{w_N^2} \left(\frac{S_i^{(KW)}}{n_i} - \frac{N+1}{2} \right)^2, \quad SC_i = \frac{n_i}{v_N^2} \left(\frac{S_i^{(AB)}}{n_i} - \mu_N \right)^2, \quad L_i = LC_i + SC_i, i = 1, 2, \ldots, k,$$

the multi-sample Lepage statistic is given as $T = \sum_{i=1}^{k} L_i, i = 1, 2, \ldots, k$. Note that the first part of L_i explains the difference in location while the second one the difference in scale. Due to the structure of the Lepage-type statistic, when the process is out-of-control, the statistic L_i is expected to take on larger values.

In order to construct the proposed Shewhart-type Lepage control chart for Phase I analysis, we need to have at hand k subgroups $X_{i1}, X_{i2}, \ldots, X_{in_i}, i = 1, 2, \ldots, k$ and for each one to calculate the plotted statistic L_i. Afterwards, the observed values are compared against the upper control limit, which for given k, n_i is determined with a prescribed *False Alarm Probability*. Note that the *False Alarm Probability* is commonly used as the design criterion while constructing and evaluating various Phase I control charts (see, e.g., Capizzi and Masarotto (2013), Jones-Farmer et al. (2009)). The *False Alarm Probability* represents simply the probability of observing at least one signal on a Phase I control chart with m plotted subgroups when the process is in-control. The in- and out-of-control performance of the proposed Phase I Lepage scheme has been studied by Li et al. (2019) through Monte-Carlo simulations.

4.13 Wilcoxon-Type Control Charts Based on Progressively Censored Reference Data

Triantafyllou (2019b) introduced a distribution−free Shewhart−type control chart implementing a modified Wilcoxon-type rank-sum statistic based on progressive Type II censoring reference data. The proposed chart seems to be an effective tool for monitoring incomplete data, because the censoring scheme applied allows the protection of experimental units at an early stage of the testing procedure. The conventional Type-I and Type-II censoring schemes do not allow the removal of units or products at a stage other than the terminal point of the process. In order to cover the abovementioned case, the so-called *progressive Type II right censoring scheme* has been applied by many authors, including Mann (1971), Thomas and Wilson (1972), Nelson (1982), Viveros and Balakrishnan (1994), Balakrishnan and Asgharzadeh (2005), Balakrishnan and Han (2007), Balakrishnan and Dembińska (2008) or Yadav et al. (2018).

Progressive Type II censoring scheme is described as follows: Suppose that a sample of m units is placed in a testing procedure. At the time of ath failure, R_a units are randomly removed from the remaining $(m - a)$ surviving units. At the $(a + 1)$th failure, R_{a+1} units are randomly removed from the remaining $(m - R_a - a - 1)$ surviving units. The testing procedure continues until the bth failure occurs, where all remaining $(m - R_a - R_{a+1} - \cdots - R_b - b)$ units are removed. The censoring scheme $(R_a, R_{a+1}, .., R_b)$ is supposed to be determined before the beginning of the failure process.

The control limits of the proposed distribution-free control chart are established following the general setup described in Sect. 4.1. The plotted statistic suggested by Triantafyllou (2019b) is defined in terms of the number M_i of observations of the test sample that fall between two successive failures of the reference sample. Suppose a progressive Type II right censoring scheme is to be adopted to the reference sample $X_1, X_2, \ldots, X_m$. The maximum value of the Wilcoxon's test statistic is achieved when all the progressive censored X-items in an interval fail before the smallest of Y-failures in the corresponding interval. The corresponding test statistic, named maximal Wilcoxon-type rank-sum statistic for the progressive Type II right censoring scheme $(R_a, R_{a+1}, .., R_b)$, is given by

$$W_{\max} = \sum_{i=a+1}^{b} M_i \left(M_0 + \sum_{j=a+1}^{i-1} M_j + (i - 1) + \sum_{j=a}^{i-1} R_j + \frac{M_i + 1}{2} M_i \right).$$

The process is declared to be in-control, if the test statistic $W_{\max}$ satisfies the condition $W_{\max} \leq W$, where the parameter W is a positive integer.

The operating characteristic function of the proposed control chart under the Lehmann alternative $G = F^\gamma, \gamma > 0$ is given as

$$oc(\gamma) = \frac{C}{\gamma^{b-a}} \sum_{m_0, m_{a+1}, \ldots, m_b}$$

$$\sum_{\xi=0}^{n-m_0-\sum\limits_{i=a+1}^{b} m_i} \binom{n - m_0 - \sum\limits_{i=a+1}^{b} m_i}{\xi} (-1)^\xi \sum_{j_a=0}^{R_a} \cdots \sum_{j_{b-1}=0}^{R_{b-1}} \binom{R_a}{j_a}\binom{R_{a+1}}{j_{a+1}} \cdots \binom{R_{b-1}}{j_{b-1}}$$

$$\times (-1)^{j_a + j_{a+1} + \cdots + j_{b-1}} B\left(m_0 + \frac{j_a + 1}{\gamma}, m_{a+1} + 1\right) \times B\left(m_0 + m_{a+1} + \frac{j_a + j_{a+1} + 2}{\gamma}, m_{a+2} + 1\right)$$

$$\times \cdots \times B\left(m_0 + m_{a+1} + m_{b-1} + \frac{j_a + j_{a+1} + \cdots + j_{b-1} + b - a}{\gamma}, m_b + 1\right)$$

$$\times B\left(\gamma(m_0 + m_{a+1} + m_b + \xi) + \frac{j_a + j_{a+1} + \cdots + j_{b-1} + b - a}{\gamma} + 1, R_b + 1\right),$$

where C is given as

$$C = \frac{n!(m - a + 1)(m - a + 1 - R_a - 1)\ldots(m - a + 1 - R_a - \ldots R_{b-1} - b + a)}{m_0! \prod\limits_{i=a+1}^{b} m_i! \left(n - m_0 - \sum\limits_{i=a+1}^{b} m_i\right)!}.$$

Based on the above expression, Triantafyllou (2019b) derived explicit formulae for the *Alarm Rate* and the *Average Run Length* for both in-control and out-of-control situations. A numerical study carried out by Triantafyllou (2019b) depicted the performance and robustness of the proposed control chart.

References

Abbasi, B., & Guillen, M. (2013). Bootstrap control charts in monitoring value at risk in insurance. *Expert Systems with Applications, 40,* 6125–6135.

Alloway, J. A., & Raghavachari, M. (1991). Control chart based on the Hodges-Lehmann estimator. *Journal of Quality Technology, 23,* 336–347.

Altukife, F. S. (2003). Nonparametric control chart based on sum of ranks. *Pakistan Journal of Statistics, 19,* 156–172.

Amin, R. W., & Widmaier, O. (1999). Sign control charts with variable sampling intervals. *Communication in Statistics: Theory and Methods, 28,* 1961–1985.

Amin, R. W., Reynolds, M. R., Jr., & Bakir, S. T. (1995). Nonparametric quality control charts based on the sign statistic. *Communication in Statistics: Theory and Methods, 24,* 1597–1623.

Antzoulakos, D. L., & Rakitzis, A. C. (2008). The revised *m*-of-*k* runs rule. *Quality Engineering, 20,* 75–81.

Asghari, S., Gildeh, B. S., Ahmadi, J., & Borzadaran, G. M. (2018). Sign control chart based on ranked set sampling. *Quality Technology & Quantitative Management, 15,* 568–588.

Bakir, S. T. (2012). A nonparametric Shewhart-type quality control chart for monitoring broad changes in a process distribution. *International Journal of Quality, Statistics and Reliability, 2012*(Article ID 147520), 10 p.

Balakrishnan, N., & Asgharzadeh, A. (2005). Inference for the scaled half-logistic distribution based on progressively type-II censored samples. *Communication in Statistics: Theory & Methods, 34,* 73–87.

Balakrishnan, N., & Dembińska, A. (2008). Progressively type-II right censored order statistics from discrete distributions. *Journal of Statistical Planning and Inference, 138,* 845–856.

Balakrishnan, N., & Han, D. (2007). Optimal progressive type-II censoring schemes for nonparametric confidence intervals of quantiles. *Communication in Statistics: Simulation and Computation, 36,* 1247–1262.

Balakrishnan, N., & Ng, H. K. T. (2006). *Precedence-type tests and applications.* Hoboken, NJ: Wiley.

Balakrishnan, N., Triantafyllou, I. S., & Koutras, M. V. (2009). Nonparametric control charts based on runs and Wilcoxon-type rank-sum statistics. *Journal of Statistical Planning and Inference, 139,* 3177–3192.

Balakrishnan, N., Triantafyllou, I. S., & Koutras, M. V. (2010). A distribution-free control chart based on order statistics. *Communication in Statistics—Theory and Methods, 39,* 3652–3677.

Balakrishnan, N., Paroissin, C., & Turlot, J. C. (2015). One-sided control charts based on precedence and weighted precedence statistics. *Quality Reliability Engineering International, 31,* 113–134.

Barabesi, L. (1998). The computation of the distribution of the sign test statistic for ranked-set sampling. *Communication in Statistics—Simulation and Computation, 27,* 833–842.

Büning, H., & Thadewald, T. (2000). An adaptive two-sample location-scale test of Lepage type for symmetric distributions. *Journal of Statistical Computation and Simulation, 65,* 287–310.

Capizzi, G., & Masarotto, G. (2009). Bootstrap-based design of residual control charts. *IIE Transactions, 41,* 275–286.

Capizzi, G., & Masarotto, G. (2013). Phase I distribution-free analysis of univariate data. *Journal of Quality Technology, 45,* 273–284.

Celano, G., Castagliola, P., Fichera, S., & Trovato, E. (2011). Shewhart and EWMA *t* charts for short production runs. *Quality Reliability Engineering International, 27,* 313–236.

Celano, G., Castagliola, P., & Trovato, E. (2012). The economic performance of a CUSUM *t* control chart for monitoring short production runs. *Quality Technology and Quantitative Management, 9,* 329–354.

Celano, G., Castagliola, P., & Chakraborti, S. (2016a). Joint Shewhart control charts for location and scale monitoring in finite horizon processes. *Computers & Industrial Engineering, 101,* 427–439.

Celano, G., Castagliola, P., Chakraborti, S., & Nenes, G. (2016b). The performance of the Shewhart sign control chart for finite horizon processes. *International Journal of Advanced Manufacturing Technology, 84,* 1497–1512.

Celano, G., Castagliola, P., Chakraborti, S., & Nenes, G. (2016c). On the implementation of the Shewhart sign control chart for low-volume production. *International Journal of Production Research, 54,* 5866–5900.

Chakraborti, S. (2000). Run length, average run length and false alarm rate of Shewhart X-bar chart: Exact derivations by conditioning. *Communications in Statistics-Simulation and Computation, 29,* 61–81.

Chakraborti, S. (2011). Nonparametric (Distribution-free) quality control charts. *Encyclopedia of Statistical Sciences,* 1–27.

Chakraborti, S., & Eryilmaz, S. (2007). A nonparametric Shewhart-type signed-rank control chart based on runs. *Communication in Statistics: Simulation and Computation, 36,* 335–356.

Chakraborti, S., & Graham, M. A. (2019b). Nonparametric (distribution-free) control charts: An updated overview and some results. *Quality Engineering, 31,* 523–544.

Chakraborti, S., & van de Wiel, M. A. (2008). A nonparametric control chart based on the Mann-Whitney statistic. In *IMS Collections* [*Beyond parametrics in interdisciplinary research*: Festschrift in Honour of Professor Pranab K. Sen], **1,** 156–172.

Chakraborti, S., van der Laan, P., & Bakir, S. T. (2001). Nonparametric control charts: An overview and some results. *Journal of Quality Technology, 33,* 304–315.

Chakraborti, S., van der Laan, P., & van de Weil, M. A. (2004). A class of distribution-free control charts. *Journal of the Royal Statistical Society, Series C-Applied Statistics, 53,* 443–462.

Chakraborti, S., Eryilmaz, S., & Human, S. W. (2009). A phase II nonparametric control chart based on precedence statistics with runs-type signaling rules. *Computational Statistics & Data Analysis, 53,* 1054–1065.

Chakraborti, S., & Graham, M. A. (2019a). *Nonparametric statistical process control.* Wiley.

Chatterjee, S., & Qiu, P. (2009). Distribution-free cumulative sum control charts using Bootstrap-based control limits. *The Annals of Applied Statistics, 3,* 349–369.

Chong, Z. L., Mukherjee, A., & Khoo, M. B. C. (2017). Distribution-free Shewhart-Lepage type premier control schemes for simultaneous monitoring of location and scale. *Computers & Industrial Engineering, 104,* 201–215.

Chowdhury, S., Mukherjee, A., & Chakraborti, S. (2014). A new distribution-free control chart for joint monitoring of unknown location and scale parameters of continuous distributions. *Quality Reliability Engineering International, 30,* 191–204.

Coehlo, M. L. I., Graham, M. A., & Chakraborti, S. (2017). Nonparametric signed-rank control charts with variable sampling intervals. *Quality Reliability Engineering International, 33,* 2181–2192.

Conover, W. J. (1999). *Practical nonparametric statistics* (3rd ed.). New York: Wiley.

Cucconi, O. (1968). *Un nuovo test non parametrico per il confronto tra due gruppi campionari* (pp. 225–248). XXVII: Giornale degli Economisti.

David, H. A., & Nagaraja, H. N. (2003). *Order Statistics* (3rd ed.), Hoboken, NJ. Wiley.

Del Castillo, E., & Montgomery, D. C, (1993). Optimal design of control charts for monitoring short production runs. *Economic Quality Control, 8,* 225–240.

Del Castillo, E., & Montgomery, D. C. (1996). A general model for the optimal economic design of $\overline{X}$ charts used to control short or long run processes. *IIE Transactions, 28,* 193–201.

Derman, C., & Ross, S. M. (1997). *Statistical Aspects of Quality Control.* San Diego: Academic Press.

Duncan, A. J. (1956). The economic design of X charts used to maintain current control of a process. *Journal of American Statistical Association, 51,* 228–242.

Figueiredo, F., & Gomes, M. I. (2016). The total median statistic to monitor contaminated normal data. *Quality Technology & Quantitative Management, 13,* 78–87.

Fligner, M. A., & Wolfe, D. A. (1976). Some applications of sample analogues to the probability integral transformation and a coverage property. *The American Statistician, 30,* 78–85.

Fu, J. C., & Lou, W. Y. W. (2003). *Distribution theory of runs and patterns and its applications: A finite markov chain imbedding approach.* Singapore: World Scientific Publishing.

Gastwirth, J. L. (1965). Percentile modifications of two-sample rank tests. *Journal of the American Statistical Association, 60,* 1127–1141.

Gibbons, J. D., & Chakraborti, S. (2010). *Nonparametric statistical inference* (5th ed.). Boca Raton: Taylor & Francis.

Graham, M., Human, S. W., & Chakraborti, S. (2010). A phase I nonparametric Shewhart-type control chart based on the median. *Journal of Applied Statistics, 37,* 1795–1813.

Hettmansperger, T. P. (1995). The ranked-set sample sign test. *Journal of Nonparametric Statistics, 4,* 263–270.

Hogg, R. V., Fisher, D. M., & Randles, R. H. (1975). A two-sample adaptive distribution-free test. *Journal of the American Statistical Association, 70,* 656–661.

Human, S. W., Chakraborti, S., & Smit, C. F. (2010). Nonparametric Shewhart-type sign control charts based on runs. *Communication in Statistics—Theory and Methods, 39,* 2046–2062.

Janacek, G. J., & Meikle, S. E. (1997). Control charts based on medians. *The Statistician, 46,* 19–31.

Jarett, R. G. (1979). A note on the intervals between coal-mining disasters. *Biometrika, 66,* 191–193.

Jensen, J. L. (1995). *Saddlepoint approximations.* New York: Oxford University Press.

Jones-Farmer, L. A., Jordan, V., & Champ, C. W. (2009). Distribution-free phase I control charts for subgroup location. *Journal of Quality Technology, 41,* 304–316.

Khilare, S. K., & Shirke, D. T. (2010). A nonparametric synthetic control chart using sign statistic. *Communication in Statistics—Theory and Methods, 39,* 3282–3293.

Khilare, S. K., & Shirke, D. T. (2012). Nonparametric synthetic control charts for process variation. *Quality Reliability Engineering International, 28,* 193–202.

Khoo, M. B. C., & Ariffin, K. N. (2006). Two improved runs rules for Shewhart $\overline{X}$ control chart. *Quality Engineering, 18,* 173–178.

Klein, M. (2000). Two alternatives to the Shewhart $\overline{X}$ control chart. *Journal of Quality Technology, 32,* 427–431.

Kossler, W. (2006). *Asymptotic power and efficiency of lepage-type tests for the treatment of combined location-scale alternatives* (Informatik-bericht nr. 200). Humboldt: Universitat zu Berlin.

Kössler, W. (2010). Max-type rank tests, U-tests and adaptive tests for the two-sample location problem—An asymptotic power study. *Computational Statistics & Data Analysis, 54,* 2053–2065.

Koti, K. M., & Jogeph Babu, G. (1996). Sign test for ranked-set sampling. *Communication in Statistics—Theory and Methods, 25,* 1617–1630.

Koutras, M. V., & Triantafyllou, I. S. (2018). A general class of nonparametric control charts. *Quality Reliability Engineering International, 34,* 427–435.

Kritzinger, P., Human, S. W., & Chakraborti, S. (2014). Improved Shewhart-type runs-rules nonparametric sign charts. *Communication in Statistics—Theory and Methods, 43,* 4723–4748.

Ladany, S. P. (1973). Optimal use of control charts for controlling current production. *Management Science, 19,* 763–772.

Lehmann, E. L. (1953). The power of rank tests. *Annals of Mathematical Statistics, 24,* 23–43.

Lepage, Y. (1971). A combination of Wilcoxon's and Ansari-Bradley's statistics. *Biometrika, 58,* 213–217.

Li, S.-Y., Tang, L.-C., & Ng, S.-H. (2010). Nonparametric CUSUM and EWMA control charts for detecting mean shifts. *Journal of Quality Technology, 42,* 209–226.

Li, C., Mukherjee, A., Su, Q., & Xie, M. (2016). Optimal design of a distribution-free quality control scheme for cost-efficient monitoring of unknown location. *International Journal of Production Research, 54,* 7259–7273.

Li, C., Mukherjee, A., & Su, Q. (2019). A distribution-free phase I monitoring scheme for subgroup location and scale based on the multi-sample Lepage statistic. *Computers & Industrial Engineering, 129,* 259–273.

Lorenzen, T. J., & Vance, L. C. (1986). The economic design of control charts: A unified approach. *Technometrics, 28,* 3–10.

Malela-Majika, J. C., & Rapoo, E. (2016). Distribution-free cumulative sum and exponentially weighted moving average control charts based on the Wilcoxon rank-sum statistic using ranked set sampling for monitoring mean shifts. *Journal of Statistical Computation and Simulation, 86,* 3715–3734.

Malela-Majika, J. C., Chakraborti, S., & Graham, M. A. (2016a). Distribution-free control charts with improved runs-rules. *Applied Stochastic Models in Business and Industry, 32,* 423–439.

Malela-Majika, J. C., Graham, M. A., & Chakraborti, S. (2016b). Distribution-free phase II Mann-Whitney control charts with runs-rules. *International Journal of Advanced Manufacturing Technology, 86,* 723–735.

Malela-Majika, J. C., Rapoo, E. M., Mukherjee, A., & Graham, M. A. (2019). Distribution-free precedence schemes with a generalized runs-rule for monitoring unknown location. *Communications on Statistics—Theory and Methods.* https://doi.org/10.1080/03610926.2019.1612914.

Mann, N. R. (1971). Best linear invariant estimation for Weibull parameters under progressive censoring. *Technometrics, 13,* 521–533.

Marozzi, M. (2009). Some notes on the location-scale Cucconi test. *Journal of Nonparametric Statistics, 21,* 629–647.

McCracken, A. K., & Chakraborti, S. (2013). Control charts for joint monitoring of mean and variance: An overview. *Quality Technology & Quantitative Management, 10,* 17–36.

McCracken, A. K., Chakraborti, S., & Mukherjee, A. (2013). Control charts for simultaneous monitoring of unknown mean and variance of normally distributed processes. *Journal of Quality Technology, 45,* 360–376.

McIntyre, G. (1952). A method for unbiased selective sampling using ranked sets. *Crop and Pasture Science, 3,* 385–390.

Montgomery, D. C. (2009). *Introduction to statistical quality control* (6th ed.). New York: Wiley.

Mood, A. M. (1954). On the asymptotic efficiency of certain nonparametric two-sample tests. *Annals of Mathematical Statistics, 25,* 514–522.

Mukherjee, A., & Chakraborti, S. (2012). A distribution-free control chart for the joint monitoring of location and scale. *Quality Reliability Engineering International, 28,* 335–352.

Mukherjee, A., & Marozzi, M. (2017). Distribution-free Lepage type circular-grid charts for joint monitoring of location and scale parameters of a process. *Quality Reliability Engineering International, 33,* 241–274.

Mukherjee, A., & Sen, R. (2015). Comparisons of Shewhart-type rank based control charts for monitoring location parameters of univariate processes. *International Journal of Production Research, 53,* 4414–4445.

Mukherjee, A., & Sen, R. (2018). Optimal design of Shewhart-Lepage type schemes and its application in monitoring service quality. *European Journal of Operational Research, 266,* 147–167.

Murakami, H. (2007). Lepage type statistic based on the modified Baumgartner statistic. *Computational Statistics & Data Analysis, 51,* 5061–5067.

Murakami, H., & Matsuki, T. (2010). A nonparametric control chart based on the mood statistic for dispersion. *International Journal of Advanced Manufacturing Technology, 49,* 757–763.

Nelson, W. (1982). *Applied life data analysis.* New York: Wiley.

Nenes, G., & Tagaras, G. (2010). Evaluation of CUSUM charts for finite-horizon processes. *Communication in Statistics: Simulation and Computation, 39,* 578–597.

Neuhäuser, M. (2011). *Nonparametric statistical tests: A computational approach.* CRC Press.

Park, H. (2009). Median control charts based on Bootstrap method. *Communications on Statistics—Simulation and Computation, 38,* 558–570.

Qiu, P. (2014). *Introduction to statistical process control.* Boca Raton: CRC Press.

Qiu, P. (2018). Some perspectives on nonparametric statistical process control. *Journal of Quality Technology, 50,* 49–65.

Qiu, P. (2019). Some recent studies in statistical process control. In *Statistical quality technologies,* 3–19.

Rakitzis, A. C., Chakraborti, S., Shongwe, S. C., Graham, M. A., & Khoo, M. B. C. (2019). An overview of synthetic-type control charts: techniques and methodology. *Quality and Reliability Engineering International, 35,* 2081–2096.

Randles, R. H., & Wolfe, D. A. (1979). *Introduction to the theory of nonparametric statistics.* New York: Wiley.

Rublík, F. (2005). The multisample version of the lepage test. *Kybernetika, 41,* 713–733.

Sabegh, M. H. Z., Mirzazadeh, A., Salehian, S., & Weber, G. W. (2014). A literature review on the fuzzy control chart; classifications & analysis. *International Journal of Supply and Operations Management, 1,* 167–189.

Su, N. C., Chiang, J. Y., Chen, S. C., Tsai, T. R., & Shyr, Y. (2014). Economic design of two-stage control charts with Skewed and dependent measurements. *The International Journal of Advanced Manufacturing Technology, 73,* 1387–1397.

Tagaras, G. (1996). Dynamic control charts for finite production runs. *European Journal of Operational Research, 91,* 38–55.

Tamura, R. (1963). On a modification of certain rank tests. *Annals of Mathematical Statistics, 34,* 1101–1103.

Thomas, D. R., & Wilson, W. M. (1972). Linear order statistic estimation for the two-parameter Weibull and extreme value distributions from type-II progressively censored samples. *Technometrics, 14,* 679–691.

Triantafyllou, I. S. (2018a). Nonparametric control charts based on order statistics: Some advances. *Communication in Statistics: Simulation and Computation, 47,* 2684–2702.

Triantafyllou, I. S. (2018b). A new distribution-free reliability monitoring scheme: Advances and applications in engineering. In *Modeling and Simulation Based Analysis in Reliability Engineering* (pp. 199–213).

Triantafyllou, I. S. (2019a). A new distribution-free control scheme based on order statistics. *Journal of Nonparametric Statistics, 31,* 1–30.

Triantafyllou, I. S. (2019b). Wilcoxon-type rank-sum control charts based on progressively censored reference data. *Communication in Statistics: Theory and Methods.* https://doi.org/10.1080/03610926.2019.1634816.

Viveros, R., & Balakrishnan, N. (1994). Interval estimation of life characteristics from progressively censored data. *Technometrics, 36,* 84–91.

Wilcoxon, F. (1945). Individual comparisons by ranking methods. *Biometrics, 1,* 80–83.

Wood, A. T. A., Booth, J. G., & Butler, R. W. (1993). Saddlepoint approximations to the CDF of some statistics with non-normal limit distributions. *Journal of the American Statistical Association, 88,* 680–686.

Xie, M., Goh, T. N., & Ranjan, P. (2002). Some effective control chart procedures for reliability monitoring. *Reliability Engineering and System Safety, 77,* 143–150.

Yadav, A. S., Singh, S. K., & Singh, U. (2018). Estimation of stress–strength reliability for inverse Weibull distribution under progressive type-II censoring scheme. *Journal of Industrial and Production Engineering, 35,* 48–55.

Yeh, L. L., Wang, F., Li, C., & Yeh, Y. M. (2011). An extension of economic design of X-bar control charts for non-normally distributed data under Weibull shock models. *Communications in Statistics—Theory and Methods, 40,* 3879–3902.

Zhang, P., Su, Q., Li, C., & Wang, T. (2014). An economically designed sequential probability ratio test control chart for short-run production. *Computers & Industrial Engineering, 78,* 74–83.

Multivariate Nonparametric Control Charts Based on Ordered Samples, Signs and Ranks

Elisavet M. Sofikitou and Markos V. Koutras

Abstract Nonparametric control charting techniques have attracted the interest of the researchers during the past decades. However, the studies in the multivariate case are not equally weighted as the ones that have already been done in the univariate case. In the present work, we firstly review the recent literature of nonparametric control charts, giving emphasis on multivariate schemes, which make use of order statistics, signs or ranks for the computation of the test statistic that is exploited for the decision-making whether the process is in- or out-of-control. In addition, we carry out a simulation study in order to evaluate the performance of these charts when compared with each other, as well as to their parametric counterparts. The numerical results take into account different shift scenarios for the location parameter, the scale parameter or both. Finally, some concluding remarks are given, as well as some ideas and directions for future work.

Keywords Joint monitoring · Nonparametric control charts · Order statistics · Ordered samples · Ranks · Signs · Statistical process control

1 Introduction

Statistical Process Control (SPC) is a very popular area of the statistical science that develops techniques for monitoring and controlling manufacturing processes over time. The main statistical–graphical tools, which are used to maintain the quality of the data at the required standards and at the same time ensure that the overall

E. M. Sofikitou
Department of Biostatistics, School of Public Health and Health Professions,
State University of New York at Buffalo, Buffalo, USA
e-mail: esofikit@buffalo.edu

M. V. Koutras (✉)
Department of Statistics and Insurance Science, School of Finance and Statistics,
University of Piraeus, Piraeus, Greece
e-mail: mkoutras@unipi.gr

© Springer Nature Switzerland AG 2020
M. V. Koutras and I. S. Triantafyllou (eds.), *Distribution-Free Methods for Statistical Process Monitoring and Control*,
https://doi.org/10.1007/978-3-030-25081-2_2

57

process operates in an efficient way, are known as Control Charts (CCs). The first CC was introduced by Shewhart (1926) in a memorandum of Bell Laboratories. A few years later, Shewhart (1931,1939) published two books where he described how statistical methods could be applied/adopted in industry; in addition, he analysed the lessons learned about their implementation from a statistical point of view. Since then, classical CCs were rapidly expanded and several wider approaches were established so that its techniques can be applied in statistical inference and decision-making in many applied fields different than industry.

The bulk of the CCs which are available in the SPC literature rely on a distributional assumption (such as normality); therefore, the test statistic as well as the Control Limits (CLs) depend on the properties of the underlying (usually completely known) distribution for which they have been designed for. However, when one is dealing with real-world data, the true distribution is unknown (at least in part) and sometimes there is not enough information about the kind and the shape of the data distribution. In these cases, the nonparametric CCs can provide useful, robust alternatives. Consequently, although the parametric CCs are expected to be more effective than their nonparametric counterparts (when there is complete knowledge of the distribution), there exist many real-life settings where they cannot be used. A detailed review of the univariate nonparametric CCs and a synopsis of their advantages has been provided by Chakraborti and his colleagues (see, for example, Chakraborti et al. 2001, 2011; Chakraborti (2011); Chakraborti and Graham (2008, 2019).

The classical approach for the construction of a nonparametric CC requires the use of two different samples: the reference or historical sample (of size m) and the test sample (of size n). The former sample is used to establish the decision rule, as well as the CLs and the most popular statistics exploited to this aid are the order statistics, ranks, signs, etc. This occurs in Phase I (also known as Retrospective Phase), during which it is very important to verify the stability of the process, by establishing reliable CLs. In SPC parlance, a process is stable or in an in-control stage, when no assignable or special causes of variation are present. It should be mentioned that the CLs are appropriately chosen to achieve a pre-specified False Alarm Rate (FAR), which is the probability to get an alarm when the process is in fact in-control, or the in-control the Average Run Length (ARL_{in}), which indicates the average number of points that are plotted in the chart before an alarm is triggered for the first time.

In Phase II (also known as Prospective Phase) the appropriately determined CLs are used to verify if the process remains in-control or shifts to an out-of-control state. For this purpose, successive test samples are drawn (independently of each other and of the reference sample of course) and using them we check whether the out-of-control signalling rule is still valid or not, e.g. whether an order statistic or a specific quartile lies between the CLs. Phase I Analysis is closely related to the efficiency of Phase II and may have a great impact on Phase II if not being adequately designed; for a detailed discussion of this topic see Chakraborti et al. (2008) and Jones-Farmer et al. (2014) as well as Jensen et al. (2006) who studied the quality and the sample size requirements of the reference sample.

The majority of CCs that have appeared in the SPC literature are usually designed for monitoring either the mean or the variability. During the past years, there is an increasing interest in the development of schemes for the joint detection of potential shifts in both the location and the scale parameter of the underlying distribution. When applying a charting technique, the critical quantity (depicted at each time point) may be based on the most recent sample (ignoring the information contained in previous samples) or it may use information associated with previous samples. In the second case, the schemes are known as CCs without memory or simply *Shewhart*-type CCs. In the first case, they are known as CCs with memory; the most popular representatives of this family of CCs are the Cumulative Sum ($CUSUM$), the Exponentially Weighted Moving Average ($EWMA$) and the Moving Average (MA) type CCs.

In the present article, we restrict our presentation to nonparametric CCs which are based on ordered samples, signs or ranks. First, in Sect. 2, we briefly review the univariate CCs; more details on this subject can be found in Chapter 1 of the present volume. In Sect. 3, we present in detail the multivariate CCs. The schemes are categorized into CCs for mean and/or variance and into CCs with or without memory. Section 4 provides numerical study of the performance of a large number of multivariate CCs presented in the literature and several comparisons between them. Finally, conclusions and comments for future work are made in Sect. 5.

2 Brief Literature Review on Univariate Charts

For reasons of completeness, we start by describing in brief the univariate CCs which are based on ordered samples, signs and ranks. We believe that this may be proved helpful for the reader since, some of these schemes were the starting point for the construction of respective charts in higher dimensions.

2.1 Monitoring of Location

It is true that the share of lion on CCs' literature goes to CCs designed for detecting potential shifts in the mean for univariate or multivariate processes. In this section, we shall cover the case of nonparametric univariate CCs.

a. *Univariate CCs without Memory*

When there is only one quality characteristic under study, the plotted statistic of a CC is usually the median. For instance, Alloway and Raghavachari (1991) suggested a CC for monitoring the median of a continuous symmetric population which used order statistics for the determination of the CLs, while Amin et al. (1995) presented a CC for the median of population that is not necessarily symmetric. The first chart is based on a (distribution-free) confidence interval that is calculated using a Hodges–Lehmann estimator and the second one uses the within sign test statistic.

A well-known CC for the median was proposed by Janacek and Meikle (1997), the CLs of which are specific order statistics from the reference sample. Chakraborti et al. (2004) extended the previous idea by developing a CC for a test sample quartile, which exploits the so-called precedence statistics. Recently, Balakrishnan et al. (2015) introduced a class of one-sided schemes that take into account the precedence and the weighted precedence statistics.

Another CC, which uses as CLs appropriately chosen order statistics from the historical sample, was developed by Albers and Kallenberg (2008). This scheme is known as $MIN\ CC$ and can be used for either individual observations or subgrouped data (in which case the subgroup minimum is computed). Moreover, Bakir (2008) proposed a CC for the median that uses signed-rank-like statistics, while Graham et al. (2010) presented a CC that makes use of the pooled median of the reference sample.

In the literature, there are available a number of schemes that use test statistics which involve signs and/or ranks. More precisely, Bakir (2004) suggested a CC based on a Wilcoxon signed-rank statistic. This chart was later modified by Chakraborti and Eryilmaz (2007) by incorporating into it several runs-rules. A class of sign CCs that is based on runs-rules was introduced by Human et al. (2010). Recently, Malela-Majika et al. (2016) also incorporated runs-rules in a CC based on the Mann–Whitney statistics, initially suggested by Chakraborti et al. (2004). The Mann–Whitney test has been applied for the construction of change-point models too, see, for example, Hawkins and Deng (2010). Non-classical, nonparametric test statistics have also been used in developing control charting techniques. In particular, Jones-Farmer et al. (2009) developed a mean-rank CC for location which exploits an adapted version of the Kruskal–Wallis test. Gadre and Kakade (2014) designed a group runs CC for the median, which considers sample group as a unit. Finally, Garde and Kakade (2019) suggested two CCs which exploit side sensitive group runs for detecting potential shifts in the process median.

b. *Univariate CCs with Memory*

Bakir and Reynolds (1979) described a $CUSUM$ procedure based on Wilcoxon (within) grouped signed-rank statistics, while Li et al. (2010) proposed $CUSUM$ CCs based on Wilcoxon rank-sum statistic. McDonald (1990) provided a $CUSUM$ technique based on sequential ranks of individual observations and Amin et al. (1995) suggested a scheme that uses a $CUSUM$ of sign tests statistics. Albers and Kallenberg (2009) improved their MIN chart (Albers and Kallenberg (2008)) by adapting a sequential approach in order to construct a $CUSUM$-like CC labelled as $CUMIN$ chart. The exceedance statistics have also been incorporated in a $CUSUM$ procedure for location monitoring, see Mukherjee et al. (2013).

Moreover, several $EWMA$ techniques have been proposed for monitoring the location. For instance, $EWMA$s of the "standardized" ranks and the sequential ranks of individual observations were, respectively, considered by Hackl and Ledolter (1991, 1992). Amin and Searcy (1991) provided an $EWMA$ which uses the Wilcoxon (grouped) signed-rank statistic, while Li et al. (2010) proposed $EWMA$ CCs based on Wilcoxon rank-sum statistic. Zhou et al. (2009) introduced an $EWMA$ procedure for detecting shifts in the mean which uses an estimator based on the Mann–Whitney statistic of the change-point.

Graham et al. (2011a, b) suggested the $EWMA$-SN and $EWMA$-SR CCs. The former scheme is for individual observations and it is based on the sign statistic, while the latter one makes use of the signed-rank statistic and can be applied only if the underlying distribution is symmetric. Working in a similar fashion to Mukherjee et al. (2013), Graham et al. (2012) considered an $EWMA$-EX control scheme based on exceedance statistics. Recently, generally weighted moving average CCs based on sign (abbrev. $GWMA$-SN) and signed-rank (abbrev. $GWMA$-SR) statistics have been developed by Lu (2015) and Chakraborty et al. (2016), respectively. It should be mentioned that the $GWMA$-SR assumes that the process distribution is symmetric.

2.2 Monitoring of Scale

The monitoring of variability might also be of interest and quite useful in a manufacturing process; yet, a few procedures are only available in the literature.

a. *Univariate CCs without Memory*

Amin et al. (1995) considered a *Shewhart*-type sign CC for variability which takes into account the interquartile range of two samples. Additionally, Bakir (2010) proposed a nonparametric procedure for testing homogeneity of several population variances (scale parameters) and he provided a graphical representation.

Das (2008) developed two CCs based on Mood's (1954) and Siegel and Tukey's (1960) nonparametric ranks tests for equality of dispersion. In a similar context, Das and Bhattacharya (2008) exploited Conover's (1980) test for monitoring variability, which is based on the joint squared ranks of squared deviations from the means.

b. *Univariate CCs with Memory*

Zombade and Ghute (2014) presented two $CUSUM$ procedures for controlling variability, the $NPCSM$-S and the $NPCSM$-M CCs. These schemes use the Mood's (1954) and Sukhamte's (1954) tests, respectively, for testing equality of variances.

2.3 Simultaneous Monitoring of Location and Scale

All the approaches presented in the previous subsections were talking about CCs capable of identifying shifts either in the mean or in the variance (alternatively in the standard deviation). As already been mentioned, the joint monitoring of both seems to be a relatively new research area that has attracted the interest of a number of researchers quite recently. A summarization of such charts is given in the overviews of Cheng and Thaga (2006) and McCracken and Chakraborti (2013), while Reynolds and Stoumbos (2006) and Ou et al. (2012) carried out comparison studies for some of these schemes. The vast majority of the CCs included in the aforementioned review articles are univariate and assume that the normality assumption is fulfilled. Below, we present univariate CCs for simultaneous monitoring that appeared in the nonparametric SPC literature during the past decade.

a. *Univariate CCs without Memory*

Some of the very first nonparametric univariate CCs capable of detecting simultaneous shifts in the mean and variability are the schemes suggested by Balakrishnan et al. (2009), which exploit run and Wilcoxon-type rank-sum statistics. The authors proposed three CCs: the test statistic of the first CC takes into account the maximum run length of the test sample observations that occur between the CLs, the second CC uses a statistic defined as the number of runs of the test sample observations (lying between the CLs) the length of which exceeds a pre-specified level and the third one considers the joint sample resulting by combining the reference and the test sample observations and the decision rule uses the sum of ranks of the observations of the latter sample that lie between the CLs (i.e. the well-known Wilcoxon-type rank-sum statistic).

Balakrishnan et al. (2010) also suggested another CC that is based on order statistics. More specifically, they improved the median CC (Janacek and Meikle (1997)) and its extensions (Chakraborti et al. (2004)) by incorporating an additional condition. The new CC uses two rules based on separate plotted statistics. The first rule, which is used to detect plausible mean shifts, checks whether a test sample order statistic (for example, the median) is placed between the CLs. The second one, which controls the variability, counts the number of the test sample observations lying between the CLs. Recently, Triantafyllou (2018) modified the first statistic of Balakrishnan et al. (2010), while the second statistic was maintained in its original form. The new modified statistic, instead of using a single-order statistic, is based on the successive number of order statistics from the test sample that have been observed between the CLs.

Of course, the CLs of the CCs described above are order statistics that have been appropriately chosen from the reference sample. It should be underlined here that, since the test rules are based not only on the location of a single observation but also incorporate a measure of dispersion (that is, the number or length of observations and number or sum of ranks), the charts become capable of detecting potential shifts in both mean and variance. This appealing property seems not to have been widely noticed, and as a result, these schemes are often been presented in the literature as CCs for monitoring location only.

Mukherjee and Chakraborti (2012) constructed the *Shewhart–Lepage CC*, the plotted statistic of which, is a statistic originally introduced by Lepage (1971) and constitutes a quadratic form that combines the Wilcoxon rank-sum statistic for location and the Ansari–Bradley statistic for scale. Motivated by this idea, Chowdhury et al. (2014) developed the *Shewhart–Cucconi CC* which exploits the Cucconi's (1968) test for equality of the scale and the location parameters of two populations.

This test is based on the sum of squares of the usual ranks of the first sample and the sum of squares of the "antiranks" of the second sample.

b. *Univariate CCs with Memory*

The Lepage statistic has also been used for the construction of univariate CCs with memory. In particular, Chowdhury et al. (2015) considered a *CUSUM-Lepage CC* for the simultaneous monitoring of the location and scale parameter of the underlying distribution. Recently, Mukherjee (2017) proposed an *EWMA-Lepage* formulation for subgroups.

3 Presentation of the Multivariate Control Charts

In this section, we present in some detail a number of multivariate CCs which use in their decision rules ordered samples, signs and ranks. Usually, in these schemes the monitoring of the mean vector and/or variance–covariance matrix of the process distribution is achieved by exploiting the multivariate versions of well-known nonparametric tests.

Before proceeding to our presentation, we give some definitions and notations that will be used in the sequel. The symbol # will indicate the *number* of observations, $||x||$ will denote the *Euclidean norm* of the vector x which is calculated by $||x|| = \sqrt{x'x}$, and I_p will stand for the $p \times p$ *identity matrix*, i.e.

$$I_p = \begin{pmatrix} 1 & 0 & 0 & \cdots & 0 \\ 0 & 1 & 0 & \cdots & 0 \\ \vdots & & & \ddots & \\ 0 & 0 & 0 & \cdots & 1 \end{pmatrix}.$$

The following three functions will also appear quite frequently in the next sections:

- the *indicator function* $I(\cdot)$, defined as

$$I(C) = \begin{cases} 1, & \text{if condition } C \text{ is true,} \\ 0, & \text{otherwise.} \end{cases}$$

- the *sign function* $sgn(\cdot)$ which is given by

$$sgn(x) = \begin{cases} -1, & \text{if } x < 0, \\ 0, & \text{if } x = 0, \\ +1, & \text{if } x > 0. \end{cases}$$

- the *multivariate spatial sign* (also called *directional vector*, since it indicates the direction of the observation by mapping it on the multidimensional unit sphere) defined as follows:

$$U(x) = \begin{cases} ||x||^{-1}x, & \text{if } x \neq 0, \\ 0, & \text{if } x = 0. \end{cases}$$

3.1 Monitoring of Location

Over the past couple of decades, several multivariate CCs with or without memory have been developed for monitoring the mean vector of a process without setting any requirement for prior knowledge of the underlying distribution.

a. Bivariate and Multivariate CCs without Memory

Let us start with the *Shewhart*-type CCs. Hayter and Tsui (1994) developed a procedure that is based on the construction of simultaneous confidence intervals for the mean. The plotted statistic computes the maximum deviation of the observations from their sample means. In particular, if $\bar{X}$ and $S = \{S_{ij}\}$ represent the sample mean vector and the sample covariance matrix of a sample $X_1, X_2, \ldots, X_N$ with empirical Cumulative Distribution Function (CDF) $\hat{F}(\cdot)$; then, Hayter and Tsui (1994) suggested using the following statistic:

$$M^{(j)} = \max_{1 \leq i \leq k} |X_i^j - \bar{X}_i| / \sqrt{S_{ii}},$$

for $1 \leq j \leq N$. The process is declared in-control as long as $M^{(j)} \leq C_a$, where a is the error rate and the critical point C_a is the solution of $\hat{F}(C_a) = 1 - a$. For the identification of the characteristic that produced the out-of-control, Hayter and Tsui (1994) suggested checking (per variable) whether the quantity $|X_i - \bar{X}_i| / \sqrt{S_{ii}}$ exceeds the critical value C_a. The sample means can be obtained from confidence intervals of the form: $\mu_i \in [X_i - C_a \cdot \sqrt{S_{ii}}, X_i + C_a \cdot \sqrt{S_{ii}}]$. However, as stated by Stoumbos et al. (2001), the M procedure described above fails in detecting possible changes in the correlation structure.

Ghute and Shirke (2012) considered a CC based on the bivariate version of the signed-rank test suggested by Bennett (1964), while Das (2009) proposed a CC which exploits the bivariate sign test of Puri and Sen (1976). Ghute (2013) also considered a bivariate CC based on the Hodges's (1955) sign test. Later, Boone and Chakraborti (2012) extended the previous ideas for monitoring the location of more than two quality characteristics. They constructed the SN^2 and SR^2 CCs which are based on the multivariate forms of the sign and the Wilcoxon signed-rank test defined in Hettmansperger (2006).

More precisely, given a bivariate sample $X_i = (X_{1i}, X_{2i}), i = 1, 2, \ldots, n$, Ghute and Shirke (2012) introduced first the following nonparametric (univariate) signed-rank statistics:

$$T_j = \sum_{i=1}^{n} I(X_{ji} \geq 0) \cdot R(X_{ji}), \quad j = 1, 2,$$

for the two characteristics, where $R(X_{ji})$ is the rank of $|X_{ji}|$ among $|X_{j1}|$, $|X_{j2}|$, ..., $|X_{jn}|$ for $j = 1, 2$. The statistics T_1 and T_2 were next combined together in a quadratic form given as follows:

$$W = (T - v)' \hat{\beta}^{-1} (T - v),$$

where $T = (T_1, T_2)'$ is the vector of coordinate-wise univariate signed-rank statistics, the variance-covariance matrix of vector T is

$$\beta = \begin{pmatrix} \beta_{11} & \beta_{12} \\ \beta_{21} & \beta_{22} \end{pmatrix} \quad \text{and} \quad v = (v_1, v_2)'$$

with v_j standing for $E(T_j | \mu = \mu_0)$ for $j = 1, 2$; μ_0 denotes the in-control sample mean vector and if the process is in-control, then $\mu = \mu_0$, $v_1 = v_2 = n(n+1)/4$, $\beta_{11} = \beta_{22} = n(n+1)(2n+1)/24$, $\beta_{12} = \beta_{21} = \sum_{i=1}^{n} \text{sgn}(X_{1i}) \text{sgn}(X_{2i}) R(X_{1i}) R(X_{2i})/4$. When the process is in-control, the quadratic form W follows asymptotically the chi-squared distribution with 2 degrees of freedom (χ_2^2). The process is declared out-of-control if W exceeds a pre-specified UCL. This CC requires the distribution of the process to be symmetric. It is worth mentioning that Ghute and Shirke (2012) improved the performance of the aforementioned CC by suggesting its synthetic version which enhances runs-rules.

Let $X_i = (X_{1i}, X_{2i})', i = 1, 2, \ldots, n$ be n independent vectors with their respective $CDFs$ $F_1(x), F_2(x), \ldots, F_n(x), x \in R^2$. Das (2009) proposed a procedure for testing whether the $CDFs$ have n specified medians. Without loss of generality, each pair of medians is assumed to be equal to $0 = (0, 0)'$, which can be achieved by an appropriate shift. In this case, the null hypothesis is described as follows:

$$H_0 : F_i(0, \infty) = F_i(-\infty, 0) = 1/2, \quad i = 1, 2, \ldots, n.$$

Defining the concordance of first and second kind by the events $(X_{1i} \leq 0, X_{2i} \leq 0)$ and $(X_{1i} \geq 0, X_{2i} \geq 0)$, the discordance of first and second kind by the events $(X_{1i} \leq 0, X_{2i} \geq 0)$ and $(X_{1i} \geq 0, X_{2i} \leq 0)$, respectively, and denoting the conditional probability of a concordance (discordance) of the first kind given that a concordance (discordance) has occurred by θ_i (τ_i) for $i = 1, \ldots, n$, then the null hypothesis takes on the form

$$H_0 : \theta_i = \tau_i = 1/2, \quad i = 1, 2, \ldots, n.$$

The CC based on this principle/test uses the following statistic:

$$T = \frac{4}{C}\left(C_1 - \frac{C}{2}\right)^2 + \frac{4}{n-C}\left(D_1 - \frac{n-C}{2}\right)^2 = \frac{(C_1 - C_2)^2}{C_1 + C_2} + \frac{(D_1 - D_2)^2}{D_1 + D_2},$$

where C_j (D_j) is the number of concordances (discordances) of the jth kind for $j = 1, 2$, $C = C_1 + C_2$, $D = D_1 + D_2$ and $C + D = n$. The null hypothesis H_0 is rejected when $T > UCL$. If $F_n(t|c)$ denotes the conditional CDF of T under H_0 given that $C = c$, then the UCL equals $t_{n,a}(c)$, a characteristic point that can be computed as the value of t for which $1 - F_n(t|c) \leq a \leq 1 - F_n(t - 0|c)$ where $0 < a < 1$. For large values of n, $F_n(t|c)$ can be approximated by the chi-square distribution and the UCL will be asymptotically expressed as $UCL = \chi_{2;a}^2$.

In order to create a sign-based CC Ghute (2013), considered the direction angle θ_i of a bivariate subgroup (X_{1i}, X_{2i}), $i = 1, 2, \ldots, n$ and he denoted the direction angle of $(-X_{1i}, -X_{2i})$ by $\theta_i^* = \theta_i + \pi (\text{mod} 2\pi)$. In addition, Ghute (2013) defined the following indicator function:

$$z_i = \begin{cases} 1, & \text{if } \theta_i' \in \{\theta_1, \theta_2, \ldots, \theta_n\}, \\ 0, & \text{if } \theta_i' \in \{\theta_1^*, \theta_2^*, \ldots, \theta_n^*\}, \end{cases}$$

where $\theta_1' < \theta_2', < \ldots, < \theta_{2n}'$ represent the ordered angles in the joint set $\{\theta_1, \ldots, \theta_n, \theta_1^*, \ldots, \theta_n^*\}$. The plotted statistic used in the CC is in fact the bivariate sign test statistic of Hodges (1955), namely

$$H = \max_{0 \leq k \leq n-1} \left| \sum_{i=1}^{n} z_{k+i} - \frac{n}{2} \right|.$$

A shift in the location parameter is declared when $H > UCL$ for an appropriately chosen Upper Control Limit (UCL).

As mentioned earlier, Boone and Chakraborti (2012) provided multivariate versions of the sign and signed-rank CCs. Particularly, for each characteristic they introduced the univariate sign statistic as follows:

$$S_j = \sum_{i=1}^{n} \text{sgn}(X_{ji} - \theta_{j0}), \quad j = 1, 2, \ldots, p,$$

where X_{ji} represents the ith observation of the jth characteristic, n is the sample size, and θ_{j0} is the pre-specified in-control median value. Manifestly, $S_j = \#(X_{ji} > \theta_{j0}) - \#(X_{ji} < \theta_{j0})$ for $j = 1, 2, \ldots, p$ and $i = 1, 2, \ldots, n$. If S is the $p \times 1$ vector of the S_j's, then the quantity $n^{-1/2}S$ follows asymptotically the multivariate normal distribution with mean vector $\mathbf{0}$ and variance–covariance matrix $n^{-1}V$. The V matrix can be estimated by $\hat{V} = (\hat{v}_{jl})$, $j, l = 1, 2, \ldots, p$ using the formulae

$$\hat{v}_{jj} = n \quad \text{and} \quad \hat{v}_{jl} = \sum_{i=1}^{n} \text{sgn}(X_{ji} - \theta_{j0}) \cdot \text{sgn}(X_{li} - \theta_{l0}).$$

Finally, the plotted statistic of the SN^2 chart is the quadratic form

$$SN^2 = S'\hat{V}^{-1}S.$$

Analogously, the univariate Wilcoxon signed-rank test statistic

$$W_j = \sum_{i=1}^{n} R(|X_{ji} - \theta_{j0}|) \cdot \text{sgn}(X_{ji} - \theta_{j0}), \quad j = 1, 2, \ldots, p,$$

is calculated as per characteristic, where $R(|X_{ji} - \theta_{j0}|)$ is the rank of $|X_{ji} - \theta_{j0}|$ among $|X_{j1} - \theta_{j0}|, \ldots, |X_{jn} - \theta_{j0}|$. If W is the $p \times 1$ vector of the W_j's, then the quantity $n^{-3/2}W_j$ follows asymptotically the multivariate normal distribution with mean vector $\mathbf{0}$ and variance–covariance matrix $n^{-3}L$, where matrix L is estimated by $\hat{L} = (\hat{\ell}_{jl})$, $j, l = 1, 2, \ldots, p$ using the formulae

$$\hat{\ell}_{jj} = n(n+1)(2n+1)/6,$$

$$\ell_{jl} = \sum_{i=1}^{n} R(|X_{ji} - \theta_{j0}|) \cdot R(|X_{li} - \theta_{l0}|) \cdot \text{sgn}(X_{ji} - \theta_{j0}) \cdot \text{sgn}(X_{li} - \theta_{l0}).$$

Finally, the plotted statistic of the SR^2 chart is given by

$$SR^2 = W'\hat{L}^{-1}W.$$

In this case, the assumption of diagonal symmetry should be met.

The limiting distribution of both SN^2 and SR^2 statistics is χ^2 with p degrees of freedom. So, for a given alarm rate, say a, two-sided CCs can be constructed exploiting the next lower and upper CLs

$$LCL = 0 \quad \text{and} \quad UCL = \chi^2_{a,p}.$$

Abu-Shawiesh and Abdullah (2001) introduced a robust CC which uses the Hodges–Lehmann's (1963) and Shamos–Bickel–Lehmann's (1976) estimators (abbrev. *HL* and *SBL*, respectively) for monitoring a bivariate, symmetric and continuous process. For a given random sample $X_1, X_2, \ldots, X_n$, let the $M = n(n+1)/2$ Walsh averages be defined by $W_r = (X_i + X_\ell)/2$, where $r = 1, 2, \ldots, M$ and $i \leq \ell = 1, 2, \ldots, n$. Then, the *HL* estimator is defined as the sample median of the Walsh averages W_r. To define the *SBL* estimator, we introduce first $U = \binom{n}{2} = n!/[2!(n-2)!]$ pairwise distances $B_r = |X_i - X_j|$, $r = 1, 2, \ldots, U$ for $i < \ell = 1, 2, \ldots, n$. The *SBL* estimator is then the sample median of the pairwise distances B_r.

Assuming that there are p variables $X_1, X_2, \ldots, X_p$ at hand, each of which consists of m subgroups of size n, a control charting procedure can be constructed for testing the hypothesis $H_0 : (\mu, v) = (\mu_0, v_0)$ versus $H_1 : (\mu, v) \neq (\mu_0, v_0)$, where (μ, v) are the location parameters.

In order to achieve that, the M Walsh averages and the U pairwise distances need to be determined first for each subgroup. Then, the entries of of the vector of HL estimators $HL = [HL_j]_{p \times 1}$ are given by

$$HL_j = \frac{1}{m} \sum_{k=1}^{m} HL_{jk}, \quad j = 1, 2, \ldots, p,$$

where

$$HL_{jk} = \underset{1 \leq i \leq \ell \leq n}{\text{median}} \left\{ \frac{X_{ijk} + X_{ljk}}{2} \right\}, \quad k = 1, 2, \ldots, m.$$

The SBL estimators are given by

$$SBL_{jk} = \underset{1 \leq i \leq \ell \leq n}{\text{median}} |X_{ijk} - X_{ljk}|,$$

for $k = 1, 2, \ldots, m$ and $j = 1, 2, \ldots, p$. In the case of $p = 2$ quality characteristics, say X_1 and X_2, the variance–covariance matrix S_{SBL} is defined as

$$S_{SBL} = \begin{pmatrix} SBL_{11} & SBL_{12} \\ SBL_{21} & SBL_{22} \end{pmatrix} = \begin{pmatrix} SBL^2(X_1) & Cov(X_1, X_2) \\ Cov(X_1, X_2) & SBL^2(X_2) \end{pmatrix},$$

where $Cov(X_1, X_2) = SBL(X_1)SBL(X_2)r(X_1, X_2), r(X_1, X_2)$ is Spearman's rank correlation and $SBL^2(X_j) = SBL_j^2 = SBL_{jj}$ for $j = 1, 2$. Finally, the plotted statistic of the chart is given by

$$T^2_{HLSBL} = n(HL - \overline{HL})' \bar{S}_{SBL}^{-1} (HL - \overline{HL}),$$

where $\bar{S}_{SBL}^{-1}$ is the inverse matrix of $\bar{S}_{SBL}$ and $\bar{S}_{SBL}$ is calculated by averaging the subgroup variance–covariance matrices over the m subgroups. The (bivariate) process is declared out-of-control if, for a pre-specified FAR, the statistic T^2_{HLSBL} falls outside the CLs which can be determined using simulation.

Holland and Hawkins (2014) used the Mann–Whitney statistic in a change-point model that is analogue to the univariate CC of Hawkins and Deng (2010). For a given sample which consists of $p \times 1$ vectors $X_1, X_2, \ldots, X_n$, the multivariate change-point model is defined as

$$X_i \sim \begin{cases} F(\boldsymbol{\mu}), & \text{if } i \leq \tau, \\ F(\boldsymbol{\mu} + \boldsymbol{\delta}), & \text{if } i > \tau, \end{cases}$$

where F is the multivariate CDF of the process, $\boldsymbol{\delta}$ is an arbitrary shift in the location vector and τ is an index indicating the point at which the change occurs. The procedure can be applied to test the hypothesis $H_0 : \delta = 0$ versus $H_1 : \delta \neq 0$ for a given fixed value $\tau = k$.

In the univariate case and in order to centre the ranks, $R(X_i)$ is set equal to $2r^{(i)} - n - 1$, where $r^{(i)}$ is the rank of X_i among the $X_1, X_2, \ldots, X_n$. Then, $R(X_i) = \sum_{j=1}^{n} \text{sgn}(X_i - X_j)$ and the Wilcoxon–Mann–Whitney statistic for the difference between the two samples $X_1, X_2, \ldots, X_k$ and $X_{k+1}, X_2, \ldots, X_n$ is given by $u_k = \sum_{i=1}^{k} R(X_i)$.

In the multivariate case, the centred rank is defined by

$$R_n(X_i) = \sum_{j=1}^{n} h(X_i, X_j), \quad 1 \leq i \leq n.$$

Choi and Marden (1997) suggested using in the above formula the kernel function $h(x, y) = (x - y)/\|x - y\|$ so that $R_n(X_i)$ represents the directional rank of X_i, i.e. the sum of the unit vectors pointing from each data point to X_i. Then, they proceeded to the introduction of a directional rank test statistic as follows. Firstly, the following notations are used to denote the within group rank vectors:

$$R_{n,k}^*(X_i) = \sum_{j=k+1}^{n} h(X_i, X_j) \quad \text{and} \quad \bar{r}_n^{(k)} = \frac{1}{k} \sum_{i=1}^{k} R_n(X_i)$$

for every potential change-point $k = 1, 2, \ldots, n - 1$. In addition, the pooled covariance matrix of the centred vectors are defined by estimating the covariance matrices of the left segments of the data $\{X_1, X_2, \ldots, X_k\}$ and the right segments of the data $\{X_{k+1}, X_{k+2}, \ldots, X_n\}$ independently, i.e.

$$\tilde{\Sigma}_{k,n} = \frac{n^2}{n - 2} \left(\frac{1}{k^2} \sum_{i=1}^{k} R_k(X_i) R_k(X_i)' + \frac{1}{(n - k)^2} \sum_{i=k+1}^{n} R_{n,k}^*(X_i) R_{n,k}^*(X_i)' \right),$$

as well as the unpooled covariance matrix, which is the sample covariance matrix of all data vectors, i.e.

$$\hat{\Sigma}_n = \frac{1}{n - 1} \sum_{i=1}^{n} R_n(X_i) R_n(X_i)'.$$

Under the null hypothesis, the quantity $[nk/(n - k)]\bar{r}_n^{(k)}{}' \tilde{\Sigma}_{k,n}^{-1} \bar{r}_n^{(k)}$ follows the χ^2 distribution with p degrees of freedom and $\hat{\Sigma}_n$ is a consistent estimator of the true covariance matrix of the rank vectors. Then, the directional rank statistic suggested by Choi and Marden (1997) to test the difference in the location vector between the left and the right data segments is given by

$$r_{k,n} = \bar{r}_n^{(k)}{}' \hat{\Sigma}_{k,n}^{-1} \bar{r}_n^{(k)},$$

where $\hat{\Sigma}_{k,n} = [(n-k)/(nk)]\hat{\Sigma}_n$. Finally, the charting procedure is constructed by maximizing the above directional rank test statistic as follows:

$$r_{max,c,n} = \max_{c<k<n-c} r_{k,n},$$

where c is the number of data points near the beginning and the end of the sequence which are not taken into account in the detection of a possible change-point. For large values of the constant c, the distribution of $r_{k,n}$ is approximated by χ_p^2 for all $c < k < n - c$. The point at which a change occurs can be estimated using $\hat{\tau} = \arg\max_{c<k<n-c} r_{k,n}$ and the process is declared out-of-control if $r_{max,c,n} > h_{a,p,c,n}$, where $h_{a,p,c,n}$ is chosen so that $P[r_{max,c,n} > h_{a,p,c,n} | r_{max,c,j} \leq h_{a,p,c,n}; j < n] = a$.

Bell et al. (2014) considered a multivariate mean-rank (abbrev. *MMR*) chart that constitutes a multivariate extension of the scheme suggested by Jones-Farmer et al. (2009). The CC uses the ranked Data Depth (DD) values focusing on Mahalanobis Depth (MD) and constitutes a Phase I scheme for subgrouped data from elliptical distributions. It goes without saying that one can estimate the Phase II parameters by using the resulting historical, in-control sample.

More precisely, assuming that an independent and identically distributed (i.i.d.) reference sample is available which consists of m subgroups of size n and is coming from a $p-$dimensional continuous process, X_{ki} denotes the $1 \times p$ vector containing info of the ith observation of the kth subgroup. To construct the *MMR CC*, for each X_{ki} a DD function $D(x_{ki}; \hat{F}_N)$ is computed (preferably satisfying the four conditions described in Zuo and Serfling (2000); the notion of DD is formally defined in Subsection 3.3), where $\hat{F}_N$ is the empirical CDF of the pooled reference sample of size $N(= m \times n)$. To each of those DD functions, integer ranks of the form $R_{ki} = 1, 2, \ldots, N$ are assigned. When the process is in-control, the mean and the variance of the random variable R_{ki} are, respectively, given by $E(R_{ki}) = (N + 1)/12$ and $Var(R_{ki}) = (N - 1)(N + 1)/12$. Eventually, the plotted statistic is the standardized subgroup mean rank

$$Z_k = \frac{\bar{R}_k - E(\bar{R}_k)}{\sqrt{Var(\bar{R}_k)}},$$

where $E(\bar{R}_k) = (N + 1)/2$, $Var(\bar{R}_k) = (N - n)(N + 1)/(12n)$ and $\bar{R}_k = \frac{1}{n}\sum_{i=1}^{n} R_{ki}$. An alarm is triggered in the process, if Z_k falls outside the CLs, which depend on the values of m, n (not on p) and can be derived empirically.

Cheng and Shiau (2015) suggested a multivariate (spatial) sign-based, *Shewhart*-type CC for subgroups (abbrev. *MSS CC*). Let X_{ki}, $k = 1, 2, \ldots, m$ and $i = 1, 2, \ldots, n$, be an i.i.d. random sample which consists of m subgroups of size n. This sample has a $p-$dimensional continuous distribution function $F(x - \theta_0)$, where θ_0 is the in-control location parameter. The estimation of the sample *spatial median* $\hat{\theta}$ can be defined as the quantity that minimizes the function $\sum_{k=1}^{m}\sum_{i=1}^{n} ||X_{ki} - \theta||$. The desired $\hat{\theta}$ satisfies $\sum_{k=1}^{m}\sum_{i=1}^{n} U(X_{ki} - \theta) = 0$. Hettmansperger and Randles (2002) suggested a procedure for estimating simultaneously the spatial median θ and Tyler's transformation matrix A, which satisfy the following conditions:

$$E(U(A(X_{ki} - \boldsymbol{\theta}))) = E\left(\frac{A(X_{ki} - \boldsymbol{\theta})}{\|A(X_{ki} - \boldsymbol{\theta})\|}\right) = 0 \quad \text{and}$$

$$E(U(A(X_{ki} - \boldsymbol{\theta}))U(A(X_{ki} - \boldsymbol{\theta}))') = E\left(\frac{A(X_{ki} - \boldsymbol{\theta})(X_{ki} - \boldsymbol{\theta})'A'}{\|A(X_{ki} - \boldsymbol{\theta})\|^2}\right) = \frac{1}{p}I_p,$$

(1)

where A is a $p \times p$ upper triangular positive-definite matrix with 1 in the upper left-hand element. Thus, the estimators of the $\boldsymbol{\theta}$ and A can be obtained by the aid of the next system of equations

$$\frac{1}{mn} \sum_{k=1}^{m} \sum_{i=1}^{n} U(A(X_{ki} - \boldsymbol{\theta})) = 0 \quad \text{and}$$

$$\frac{1}{mn} \sum_{k=1}^{m} \sum_{i=1}^{n} U(A(X_{ki} - \boldsymbol{\theta}))U(A(X_{ki} - \boldsymbol{\theta}))' = \frac{1}{p}I_p.$$

(2)

It should be mentioned that the Hettmansperger–Randles estimators can be computed in high dimensions by using an iterative algorithm, which provides estimates for the spatial median $\hat{\boldsymbol{\theta}}$ and Tyler's scatter matrix S_T. Then, the corresponding Tyler's transformation matrix A_T is such that $A_T'A_T = S_T^{-1}$. In this case, the multivariate sign vector for each X_{ki} is $U_{ki} = U(A_T(X_{ki} - \hat{\boldsymbol{\theta}}))$ and the plotted statistic is given by

$$Q_k = np\bar{U}_k'\bar{U}_k$$

for $k = 1, 2, \ldots, m$, where $\bar{U}_k = \sum_{i=1}^{n} U_{ki}/n$. The process is declared out-of-control when the quantity Q_k takes values larger than a critical value.

Combining the last two procedures, Capizzi and Masarotto (2017) came up with a scheme based on multivariate signed-ranks which integrate both the spatial signs and the ranks of MDs for detecting location shifts. Their procedure suggests calculating first the scale and location estimates, which are given by

$$S = \begin{cases} \dfrac{1}{2(m-1)} \sum_{k=2}^{m} (X_{k,1} - X_{k-1,1})(X_{k,1} - X_{k-1,1})', & \text{if } n = 1, \\[3mm] \dfrac{1}{m(n-1)} \sum_{k=1}^{m} \sum_{i=1}^{n} (X_{k,i} - \bar{X}_k)(X_{k,i} - \bar{X}_k)', & \text{if } n > 1, \end{cases}$$

and

$$\ell = S^{1/2}(\text{spatial median of } S^{1/2}\bar{X}_1, \ldots, S^{1/2}\bar{X}_m),$$

respectively, where $\bar{X}_{k,i} = (1/n) \sum_{i=1}^{n} X_{k,i}$ and $S^{1/2}$ is the square root of S, i.e. a matrix such that $S = S^{1/2}(S^{1/2})'$. Using the above estimates, the observed data are standardized as $z_{k,i} = S^{1/2}(X_{k,i} - \ell)$ and the multivariate signed-ranks of the $z_{k,i}$'s are computed by

$$
u_{k,i} = \begin{cases} \mathbf{0}, & \text{if } z_{k,i} = \mathbf{0}, \\[2ex] \dfrac{\sqrt{F_{\chi_p^2}^{-1}\left(\dfrac{r_{k,i}}{1+mn}\right)}}{||z_{k,i}||} z_{k,i}, & \text{if } z_{i,j} \neq \mathbf{0}, \end{cases}
$$

where $r_{k,i}$ is the rank of $||z_{k,i}||$ among the $||z_{1,1}||, \ldots, ||z_{m,n}||$ and $F_{\chi_p^2}$ is the CDF of a χ^2 random variable with p degrees of freedom.

Then, the next steps involve fitting the following multivariate linear regression model:

$$
u_{k,i} = \boldsymbol{\beta}_{common} + \sum_{\tau=2}^{m-1} \boldsymbol{\beta}_{step,\tau} I(k \geq \tau) + \sum_{\tau=1}^{m} \boldsymbol{\beta}_{isolated,r} I(k = \tau) + (residual)_{k,i},
$$

where the $\boldsymbol{\beta}$'s are unknown $p-$dimensional parameter vectors. In particular, $\boldsymbol{\beta}_{common}$ is the "stable" level of the signed-ranks, $\boldsymbol{\beta}_{step,\tau}$ is a level change starting from time τ and affecting all the subsequent observations, $\boldsymbol{\beta}_{isolated,\tau}$ affects only the observations at time τ and it corresponds to an isolated outlier. Then, a procedure is established to check the hypothesis H_0 : all the $\boldsymbol{\beta}_{step,\cdot}$ and $\boldsymbol{\beta}_{isolated,\cdot}$ are zero. The next step involves fitting a model with an increasing number of parameters (isolated or step shifts) to the signed-ranks using a forward search algorithm. At the jth step, the fitted values are given by $\hat{u}_k^{(j)} = \hat{\boldsymbol{\beta}}_0^{(j)} + \hat{\boldsymbol{\beta}}_1^{(j)} \xi_k^{(1)} + \ldots + \hat{\boldsymbol{\beta}}_j^{(j)} \xi_k^{(j)}$ for $k = 1, 2, \ldots, m$, where $\hat{\boldsymbol{\beta}}_r^{(j)}$ or $r = 0, 1, \ldots, j$ are the $j-$dimensional parameter vectors and $\xi_k^{(j)}$ is a scalar sequence which corresponds to either an isolated ($\xi_k^{(j)} = I(k = \tau^{(j)})$) or a step shift ($\xi_k^{(j)} = I(k \geq \tau^{(j)})$) for some $\tau^{(j)}$. Since, this is a regression model, the type (either isolated or step) and the time ($\tau^{(j)}$) of the shift $\xi_k^{(j)}$ and the parameters $\hat{\boldsymbol{\beta}}_r^{(j)}$ at the jth step can be determined by minimizing the residual sum of squares $\sum_{k=1}^{m} \sum_{i=1}^{n} ||u_{k,i} - \hat{u}_i^{(k)}||^2$ on conditioning to the type and time shifts of the previous $(j-1)$ steps. Finally, at every step, the explained variance

$$
T_j = n \sum_{k=1}^{m} ||\hat{u}_k^{(j)}||^2 - mn||\bar{\bar{u}}||^2,
$$

where $\bar{\bar{u}} = \sum_{k=1}^{m} \sum_{i=1}^{n} u_{k,i}/mn$ and the following (overall) testing statistic are computed

$$
W_{OBS} = \max_{j=1,2,\ldots,J} \frac{T_j - E_0(T_j)}{\sqrt{Var_0(T_j)}}.
$$

It should be mentioned that Capizzi and Masarotto (2017) provided a procedure to compute the $p - value$ of the process, as well as a post-signal diagnostic method (based on the adaptive $LASSO$) which can be exploited to identify the time points at which shifts occurred and the variables that produce the out-of-control signal.

Koutras and Sofikitou (2017a) suggested a multivariate nonparametric control chart, the $O2$ (Order–Order) chart, which makes use of order statistics only as a decision criterion. Their chart is an extension to the bivariate case of the univariate median CC (Janacek and Meikle (1997)) and its variants (Chakraborti et al. (2004)). To describe the procedure, let $(X_1^{(R)}, Y_1^{(R)})$, $(X_2^{(R)}, Y_2^{(R)})$, ..., $(X_m^{(R)}, Y_m^{(R)})$ denote the observations of a historical sample of size m collected from an unknown, at least in part, continuous bivariate distribution. The associated joint CDF will be given by $F_{X,Y}^{(R)}(x, y) = F(x, y)$, while $F_X(x)$ and $F_Y(y)$ will stand for its marginal distributions. Subsequently, successive test samples of size n are drawn, which are independent of each other and of the reference sample as well. These samples are denoted by $(X_1^{(T)}, Y_1^{(T)})$, $(X_2^{(T)}, Y_2^{(T)})$, ..., $(X_n^{(T)}, Y_n^{(T)})$, while their joint CDF and the corresponding marginals are given by $F_{X,Y}^{(T)}(x, y) = G(x, y)$ and $G_X(x)$, $G_Y(y)$. In addition, let $C(u, v)$ and $D(u, v)$ stand for the bivariate copulas, associated with the respective in-control and out-of-control joint continuous CDFs $F(x, y)$ and $G(x, y)$. According to Sklar's (1959) theorem, the joint CDF of the bivariate random variable (X, Y) can be expressed as $F(x, y) = C(F_X(x), F_Y(y))$ when the process is in-control and $G(x, y) = D(G_X(x), G_Y(y))$ when it shifts out-of-control. Manifestly, $F(F_X^{-1}(u), F_Y^{-1}(v)) = C(F_X(F_X^{-1}(u)), F_Y(F_Y^{-1}(v))) = C(u, v)$, $G(G_X^{-1}(u), G_Y^{-1}(v)) = D(G_X(G_X^{-1}(u)), G_Y(G_Y^{-1}(v))) = D(u, v)$. When applying the $O2$ chart, the main goal is to ascertain whether the underlying process remains in-control or shifts from its in-control (null) distribution $F(x, y)$ to an out-of-control distribution, say $G(x, y)$; typically, this is equivalent to testing the null hypothesis $H_0 : F(x, y) = G(x, y)$ versus its two-sided alternative $H_1 : F(x, y) \neq G(x, y)$.

To establish the $O2$ chart, the authors used as plotted statistic the pair $(X_{r:n}^{(T)}, Y_{s:n}^{(T)})$ of bivariate order statistics obtained from the test sample for $1 \leq r, s \leq n$. The CLs were four appropriately chosen order statistics from the reference sample, namely

$$\text{LCL} = (LCL_X, LCL_Y) = (X_{a:m}^{(R)}, Y_{c:m}^{(R)}),$$
$$\text{UCL} = (UCL_X, UCL_Y) = (X_{b:m}^{(R)}, Y_{d.m}^{(R)}), \tag{3}$$

where $1 \leq a < b \leq m$ and $1 \leq c < d \leq m$. Aiming at a FAR that does not exceed a pre-specified level α, the values of the design parameters a, b, c, d are determined through the condition $FAR = 1 - P(\text{LCL} \leq \delta \leq \text{UCL}|H_0) \leq \alpha$. Consequently, the process is declared in-control if the following inequalities hold true:

$$X_{a:m}^{(R)} \leq X_{r:n}^{(T)} \leq X_{b:m}^{(R)} \text{ and } Y_{c:m}^{(R)} \leq Y_{s:n}^{(T)} \leq Y_{d:m}^{(R)}. \tag{4}$$

It should be stressed that when the medians are used, it is plausible to choose symmetric CLs, i.e. $b = m - a + 1$ and $d = m - c + 1$. Moreover, if one wishes to detect equal mean shifts ($\mu_1 = \mu_2$), then he/she should set $a = c$; otherwise, if, for example, there is an indication that larger shifts occur in the first characteristic (say $\mu_1 > \mu_2$), then it is reasonable to use $a > c$ because the CLs which correspond to the $X-$variate become more narrow and the first condition of Rule (4) becomes more sensitive. The authors provided design tables (for different values of the design parameters along with different levels of FAR) to help the practitioner choose the CLs according to his/her needs. They also gave formulae for the calculation of the AR and the FAR. In the first case, the characteristics share the same in-and out-of-control marginals (i.e. $F_X(x) = G_X(x)$ and $F_Y(y) = G_Y(y)$) but they do not have the same copula (i.e. $C(u, v) \neq D(u, v)$); while in order to calculate the FAR both characteristics have the same marginals and the same copula (i.e. $F_X(x) = G_X(x)$, $F_Y(y) = G_Y(y)$ and $C(u, v) = D(u, v)$). This means that the marginal effect is removed in both cases, so that the formulae do not depend on the choice of the marginal distributions. The performance of the $O2$ CC is typically affected by the dependence structure of the monitored characteristics (as reflected through the copula) and not by their marginal distributions. Therefore, the $O2$ CC should be formally characterized as a semiparametric CC and not a nonparametric one. However, as evidenced by the extensive numerical experimentation carried out by Koutras and Sofikitou (2017a), the values of the FAR and the ARL_{in} are almost the same when different distributions/copulas are used, therefore it can be used in practice as a fully nonparametric CC.

Koutras and Sofikitou (2017b) also proposed the OC (Order statistics and Concomitants) scheme, which is a modification of the $O2$ so that the dependence structure of the initial characteristics has a direct impact on the decision rule. To achieve that the CLs involve now two particular pairs from the reference sample, namely $(X_{a:m}^{(R)}, Y_{[a:m]}^{(R)})$ and $(X_{b:m}^{(R)}, Y_{[b:m]}^{(R)})$, where $1 \leq a < b \leq m$ and $Y_{[a:m]}^{(R)}, Y_{[b:m]}^{(R)}$ denote the concomitants of $X_{a:m}^{(R)}, X_{b:m}^{(R)}$, respectively. We recall that, if the X-variates are arranged in ascending order as $X_{1:m}^{(R)} \leq X_{2:m}^{(R)} \leq \ldots \leq X_{m:m}^{(R)}$, then their paired Y-variates $Y_{[1:m]}^{(R)}, Y_{[2:m]}^{(R)}, \ldots, Y_{[m:m]}^{(R)}$ have been termed concomitants of $X_{1:m}^{(R)}, X_{2:m}^{(R)}, \ldots, X_{m:m}^{(R)}$. In this case, the process is assumed to be in-control if the following conditions hold true:

$$X_{a:m}^{(R)} \leq X_{r:n}^{(T)} \leq X_{b:m}^{(R)} \text{ and } \min\left\{Y_{[a:m]}^{(R)}, Y_{[b:m]}^{(R)}\right\} \leq Y_{s:n}^{(T)} \leq \max\left\{Y_{[a:m]}^{(R)}, Y_{[b:m]}^{(R)}\right\}. \quad (5)$$

Notice that the use of min and max functions is unavoidable, since the true order of $Y_{[a:m]}^{(R)}, Y_{[b:m]}^{(R)}$ is not known. It should be underlined that the OC chart is capable of identifying shifts in location as well as in correlation (because of the use of concomitants). The chart does not require any distributional assumption about the marginals; however, the values of FAR and ARL_{in} do vary for different choices of the correlation level. So, the performance of the OC chart is affected by the dependence structure and it is to expect that the chart becomes more robust as the correlation becomes stronger.

b. Multivariate CCs with Memory

Kapatou and Reynolds (1994, 1998) (see also Kapatou (1996)) proposed $EWMA$ schemes for small samples based on sign and signed-rank statistics. Let us assume that a small sample $X_k = [X_{kji}]$ of size n is collected at equal time intervals, where $k = 1, 2, \ldots$ denotes the time at which the sample was collected, and $j = 1, 2, \ldots, p$ is used to describe the p variables of the sample. X_k is also assumed to come from a p-dimensional CDF the marginal distributions of which are continuous and symmetric. The $EWMA$ sign statistic introduced by Kapatou and Reynolds is given by

$$Y_{kj} = (1 - r)Y_{(k-1)j} + rS_{kj} = (1 - r)^k Y_{0j} + \sum_{l=0}^{k-1}(1 - r)^l r S_{(k-l)j},$$

for $j = 1, 2, \ldots, p$ and $k = 1, 2, \ldots$. The starting value for Y_{kj}'s is defined as $Y_{0j} = E(S_{kj}|\mu = \mu_0)$, where μ_0 is the vector with the in-control (target) values and

$$S_{kj} = \sum_{i=1}^{n} I_{kij} = \#(X_{kji} > \mu_{0j})$$

is the sign-based statistic, where $I_{kji} = I(X_{kji} > \mu_{0j})$. Moreover, the $EWMA$ signed-rank statistic is given by

$$Y_{kj} = (1 - r)Y_{(k-1)j} + rT_{kj}^{+} = (1 - r)^k Y_{0j} + \sum_{i=0}^{k-1}(1 - r)^j r T_{(k-i)j}^{+},$$

for $j = 1, 2, \ldots, p, i = 1, 2, \ldots, n$ and $k = 1, 2, \ldots$. The starting value is now provided by $Y_{0j} = E(T_{kj}^{+}|\mu = \mu_0)$, where μ_0 is the vector with the in-control (target) values and

$$T_{kj}^{+} = \sum_{i=1}^{n} I_{kji} R_{kji}$$

is the sign-based statistic, which makes use of the rank of $|X_{kji} - \mu_{0j}|$ among $|X_{kj1} - \mu_{0j}|, \ldots, |X_{kjn} - \mu_{0j}|$ a quantity denoted by R_{kji} in the last formula. It should be mentioned that the aforementioned CCs lack of the affine invariance property, so any kind of linear transformation or rotation could alter the results (Beltran (2006)).

Zou and Tsung (2011) suggested a CC for monitoring process mean which combines the spatial sign test of Randles (2000) with the $EWMA$ procedure. Let us assume that we have at hand m_0 i.i.d., reference sample observations of dimension $p \geq 1$, say $X_{-m_0+1}, \ldots, X_0$, and that the tth observation from the multivariate change-point model below

$$X_t \sim \begin{cases} F_0(x - \mu_0), & \text{for } t = -m_0 + 1, \ldots, 0, 1, \ldots, \tau, \\ F_0(x - \mu_1), & \text{for } t = \tau + 1, \ldots, \end{cases}$$

where τ denotes the (unknown) point at which a change occurs and $\mu_0 \neq \mu_1$; the primary goal is to obtain a multivariate median θ_0 and a transformation matrix A_0 from the reference sample. The authors suggest using the affine equivariant multivariate median introduced by Hettmansperger and Randles (2002) along with the corresponding A_0, which can be constructed by solving the system of equations defined in (1) if X_{ki} is replaced by X_t. The sample estimates $(\hat{\boldsymbol{\theta}}_0, \hat{\boldsymbol{A}}_0)$ can be obtained from the historical sample by the aid of the system of equations of (2) if we set $n = 1$, m and X_{ki} are replaced by m_0 and X_t, respectively, and the outer summation is performed over $t \in \{-m_0 + 1, \ldots, 0\}$.

The observation collected for future monitoring is standardized, i.e. $V_t = U(A_0(X_t - \boldsymbol{\theta}_0))$, and a multivariate $EWMA$ procedure is established by the recursive scheme $W_t = (1 - \lambda)W_{t-1} + \lambda V_i$ is defined, where $V_0 = E(V_t) = \boldsymbol{0}$. The plotted statistic is given by

$$Q_t = \frac{2 - \lambda}{\lambda} p W_t' W_t$$

and the resulting CC triggers an alarm when the quantity Q_t takes values larger than a critical value $UCL > 0$ (chosen so as a pre-specified ARL_{in} is attained). It should be mentioned that in the case of subgroups ($n > 1$), i.e. when a group of observations $\{X_{t1}, X_{t2}, \ldots, X_{tn}\}$ is available for each t, then the standardization is made by using $V_t = (1/n) \sum_{i=1}^{n} U(A_0(X_{ti} - \boldsymbol{\theta}_0))$.

Motivated by the previous idea, Zou et al. (2012) also proposed self-starting techniques which, instead of being based on spatial sign, are based on spatial ranks. In the multivariate case, the average of the spatial signs of the pairwise differences is given by $r_t = R_E(X_t) = (1/n) \sum_{i=1}^{n} U(X_t - X_i)$, where $R_E(\cdot)$ represents the empirical spatial rank function; while the theoretical spatial rank function of the vector X (with respect to F) is obtained by replacing the average with the expectation in $R_E(\cdot)$, i.e. $R_F(X) = E_Y[U(X - Y)]$, where $Y \sim F$ and $E_Y[\cdot]$ denotes the expectation with respect to the random vector Y.

Firstly, Zou et al. (2012) suggested a theoretical spatial-based $EWMA$ CC (abbrev. $TREWMA$), which can be used to test the null hypothesis $H_0 : \boldsymbol{\mu} = \boldsymbol{\mu}_0$ versus the alternative $H_0 : \boldsymbol{\mu} \neq \boldsymbol{\mu}_0$. According to the authors, a reasonable test statistic to be used is $R_F'(X)[Cov(R_F(X))]^{-1}R_F(X)$. An affine-invariant modification of the aforementioned statistic can then be established through the formula $Q^{R_F} = R_F'(MX)[Cov(R_F(MX))]^{-1}R_F(MX)$, where $S = (M'M)^{-1}$. The suggested plotted statistic of the $TREWMA$ CC is given by

$$Q_t^{R_F} = \frac{2 - \lambda}{\lambda} W_t'[Cov(R_F(MX))]^{-1} W_t,$$

where W_t is the $EWMA$ sequence produced by the recursive scheme $W_t = (1 - \lambda)W_{t-1} + \lambda R_F(MX_t)$ with $W_0 = \boldsymbol{0}$.

Secondly, Zou et al. (2012) proposed an empirical spatial-based $EWMA$ CC (abbrev. $SREWMA$). In this CC, the empirical spatial rank function of the tth future observation is obtained from the expression $R_E(\hat{M}_{t-1}X_t) = [1/(m_0 + t - 1)] \sum_{j=-m_0+1}^{t-1} U(\hat{M}_{t-1}(X_t - X_j))$, where $\hat{S}_{t-1} = (\hat{M}_{t-1}'\hat{M}_{t-1})^{-1} = \sum_{j=-m_0+1}^{t-1} (X_j - \bar{X}_{t-1})(X_j - \bar{X}_{t-1})'$ is the sample covariance matrix obtained from the $m_0 + t - 1$ historical observations and $\bar{X}_{t-1}$ denotes the corresponding sample mean. Provided that, when the process is in-control, the equality $Cov(R_F(Mx)) = E[||R_F(MX_t)||^2]I_p/p$ holds true, the authors suggest using

$$\widehat{Cov}(R_E(\hat{M}_{t-1}X_t)) \approx \hat{E}[||R_F(MX_t)||^2]I_p/p \quad \text{and}$$

$$\hat{E}[||R_F(MX_t)||^2] \approx \frac{1}{m_0 + t - 1}\left[\sum_{j=-m_0+1}^{0}||\tilde{R}_E(\hat{M}_0X_j)||^2 + \sum_{j=1}^{t-1}||R_E(\hat{M}_{j-1}X_j)||^2\right],$$

where $\tilde{R}_E(\hat{M}_0X_j) = (1/m_0)\sum_{k=-m_0+1}^{0}U(\hat{M}_0(X_j - X_k))$. The plotted statistic of the *SREWMA CC* is given by

$$Q_t^{R_E} = \frac{2-\lambda}{\lambda\xi_t}||V_t||^2,$$

where $V_t = (1-\lambda)V_{t-1} + \lambda R_E(\hat{M}_{t-1}X_t), v_0 = 0$ and $\xi_t \equiv \hat{E}[||R_F(MX_t)||^2]$. The process is declared out-of-control in the event that $Q_t^{R_E} > UCL$. The authors provided some guidance on how one should choose the smoothing weight λ, the parameter m_0, as well as the UCLs.

Finally, Zi et al. (2013) integrated a multivariate spatial sign test and an *EWMA* procedure to on-line sequential monitoring. The goal here is to test the hypothesis $H_0 : \theta = 0$ versus its alternative $H_1 : \theta = \delta d_1$ or $\theta = \delta d_2$ or ... or $\theta = \delta d_r$, where $d_1, d_2, \ldots, d_r$ are plausible different (known) directions and δ is an unknown constant. The procedure on estimating the multivariate centre θ_0 and the transformation matrix A_0 remains the same as before. The plotted statistic of the proposed multivariate directional sign *EWMA* (abbrev. *MDSE*) *CC* is given by

$$M_t = \frac{(2-\lambda)p}{\lambda}\max_{1\le j\le r}\left[(d_j'A_0'W_t)^2/||d_j||^2_{A_0'A_0}\right],$$

where $W_t = (1-\lambda)W_{t-1} + \lambda V_t$, $W_0 = 0$. The process is out-of-control if $M_t > UCL$. If each observation consists of n subgroups, then $V_t = (1/n)\sum_{j=1}^{n}U(A_0(X_{tj} - \theta_0))$. The authors also suggested the use of formula $\zeta^* = \arg\max_{1\le j\le r}[(d_j' A_0'W_i)^2/||d_j||^2_{A_0'A_0}]$ to detect the component where the shift has occurred.

Multivariate nonparametric *CUSUM* schemes have also been developed. For example, Qiu and Hawkins (2001) constructed a *CUSUM* procedure, which is based on the "antiranks" (i.e. the indices of order statistics) of the measurements under study. More precisely, denoting by $X(t) = (X_1(t), X_2(t), \ldots, X_p(t))'$ are p measurements taken at the t-th time point, then the antirank vector of $X(t)$, say $A(t) = (A_1(t), A_2(t), \ldots, A_p(t))'$, is a permutation of $(1, 2, \ldots, p)'$ such that $X_{A_1(t)} \le X_{A_2(t)} \le \ldots X_{A_p(t)}$ are the order statistics of $\{X_j(t), j = 1, 2, \ldots, p\}$. In addition, let us define $\xi_1(t) := (\xi_1(t), \xi_2(t), \ldots, \xi_p(t))'$, where $\xi_j(t) = I(A_1(t) = j)$ is the indicator variable of the event that the j-th $(1 \le j \le p)$ measurement takes on the smallest value among the p measurements at the t-th time point. According to Qiu and Hawkins (2001), in a multivariate setting, testing the null hypothesis $H_0 : \mu_1 = \mu_2 = \ldots = \mu_p = 0$ is equivalent to testing the following combined hypotheses $H_0^{(1)} : \mu_1 = \mu_2 = \ldots = \mu_p$ and $H_0^{(2)} : \sum_{j=1}^{p}\mu_j = 0$. So, a violation at

H_0 indicates that at least one of the $H_0^{(1)}$, $H_0^{(2)}$ are violated. Hypothesis $H_0^{(1)}$ can be tested using a multivariate $CUSUM$ procedure, while hypothesis $H_0^{(2)}$ can be tested using a univariate $CUSUM$ procedure based on $\sum_{j=1}^{p} X_j(t)$. The authors prove that instead of testing $H_0^{(1)}$, one can alternatively test the hypothesis $H_0^{(1)*}$: the distribution of $A_1(t)$ is $\{g_j, \; j = 1, 2, \ldots, p\}$. In order to set up the (multivariate) $CUSUM$ procedure, the following quantities need to be calculated first:

$$\begin{cases} S_n^{(1)} = \mathbf{0}, \\ S_n^{(2)} = \mathbf{0}, & \text{if } C_n \leq k_1, \\ S_n^{(1)} = (S_{n-1}^{(1)} + \boldsymbol{\xi}_1(n))(C_n - k_1)/C_n, \\ S_n^{(2)} = (S_{n-1}^{(2)} + \boldsymbol{g}_1)(C_n - k_1)/C_n, & \text{if } C_n > k_1, \end{cases}$$

and

$$\begin{aligned} C_n = {} & [(S_{n-1}^{(1)} - S_{n-1}^{(2)}) + (\boldsymbol{\xi}_1(n) - \boldsymbol{g}_1)]' \\ & \times \text{diag}((S_{n-1,1}^{(2)} + g_1)^{-1}, \ldots, (S_{n-1,p}^{(2)} + g_p)^{-1}) \\ & \times [(S_{n-1}^{(1)} - S_{n-1}^{(2)}) + (\boldsymbol{\xi}_1(n) - \boldsymbol{g}_1)], \end{aligned}$$

where $\boldsymbol{g}_1 = (g_1, g_2, \ldots, g_p)'$, $S_0^{(1)} = S_0^{(2)} = \mathbf{0}$ and $k_1 \geq 0$ denotes a constant. Next, the quantity

$$y_n = (S_n^{(1)} - S_n^{(2)})' \times \text{diag}(1/S_{n,1}^{(2)}, \ldots, 1/S_{n,p}^{(2)}) \times (S_n^{(1)} - S_n^{(2)})$$

should be evaluated. Then, the process is declared out-of-control if $y_n > h_1$, where h_1 is a CL that can be determined using simulation. The authors showed that (a) $y_n = \max(0, C_n - k_1)$ and (b) a pre-specified ARL_{in} can be achieved if k_1 takes values in the interval $\left[0, \max_{1 \leq \ell \leq p} \frac{\sum_{j \neq \ell} g_j}{g_\ell} \right)$.

Later on, Qiu and Hawkins (2003) came up with another $CUSUM$ procedure, which is based on the order information among the measurement components as well as on the order information between the measurement components and their nominal means. They denoted by $\boldsymbol{B}(t) = (B_1(t), B_2(t), \ldots, B_p(t), B_{p+1}(t))'$ the antirank vector of $\boldsymbol{Y}(t) := (X_1(t), X_2(t), \ldots, X_p(t), 0)'$, that is, a permutation of $(1, 2, \ldots, p, p + 1)'$ such that $Y_{B_1(t)} \leq Y_{B_2(t)} \leq \ldots Y_{B_{p+1}(t)}$ are the order statistics of $\boldsymbol{Y}(t)$. Moreover, they defined $\boldsymbol{\eta}_1(t) := (\eta_1(t), \eta_2(t), \ldots, \eta_{p+1}(t))'$, where $\eta_j(t) = I(B_1(t) = j)$ for $1 \leq j \leq p + 1$. Now, in order to test the hypothesis $H_0^{(1)*}$, the authors suggest calculating first the following vector sequence:

$$\begin{cases} S_n^{(1)} = \mathbf{0}, \\ S_n^{(2)} = \mathbf{0}, & \text{if } C_n \leq k, \\ S_n^{(1)} = (S_{n-1}^{(1)} + \boldsymbol{\eta}_1(n))(C_n - k)/C_n, \\ S_n^{(2)} = (S_{n-1}^{(2)} + \boldsymbol{d})(C_n - k)/C_n, & \text{if } C_n > k, \end{cases}$$

and

$$C_n = [(S_{n-1}^{(1)} - S_{n-1}^{(2)}) + (\eta_1(n) - d)]'$$

$$\times \text{diag}\left(\frac{1}{S_{n-1,1}^{(2)} + d_1}, \dots, \frac{1}{S_{n-1,p+1}^{(2)} + d_{p+1}}\right)$$

$$\times [(S_{n-1}^{(1)} - S_{n-1}^{(2)}) + (\eta_1(n) - d)],$$

where $d = (d_1, d_2, \dots, d_{p+1})'$, $S_0^{(1)} = S_0^{(2)} = 0$ and $k \geq 0$ denotes a constant. Next, we compute the quantity

$$y_n = (S_n^{(1)} - S_n^{(2)})' \times \text{diag}(1/S_{n,1}^{(2)}, \dots, 1/S_{n,p+1}^{(2)}) \times (S_n^{(1)} - S_n^{(2)})$$

and the process is declared out-of-control if $y_n > h$, where h is a CL determined via simulation. Similarly, the authors showed that (a) $y_n = \max(0, C_n - k_1)$ and (b) a pre-specified ARL_{in} can be achieved if k_1 takes values in the interval $\left[0, \max_{1 \leq \ell \leq p+1} \frac{\sum_{j \neq \ell} d_j}{d_\ell}\right).$

3.2 Monitoring of Scale

It seems that, in the literature of nonparametric CCs, not much work has been done for the construction of CCs based on order statistics, ranks and signs, which are capable of monitoring the scale parameter of a multivariate process. In what follows, we present in some detail the few schemes that are currently available in the literature.

a. *Multivariate CCs without Memory*

To the best of our knowledge, only Osei-Aning et al. (2017) have considered nonparametric *Shewhart*-type CCs for monitoring the covariance matrix of a bivariate process. The plotted statistics used for these charts are the maximum of variance measures computed separately for each of the two characteristics for the available sample. Four different variance measures were practised, namely the standard deviation, the interquartile range, the absolute deviation from the sample median and the median absolute deviation. Assuming that $(X_1, Y_1), (X_2, Y_2), \ldots, (X_n, Y_n)$ is a bivariate sample of size n from the monitored process, the plotted statistics are

$$SMAX = \max\{S_X, S_Y\},$$
$$QMAX = \max\{Q_X, Q_Y\},$$
$$MDMAX = \max\{MD_X, MD_Y\},$$
$$MADMAX = \max\{MAD_X, MAD_Y\},$$

where

$$S_X = \sqrt{\frac{1}{n-1}\sum_{i=1}^{n}(X_i - \bar{X})^2}, \qquad S_Y = \sqrt{\frac{1}{n-1}\sum_{i=1}^{n}(Y_i - \bar{Y})^2},$$

$$Q_X = (Q_X^{(3)} - Q_X^{(1)})/1.34898, \qquad Q_Y = (Q_Y^{(3)} - Q_Y^{(1)})/1.34898,$$

$$MD_X = \frac{1}{n}\sum_{i=1}^{n} |X_i - \delta_X|, \qquad MD_Y = \frac{1}{n}\sum_{i=1}^{n} |Y_i - \delta_Y|,$$

$$MAD_X = 1.4826 \, \text{med} \, |X_i - \delta_X|, \qquad MAD_Y = 1.4826 \, \text{med} \, |Y_i - \delta_Y|,$$

and δ_X, δ_Y are the sample medians. In each case, an alarm is triggered when the statistic takes values higher than an upper CL, which is selected so as a the desirable ARL_{in} level is achieved.

b. *Multivariate CCs with Memory*

Haq and Khoo (2018) suggested an $EWMA$ sign CC for monitoring process dispersion. Let $\mathbf{X} = (X_1, X_2, \ldots, X_p)'$ be a $p-$dimensional vector with mean $\boldsymbol{\mu}_X = E(\mathbf{X}) = (E(X_1), E(X_2), \ldots, E(X_k))' = (\mu_{X_1}, \mu_{X_2}, \ldots, \mu_{X_p})'$ and variance–covariance matrix $\boldsymbol{\Sigma}_X = Cov(\mathbf{X}) = E[(\mathbf{X} - \boldsymbol{\mu}_X)(\mathbf{X} - \boldsymbol{\mu}_X)']$, where p indicates the number of characteristics under study. It is known that when the process is in-control, the quantity $(\mathbf{X} - \boldsymbol{\mu}_X)(\mathbf{X} - \boldsymbol{\mu}_X)'$ is an unbiased estimator of $\boldsymbol{\Sigma}_X$, i.e. $E[(\mathbf{X} - \boldsymbol{\mu}_X)(\mathbf{X} - \boldsymbol{\mu}_X)'] = \boldsymbol{\Sigma}_X$. Moreover, it holds true that $E[\text{tr}\{(\mathbf{X} - \boldsymbol{\mu}_X)(\mathbf{X} - \boldsymbol{\mu}_X)'\boldsymbol{\Sigma}_X^{-1}\}] = \text{tr}(\mathbf{I}_p) = p$, where $\text{tr}(\cdot)$ is the trace of the matrix $(\cdot)$.

Suppose now that $\mathbf{X}_t = (X_{1t}, X_{2t}, \ldots, X_{nt})$, $t = 1, 2, \ldots$ is a random sample (of size n) drawn from the process at the $t-$th time point, where the random vectors $\mathbf{X}_{it} = (X_{1it}, X_{2it}, \ldots, X_{pit})'$ are independent of each other for $i = 1, 2, \ldots, n$. It goes without saying that if the process is in-control, then $E(Y_{it}) = E(\mathbf{X}_{it} - \boldsymbol{\mu}_X)'\boldsymbol{\Sigma}_X^{-1}(\mathbf{X}_{it} - \boldsymbol{\mu}_X) = p$ for $i = 1, 2, \ldots, n$. The authors considered the random variable $A_t = \sum_{i=1}^{n} I(Y_{it} > p)$, which is the total number of instances in which $Y_{it} > k$. A_t follows a Bernoulli distribution with parameters n and p. If the process is in-control, then $p = p_0 = P(Y_{it} > k)$ and the value of p depends on the probability distribution of Y_{it}. Then, they used the transformation $B_t = \sin^{-1}(\sqrt{A_t/n})$, because it asymptotically follows a Normal distribution with mean $\sin^{-1}(\sqrt{p})$ and variance $1/(4n)$. The last quantity is exploited to determine the $EWMA$ statistic of the CC, which is given by

$$C_t = \lambda \cdot B_t + (1 - \lambda) \cdot C_{t-1},$$

where $C_0 = \sin^{-1}(\sqrt{p_0})$ and $\lambda \in (0, 1]$ is the smoothing constant.

Notice that when the process is in-control, $E(C_t) = E(B_t) = \sin^{-1}(\sqrt{p_0})$ and $Var(C_t) = (\lambda/(2 - \lambda)) \cdot (1 - (1 - \lambda)^{2t}) \cdot (1/(4n)) \overset{\text{asympt.}}{\approx} (\lambda/(2 - \lambda)) \cdot (1/(4n))$. Therefore, the CLs of the $EWMA$ sign CC will be calculated by

$$UCL = \sin^{-1}(\sqrt{p_0}) + \frac{w}{2\sqrt{n}} \cdot \sqrt{\frac{\lambda}{2 - \lambda}},$$

$$Centerline = \sin^{-1}(\sqrt{p_0}),$$

$$LCL = \sin^{-1}(\sqrt{p_0}) - \frac{w}{2\sqrt{n}} \cdot \sqrt{\frac{\lambda}{2 - \lambda}},$$

where $w(>0)$ is appropriately chosen so as a pre-specified value of ARL_{in} is achieved.

It should be mentioned that both the values of λ and w affect the ARL. Haq and Khoo (2018) gave formulae for the ARL and the Standard Deviation of the Run Length ($SDRL$), which can be calculated by the aid of Markov chain. It is worth mentioning that this chart was initially suggested for monitoring the process variability; however, the numerical study revealed that it is also capable of detecting shifts in both mean and variance.

3.3 Simultaneous Monitoring of Location and Scale

When monitoring a manufacturing process, it is of special interest to have a tool that is capable of detecting the presence of assignable causes that trigger simultaneous shifts to both scale and location and not separately in either mean or variability. In the SPC literature, two different procedures have been proposed to construct CCs for joint monitoring of both mean and variance. One technique makes use of a single, combined plotted statistic, which is capable of monitoring both the mean and the variance. The other technique makes use of two distinct statistics and creates a two-chart monitoring scheme which consists of two graphs: one for monitoring the scale and one for monitoring the location. Usually an increase in variability is related to deterioration of the process, so in the former case it is common that the CCs have only one CL (upper CL).

a. *Bivariate and Multivariate CCs without Memory*

Liu (1995) was the first who considered CCs for the joint monitoring of mean and variability using one-chart monitoring schemes with combined statistics. She proposed two *Shewhart*-type CCc (abbrev. r and Q), which are based on the notion of DD and constitute a generalization of the classical univariate parametric X and $\bar{X}$ charts.

To present the proposed CCs, let us first assume that $X_1, X_2, \ldots, X_m$ is the reference sample coming from an $p-$dimensional distribution F, and denote by $\hat{F}_m(\cdot)$ the empirical distribution induced by the reference sample. In the sequel, a test sample $Y_1, Y_2, \ldots, Y_n \sim G$ is collected, which should be compared to the reference sample in order to ascertain whether or not the process has remained is in-control. This can be translated to a hypothesis problem, where

$H_0 : F = G$ with a false alarm rate a,

H_1 : there is a shift in location and/or an increase in the variance.

The statistics, which are used for determining the plausible differences between the in-and out-of-control distributions F and G, are based on the notion of DD. For each point $x \in R^p$, the *simplicial depth* of x with respect to F is provided by

$$SD_F(x) = P_F\{x \in s[X_1, X_2, \ldots, X_{p+1}]\},$$

where $s[X_1, X_2, \ldots, X_{p+1}]$ is a polygon whose vertices $X_1, X_2, \ldots, X_{p+1}$ are $p + 1$ random observations from the distribution F. The quantity $SD_F(x)$ measures how deep or central is the point x with respect to F distribution. When the distribution F is not known and only the sample $X_1, X_2, \ldots, X_m$ is available, the *sample simplicial depth* of x is defined as

$$SD_{\hat{F}_m}(x) = \binom{m}{p+1}^{-1} \sum I(x \in s[X_{i_1}, X_{i_2}, \ldots, X_{i_{p+1}}]),$$

where the summation is carried over all the possible subsets of $X_1, X_2, \ldots, X_m$ of size $p + 1$. The quantity $SD_{\hat{F}_m}(x)$ measures how deep is the point x is with respect to the sample $X_1, X_2, \ldots, X_{p+1}$.

Without loss of generality, the notion of DD can be based on the Mahalanobis distance. In this case, how deep a point $\mathbf{x}$ is with respect to a given distribution F can be measured by how small is its quadratic distance to the mean. The *Mahalanobis Depth* is given by

$$\mathbf{MD}_F(x) = \frac{1}{1 + (x - \mu_F)'\Sigma_F^{-1}(x - \mu_F)},$$

where μ_F and Σ_F are the mean vector and the variance–covariance matrix of the distribution F, respectively. The empirical form of the quantity $\mathbf{MD}_F(x)$ is given by

$$\mathbf{MD}_{\hat{F}_m}(x) = \frac{1}{1 + (x - \bar{X})'S_F^{-1}(x - \bar{X})},$$

where $\bar{X}$ is the sample mean of $X_1, X_2, \ldots, X_m$, and S_F is the sample variance–covariance matrix. It should be stressed that $\mathbf{MD}_F(\cdot)$ is affine invariant.

If the depths of $X_1, X_2, \ldots, X_m$ are arranged in ascending order, i.e. $X_{[1]}, X_{[2]}, \ldots, X_{[m]}$, then $X_{[m]}$ represents the most central point. The smallest the order of this point, the more distant the point is from the F distribution.

Liu (1995) set

$$r_F(x) = P\left(D_F(X) \le D_F(x) \mid X \sim F\right), \tag{6a}$$

$$r_{\hat{F}_m}(x) = \frac{\#\{X_k \mid D_{\hat{F}_m}(X_i) \le D_{\hat{F}_m}(x), \, k = 1, 2, \ldots, m\}}{m}. \tag{6b}$$

The quantities $\{r_{\hat{F}_m}(X_1), r_{\hat{F}_m}(X_2), \ldots\}$ are computed only if the $X_1, X_2, \ldots, X_m$ are available, but not F.

Supposing that each subset is of size n, the means of $r_F(Y_i)$ or of $r_{\hat{F}_m}(Y_i)$ are computed by the formulae $Q(F, \hat{G}_n^k)$ and $Q(\hat{F}_m, \hat{G}_n^k)$, where $\hat{G}_n^k$ is the empirical distribution of the Y_i which belongs to the k−th subset for $k = 1, 2, \ldots$. Denoting the empirical distribution of the sample $Y_1, Y_2, \ldots, Y_n \sim G$ by $\hat{G}_n(\cdot)$, the following quantities can be defined:

$$Q(F, G) = P\{D_F(X) \le D_F(Y) \mid X \sim F, Y \sim G\},$$

$$Q(F, \hat{G}_n) = \frac{1}{n} \sum_{i=1}^{n} r_F(Y_i), \tag{7a}$$

$$Q(\hat{F}_m, \hat{G}_n) = \frac{1}{n} \sum_{i=1}^{n} r_{\hat{F}_m}(Y_i). \tag{7b}$$

For the construction of the r chart, Liu (1995) suggests plotting either the quantities $\{r_F(Y_1), r_F(Y_2), \ldots\}$ or $\{r_{\hat{F}_m}(Y_1), r_{\hat{F}_m}(Y_2), \ldots\}$ against time $t = 1, 2, \ldots$ by the aid of Formulae (6a) or (6b). The respective CLs will be

$$Centreline = 1/2 \quad \text{and} \quad LCL = a,$$

where a is the FAR. In each case, the process is declared out-of-control if the $r_F(Y_t)$'s or $r_{\hat{F}_m}(Y_t)$'s fall below the LCL.

For the construction of the Q chart, Liu (1995) suggests computing the means of the quantities $\{r_F(Y_1), r_F(Y_2), \ldots\}$ or $\{r_{\hat{F}_m}(Y_1), r_{\hat{F}_m}(Y_2), \ldots\}$, i.e. $Q(F, \hat{G}_n^k)$ or $Q(\hat{F}_m, \hat{G}_n^k)$ with respect to the kth subset ($k = 1, 2, \ldots$), by exploiting Formulae (7a) and (7b). The points plotted in the chart can be $\{Q(F, \hat{G}_n^1), Q(F, \hat{G}_n^2), \ldots, \}$ or $\{Q(F, \hat{G}_n^1), Q(F, \hat{G}_n^2), \ldots, \}$ if only $X_1, X_2, \ldots, X_m$ are available, against time $t = 1, 2, \ldots$. The CLs that correspond to the $Q(F, \hat{G}_n^k)$'s and $Q(\hat{F}_m, \hat{G}_n^k)$'s are respectively given by

$$LCL = \frac{1}{2} - z_a \sqrt{\frac{1}{12n}} \quad \text{and} \quad LCL = \frac{1}{2} - z_a \sqrt{\frac{1}{12}\left(\frac{1}{m} + \frac{1}{n}\right)},$$

which obviously depend on the choice of n and m. In each case, there is a $Centreline = 1/2$ and the CLs are applicable only for large values of n (i.e. $n \geq 5$). If n takes small values and at the same time $a \leq 1/n!$, then the CLs are computed by

$$Centreline = 1/2 \quad \text{and} \quad LCL = (n!a)^{1/n}/n,$$

where a is the FAR.

It is of interest to note here that Hamurkaroğlu et al. (2004) constructed the r and Q CCs of Liu's (1995) using the Mahalanobis depth to obtain the ranks of the observations. Moreover, a number of researchers—motivated by the work of Liu—exploited Principal Component Analysis (PCA)in order to reduce the number of dimensions. In particular, Zarate (2004) extended Liu's idea by computing first the Mahalanobis depth ranks of the principal components and then plotting the afore-mentioned ranks on the r chart. Working in a similar fashion, Beltran (2006) used again the r chart, but this time, the simplicial depth ranks of the first and the last principal components were exploited.

In addition, Li et al. (2014) provided a change-point control scheme (abbrev. $CPDP$) for individual observations based also on DD. Let $X_1, X_2, \ldots, X_m$ be m independent observations coming from a $p-$dimensional distribution F. When the process is in-control, all the observations share the same mean vector μ and variance–covariance matrix Σ. If we assume that a shift (in the mean, variance or both) occurs after the m_1th observation, then the parameters of the first m_1 observations are (μ, Σ) and the parameters of the remaining $m_2(= m - m_1)$ ones are (μ_1, Σ_1). In this case, a departure from distribution F to an out-of-control distribution G is reflected by a decrease in DD. Dai et al. (2004) defined the statistic

$$Q(m_1) = \sum_{j=m_1+1}^{m} R_{m_1}(j),$$

where

$$R_{m_1}(j) = \#\{X_i | D_{F_{m_1+1}}(X_i) < D_{F_{m_1+1}}(X_j), \ i = 1, 2, \ldots, m_1\}$$
$$+ \frac{1}{2}\#\{X_i | D_{F_{m_1+1}}(X_i) = D_{F_{m_1+1}}(X_j), \ i = 1, 2, \ldots, m_1\},$$

as well as its standardized form

$$SQ(m_1) = \frac{Q(m_1) - E(SQ(m_1))}{\sqrt{Var(SQ(m_1))}} = \frac{Q(m_1) - m_1(m - m_1)/2}{\sqrt{m_1(m - m_1)(m + 1)/12}}, \qquad (8)$$

which follows a standard Normal distribution as $m_1 \to \infty$ and $m_2 = m - m_1 \to \infty$. The $CPDP$ CC can be constructed by plotting the statistics $SQ(i)$ versus i for $1 \leq i < m$. The process will be considered out-of-control if $\max_{1 \leq i < m} SQ(i) > h_{m,a}$, where $h_{m,a}$ is the CL and a is the FAR. The authors proved also that the maximum likelihood estimator of the change point is given by $\hat{\tau} = \arg \max_{1 < t < m} \{SQ_t\}$.

Recently, Koutras and Sofikitou (2019) incorporated measures of dispersion in the $O2$ CC (Koutras and Sofikitou (2017a)) and they came up with two new semiparametric schemes: the $O2$ $N2$ and the $O4$ CCs. These charts have similar properties with the $O2$ CC, and can also be considered as fully nonparametric CCs. To formulate the rules of these charts, we recall the notation used in (3) and define the following enumerating function:

$$L(z_1, z_2, \ldots, z_n; w_1, w_2) = |\{i \in \{1, 2, \ldots, n\} : w_1 \leq z_i \leq w_2\}|,$$

which counts the number of the z_i's located between the values w_1 and w_2 for given $z_1, z_2, \ldots, z_n \in \mathbb{R}$ and $w_1, w_2 \in \mathbb{R}$. The decision rule of $O2$ $N2$, exploits the rth and sth order statistics of the test sample $X_{r:n}^{(T)}$ and $Y_{s:n}^{(T)}$ for $1 \leq r, s \leq n$, along with the following enumerating statistics:

$$L_X = L(X_1^{(T)}, \ldots, X_n^{(T)}; Y_{a:m}^{(R)}, X_{b:m}^{(R)}) = |\{i \in \{1, \ldots, n\} : X_{a:m}^{(R)} \leq X_i^{(T)} \leq X_{b:m}^{(R)}\}|,$$

$$L_Y = L(Y_1^{(T)}, \ldots, Y_n^{(T)}; Y_{c:m}^{(R)}, Y_{d:m}^{(R)}) = |\{j \in \{1, \ldots, n\} : Y_{c:m}^{(R)} \leq Y_j^{(T)} \leq Y_{d:m}^{(R)}\}|.$$

The statistics L_X and L_Y return the number of $X-$and $Y-$observations, from the whole range of the test sample, which lie between the CLs. According to the $O2$ $N2$ (Order–Order and Number–Number) chart, the process is declared in-control if the next conditions hold true:

$$LCL_X \leq X_{r:n}^{(T)} \leq UCL_X, \ LCL_Y \leq Y_{s:n}^{(T)} \leq UCL_Y \text{ and } L_X \geq \ell_X, \ L_Y \geq \ell_Y \quad (9)$$

or equivalently

$$X_{a:m}^{(R)} \leq X_{r:n}^{(T)} \leq X_{b:m}^{(R)} \quad (9a)$$
$$Y_{c:m}^{(R)} \leq Y_{s:n}^{(T)} \leq Y_{d:m}^{(R)} \quad (9b)$$

and

$$L_X \geq \ell_X \quad (9c)$$
$$L_Y \geq \ell_Y. \quad (9d)$$

An appealing property of the aforementioned CC is that it can discriminate whether the observed shifts are due to mean and/or variance; at the same time it is very easy to detect the out-of-control characteristic(s). More precisely, if Condition (9a) or (9b) (resp. (9c) or (9d)) is violated then the alarm is triggered because of a means (resp. variance) shift. On the other hand, a violation only on the Conditions (9a) and/or (9c) (resp. (9b) and/or (9d)) indicates that a mean and/or a variance shift has occurred in the first (resp. second) characteristic.

The construction of the $O4$ CC is based on the idea that the process is in-control if consecutive test sample observations (belonging to the intervals $I_X = [X_{r_1:n}^{(T)}, X_{r_2:n}^{(T)}]$ and $I_Y = [Y_{s_1:n}^{(T)}, Y_{s_2:n}^{(T)}]$) lie between the CLs. This leads to checking whether or not four order statistics from the test sample belong to the CLs. Thus, the process is considered to be out-of-control if at least one of the following conditions is violated:

$$LCL_X \leq X_{r_1:n}^{(T)} \leq X_{r_2:n}^{(T)} \leq UCL_X \text{ and } LCL_Y \leq Y_{s_1:n}^{(T)} \leq Y_{s_2:n}^{(T)} \leq UCL_Y \quad (10)$$

or equivalently

$$X_{a:m}^{(R)} \leq X_{r_1:n}^{(T)} \leq X_{r_2:n}^{(T)} \leq X_{b:m}^{(R)} \quad (10a)$$

$$Y_{c:m}^{(R)} \leq Y_{s_1:n}^{(T)} \leq Y_{s_2:n}^{(T)} \leq Y_{d:m}^{(R)} \quad (10b)$$

where $1 \leq r_1 \leq r_2 \leq n$ and $1 \leq s_1 \leq s_2 \leq n$. This CC is capable of detecting the variable(s) that triggered the alarm but it cannot identify whether the alarm is due to a means shift, a variance shift or both. In particular, if only Condition (10a) (resp. (10b) is violated, we may infer that the mean and/or the variance of the first (resp. second) characteristic has shifted out-of-control. If both Conditions (10a) and (10b) are simultaneously violated, then the shift should be attributed to both variables.It should be stressed that if the ranges of the intervals I_X and I_Y are narrow, the rule becomes more sensitive to mean shifts, while the use of wide interval ranges makes the CC more efficient in the detection of variance shifts.

It is of interest to notice that if $\ell_X = \ell_Y = 1$ and $r_1 = r_2$, $s_1 = s_2$, Rules (9) and (10) of the respective $O2N2$ and $O4$ CCs coincide with Rule (4) of the $O2$ chart. As a result, the latter chart can be viewed as a special case of the other two schemes. A key advantage of all these three schemes is that non-symmetric CLs can be used and thereof the CCs can be more effective in detecting specific shifts, like, for example, shifts for a specific characteristics or shifts in either location and/or scale parameter.

b. *Multivariate CCs with Memory*

Up to date, the multivariate CCs with memory that have been proposed in the literature are basically couched on the notion of DD. Liu (1995) introduced the S chart that is a DD-based generalization of the univariate $CUSUM$, capable of detecting potential increasing shifts in the mean and/or variability. Let us recall the notation used earlier to describe Liu's r and Q charts. For the construction of the S chart, Liu (1995) suggested to plot one of the following statistics that exploite Formulae (6a)–(6b) or (7a)–(7b)

$$S_n(F) = \sum_{i=1}^{n} \left[r_F(X_i) - \frac{1}{2} \right] = n \left[Q(F, \hat{G}_n) - \frac{1}{2} \right],$$

$$S_n(\hat{F}_m) = \sum_{i=1}^{n} \left[r_{\hat{F}_m}(X_i) - \frac{1}{2} \right] = n \left[Q(\hat{F}_m, \hat{G}_n) - \frac{1}{2} \right],$$

and using as $LCLs$ the quantities

$$LCL = -z_a \sqrt{\frac{n}{12}} \quad \text{and} \quad LCL = -z_a \sqrt{\frac{n^2}{12}\left(\frac{1}{m}+\frac{1}{n}\right)}.$$

In order to have a constant LCL, one may standardize the plotted statistics as follows:

$$S_n^*(F) = S_n(F)\Big/\left(\sqrt{n/12}\right) \quad \text{and} \quad S_n^*(F_m) = S_n(F_m)\Big/\sqrt{\frac{n^2}{12}\left(\frac{1}{m}+\frac{1}{n}\right)},$$

for $n = 1, 2, \ldots$ and use the simplified CLs (this scheme is known as S^* chart)

$$Centreline = 0 \quad \text{and} \quad LCL = -z_a.$$

Dai et al. (2004) considered also a $CUSUM$ CC based on DD for Phase I. The rationale is very similar to the procedure described in the change-point model of Li et al. (2014), the difference lying in the formulation of the plotted statistic. The $CUSUM$ statistic has the following form:

$$S_i = \max\{0, S_{i-1} - SQ_i - k\},$$

where the quantity SQ_i is calculated by Equation (8), $S_0 = 0$ and k is a design parameter which, as mentioned by the authors, has an optimal value equal to 2. An alarm is triggered if S_i exceeds a threshold $h_{m,p}$ for which the authors suggested the approximate formula $h_{m,p} = 1.0936n - 1.4746p$. They also gave an estimate of the position of shift, that is, $\hat{\tau} = \arg \max_{1<t<m}\{|SQ_t|\}$.

Li et al. (2013) presented two $CUSUM$ procedures based on spatial sign and DD (abbrev. SS-CUSUM, DD-CUSUM) which are affine invariant under rotation-scale transformations on all components. They employed the transformation method proposed by Hettmansperger and Randles (2002). Assuming that $X_1, X_2, \ldots, X_m$ represents the reference sample and $Y_1, Y_2, \ldots$ the test samples, then the observations X_i and Y_i are transformed as follows:

$$X_i^* = \hat{A}_m(X_i - \hat{\theta}_m) \quad \text{and} \quad Y_i^* = \hat{A}_m(Y_i - \hat{\theta}_m). \tag{11}$$

The parameters $(\hat{\theta}_m, \hat{A}_m)$ are obtained by solving the system of Eq. (2) for $n = 1$.

To construct the nonparametric SS-CUSUM chart, Li et al. (2013) extended the multivariate (parametric) $CUSUM$ procedure of Crosier (1988) by replacing the original observations with their spatial signs. More specifically, they introduced the following statistics:

$$C_n = [(S_{n-1} + U_n)'(S_{n-1} + U_n)]^{1/2},$$

$$S_n = \begin{cases} 0, & \text{if } C_n \leq k, \\ S_n = (S_{n+1} + U_n)(1 - k/C_n), & \text{if } C_n > k, \end{cases}$$

where $U_n = U(Y_n^*)$ is the spatial sign of Y_n^*, $k > 0$ and $S_0 = 0$. In this case, an alarm is triggered if the quantity $L_n = (S_n' S_n)^{1/2} > h$, where h is the CL that can be predetermined on the basis of k and a desired level of ARL_{in}.

Motivated by Crosier's (1988) procedure, Li et al. (2013) replaced the squared root of Hotelling's T^2 appearing there by $1 - r_{\hat{F}_m}(Y_n)$ and created the following $CUSUM$ procedure:

$$S_n = \max(0, S_{n-1} + (1 - r_{\hat{F}_m}(Y_n)) - k) \overset{asympt.}{\approx} \max(0, S_{n-1} + (0.5 - r_{\hat{F}_m}(Y_n)) - k);$$

the quantity $r_{\hat{F}_m}(Y_n)$ is computed by the aid of Equation (6a). Finally, in order to attain the affine invariance property, the authors suggested using the transformed data X_n^*, Y_n^* (defined in (11)) instead of X_n, Y_n. Consequently, the plotted statistic of the DD-$CUSUM$ chart takes on the form

$$S_n = \max(0, S_{n-1} + (0.5 - r_{\hat{F}_m^*}(Y_n^*)) - k),$$

where $S_0 = 0$, $k > 0$ and $r_{\hat{F}_m^*}(Y_n^*)$ is computed via Formula (6b). The process is considered out-of-control if S_n takes values greater than h, which depends on k and ARL_{in}.

Messaoud et al. (2004, 2008) generalized the (univariate) $EWMA$ scheme for individual observations of Hackl and Ledolter (1992) by considering a multivariate $EWMA$ chart which takes into account the sequential ranks of DD. We will begin by introducing first some new definitions. According to Hackl and Ledolter (1992), if $X_t, t = 1, 2, \ldots$, are independent (univariate) samples coming from a continuous distribution $F(x)$, the *sequential rank* that is the rank of X_t among the most recent m ($m > 1$) observations $X_t, X_{t-1}, \ldots, X_{t-m+1}$, is defined as

$$R_t^* = 1 + \sum_{i=t-m+1}^{t} I(X_t > X_i). \tag{12}$$

The *standardized sequential rank* is then defined by

$$R_t^{(m)} = \frac{2}{m}\left(R_t^* - \frac{m+1}{2}\right).$$

For the formulation of the chart, let us assume that the reference consists of the m most recent observations $X_{t-m+1}, X_{t-m+2}, \ldots, X_t$, where $X_t = (X_{t1}, X_{t2}, \ldots, X_{tp})'$ stands for a vector with p characteristics. The depth of X_t is calculated with respect to the aforementioned sample and the sequential rank (R_t^*) of $D_m(X_t)$ among $D_m(X_{t-m}), \ldots, D_m(X_{t-1})$ by the aid of Eq. (12), i.e.

$$R_t^* = 1 + \sum_{i=t-m+1}^{t} I(D_m(X_t) > D_m(X_i)).$$

The plotted statistics of the $EWMA$ scheme is given by

$$T_t = \min\{B, (1 - \lambda)T_{t-1} + \lambda R_t^{(m)})\},$$

for $t = 1, 2, \ldots$, where $R_t^{(m)}$ is the standardized sequential rank of $D_m(X_t)$ among $D_m(X_{t-m}), \ldots, D_m(X_{t-1})$ and $0 < \lambda \leq 1$ is a smoothing parameter; B is a reflection boundary and $T_0 = u$ with $h \leq u \leq B$. The process is declared out-of-control when $T_t > h$, where h is an appropriate LCL ($h < 0$). Since the choice of the parameters B, λ and u depends on the ARL of the lower sided $EWMA$, Messaoud et al. (2004) provided a formula for the determination of the ARL as a function of u when h, B and λ are fixed.

As mentioned earlier in the present section, Haq and Khoo (2018) introduced a multivariate $EWMA$ sign CC for scale monitoring which was subsequently proved capable of detecting potential shifts in both process mean and/or variance. Therefore, that CC could be used for the joint monitoring of location and scale.

Finally, Liu et al. (2004) studied a DD-based MA chart (abbrev. $DDMA$) for detecting process shifts in both location and scale. If $X_1, X_2, \ldots, X_m \sim F$ and $Y_1, Y_2, \ldots, Y_n \sim G$ represent the reference and the test sample, respectively, then, then MA chart uses the following moving averages of length q:

$$\tilde{Y}_q = (Y_1 + \ldots + Y_q)/q,$$
$$\tilde{Y}_{q+1} = (Y_2 + \ldots + Y_{q+1})/q,$$
$$\vdots$$
$$\tilde{Y}_n = (Y_{n-q+1} + \ldots + Y_n)/q.$$

Let $\tilde{Y}_i \in \tilde{Y} = \{\tilde{Y}_q, \ldots, \tilde{Y}_n\}$. The corresponding reference sample used for monitoring $\tilde{Y}_i \in \tilde{Y}$ is $\tilde{X} = \{\tilde{X}_q, \ldots, \tilde{X}_m\}$, where

$$\tilde{X}_q = (X_1 + \ldots + X_q)/q,$$
$$\tilde{X}_{q+1} = (X_2 + \ldots + X_{q+1})/q,$$
$$\vdots$$
$$\tilde{X}_m = (X_{m-q+1} + \ldots + X_m)/q.$$

The plotted statistic of the $DDMA$ CC is simply

$$r_{\hat{F}_{m-q+1}}(\tilde{Y}_i) = \frac{\#\{\tilde{X}_k \mid D_{\hat{F}_{m-q+1}}(\tilde{X}_k) < D_{\hat{F}_{m-q+1}}(\tilde{X}_i), \ k = q, \ldots, m\}}{m - q + 1}, \tag{13}$$

where $D_{\hat{\tilde{F}}_{m-q+1}}(\cdot)$ is the empirical depth with respect to $\hat{\tilde{F}}_{m-q+1}$, i.e. the empirical distribution of $\tilde{X}$. Once again, the CLs are

$$Centreline = 1/2 \quad \text{and} \quad LCL = a,$$

where a is the FAR we wish to reach.

4 Comparison Study

The purpose of the present Section is to examine the performance of some of the multivariate nonparametric CCs presented earlier under different shift scenarios. To set up fair comparisons, we do not contrast charts which do not belong in the same family/category (i.e. CCs with memory versus CCs without memory; CCs for individual observations versus CCs for subgroups) or charts which are not capable of detecting the same kind of shift (i.e. mean shifts, variance shifts or both). Therefore, we focus on comparing the *Shewhart*-type CCs between each other, excluding those which are based on a change-point formulation. In addition, we do not provide any numerical calculations for CCs with memory given that there are only a few schemes available per category ($EWMA, CUSUM, MA$) which may not be built to test exactly the same hypotheses.

The assessment of the performance of the CCs under comparison is based on the ARL. The ARL_{in} of all the charts of our numerical experimentation is set to 200 (approximately) and then the out-of-control ARL (ARL_{out}) is computed for three different shift scenarios in location and/or scale. In every case, semiparametric or nonparametric CCs are compared to each other, as well as to their popular parametric counterparts.

Scenario I: *Shifts in Mean*

Firstly, we generate data from a bivariate normal and a student's $t-$distribution with 5 degrees of freedom, utilizing the following in-control distribution parameters:

$$\mu_{in} = \begin{pmatrix} 1 \\ 1 \end{pmatrix}, \quad \Sigma_{in} = \begin{pmatrix} 1.0 & 0.7 \\ 0.7 & 1.0 \end{pmatrix} \quad \text{and} \quad \Sigma_{t(5)} = \begin{pmatrix} 0.60 & 0.42 \\ 0.42 & 0.60 \end{pmatrix} = \frac{3}{5}\Sigma_{in}.$$

Note that the matrix $\Sigma_{t(5)}$ is chosen so that the same covariance matrix is achieved for both distributions. In this way, it is secured that shifts occur only in the mean vector μ of each distribution and not in the variance–covariance matrix or in the correlation. It should be mentioned that the standard deviations in both characteristics are assumed to remain equal to 1 ($\sigma_X = \sigma_Y = 1$) and the correlation was chosen to be moderate/relatively strong ($\rho = 0.7$).

Under this scenario, we compare the parametric χ^2 chart, the nonparametric SN^2 and SR^2 charts proposed by Boone and Chakraborti (2012) and the semiparametric

$O2$, OC, $O2N2$, $O4$ charts proposed by Koutras and Sofikitou (2019, 2017a, b). In the case that the asymptotic CLs of the χ^2, SN^2 and SR^2 CCs do not yield an $ARL_{in} \approx 200$, the CLs are estimated so as the desired ARL_{in} value is achieved. In the $O2$, OC and $O2N2$ charts, the test sample medians are used so that the number of tabulated parameters be reduced. For this purpose, the parameters r, s appearing in (4), (5) and (9) are set equal to $(n + 1)/2$. In addition, the design parameters of the $O4$ chart are given by $r_1 = (n + 1)/2 - t_X$, $s_1 = (n + 1)/2 - t_Y$, $r_2 = (n + 1)/2 + t_X$, $s_2 = (n + 1)/2 - t_Y$. It is of interest to note that, should one use the same values for t_X, t_Y, i.e. $t_X = t_Y = t$, the decision rule of the resulting CC will make use of the most central pairs of the test sample.

As a consequence, when the medians or the most central pairs are used then the limits of the $O2$, $O2N2$ and $O4$ charts (as defined in (3)) are symmetrically placed, i.e. $b = m - a + 1$, $d = m - c + 1$. The corresponding CLs are determined in two ways: when equal mean shifts occur per characteristic, then the choice $a = c$ (and $b = d$) is made; should larger shifts be expected to occur in the mean of the first characteristic ($\mu_X > \mu_Y$), then it is intuitively obvious that one should select $a > c$. Finally, in the case of the OC scheme, both symmetric and asymmetric CLs are exploited, which are, respectively, denoted by a, b $(= m - a + 1)$ and a^*, b^*. Of course, in each case the CLs with respect to the second characteristic are just the corresponding concomitants. The reference and test sample sizes used were set equal to $m = 1000$ and $n = 15$.

It should be stressed that although the $O2N2$ and $O4$ CCs may have originally been constructed for detecting possible simultaneous shifts in the mean and variability, they can also be applied when only mean shifts occur. This can be achieved by making Rules (9) and (10) much more sensitive to mean shifts than to shifts in variability. Since under Scenario I there are no shifts in the variance, we set $\ell_X = \ell_Y = \ell$ and $t_X = t_Y = t$. Apparently, when it comes to the conditions which control the variability, only a few test sample observations should be expected to lie between the CLs. Hence, a natural choice would be the use of relatively small values for the parameters ℓ and t, in which case the performance of the $O2N2$ and $O4$ schemes will be very similar to the one of the $O2$ chart.

All the numerical results are summarized in Tables 1 and 2, where the values in bold indicate the smaller ARL_{out} and, as a result, the chart with the fastest detection capability. The underlined ARL_{out} values indicate the second-fastest chart. The numbers, presented in Table 1, were calculated using bivariate normal data. As expected, the parametric χ^2 CC overperforms the semiparametric and nonparametric counterparts except for the OC chart with non-symmetric CLs, which has uniformly the best performance for small shifts. Table 2 illustrates the behaviour of the CCs when bivariate $t_{(5)}$ data are used. In the event of equal shifts in the means of both characteristics ($\mu_X = \mu_Y = 1.0(0.3)1.9$), the semiparametric $O2$, OC, $O2N2$, $O4$ charts and the nonparametric SN^2, SR^2 charts have almost the same performance.

When larger shifts occur in the first characteristic ($\mu_X = 1.0(0.1)1.5$ and $\mu_Y = 1.0$), the CLs are adjusted so that the detection rule of the semiparametric CCs —corresponding to the mean of characteristic X (i.e. first condition of Rules (4), (5) and Conditions (9a), (10a)— becomes more sensitive. In this case, $O2$, $O4$ and

Table 1 ARL_{out} comparison of the χ^2, SN^2, SR^2, $O2$, OC, $O2N2$, $O4$ charts with common $ARL_{in} \approx 200$ for bivariate normal data

St. Dev. Shifts	Parametric chart χ^2	Nonparametric charts SN^2	SR^2	Semiparametric charts $O2$	OC		$O2N2$	$O4$
	$UCL = 10.596$	$UCL = 8.60$	$UCL = 8.60$	$a = c = 166$ $b = d = 835$	$a = 139$ $b = 862$	$a^* = 54$ $b^* = 835$	$a = c = 165$ $b = d = 836$ $\ell = \ell_X = \ell_Y = 4$	$a = c = 89$ $b = d = 912$ $t = t_X = t_Y = 2$
$\delta_X = \delta_Y$								
1.00	200.42	216.98	203.02	203.88	207.48	203.71	203.61	199.70
1.30	25.04	34.25	33.74	31.23	35.91	**19.16**	32.49	36.88
1.60	**3.45**	6.07	5.86	5.12	5.96	3.76	5.19	5.98
1.90	**1.33**	2.16	1.96	1.77	1.96	1.56	1.79	1.95
	$UCL = 10.596$	$UCL = 8.60$	$UCL = 8.78$	$a = c = 166$ $b = d = 835$	$a = 139$ $b = 862$	$a^* = 54$ $b^* = 835$	$169 = a > c = 165$ $832 = b < d = 836$ $\ell = \ell_X = \ell_Y = 3$	$100 = a > c = 70$ $901 = b < d = 931$ $t = t_X = t_Y = 2$
$\delta_X = \delta_Y$								
1.00	200.42	216.98	203.02	206.33	207.48	203.71	197.69	200.52
1.10	107.12	180.20	140.11	167.37	148.57	**93.61**	159.68	152.53
1.20	**35.00**	111.02	64.40	98.75	82.08	45.23	95.70	85.15
1.30	**12.57**	59.84	28.11	50.52	41.18	24.78	48.59	42.45
1.40	**5.36**	32.70	12.78	26.11	21.49	13.61	25.59	22.29
1.50	**2.86**	18.55	6.67	14.24	12.56	8.20	13.80	12.42

Table 2 ARL_{out} comparison of the χ^2, SN^2, SR^2, $O2$, OC, $O2N2$, $O4$ charts with common $ARL_{in} \approx 200$ for bivariate $t(5)$ data

St. Dev. Shifts	Parametric chart	Nonparametric charts		Semiparametric charts				
	χ^2	SN^2	SR^2	$O2$	OC		$O2N2$	$O4$
$\delta_X = \delta_Y$	$UCL = 12.22$	$UCL = 8.60$	$UCL = 8.73$	$a = c = 166$ $b = d = 835$	$a = 139$ $b = 862$	$a^* = 54$ $b^* = 835$	$a = c = 165$ $b = d = 836$ $\ell = \ell_X = \ell_Y = 4$	$a = c = 89$ $b = d = 912$ $t = t_X = t_Y = 2$
1.00	199.37	217.98	204.25	206.43	205.46	203.39	201.76	204.60
1.30	39.49	22.37	22.17	26.33	31.85	**15.94**	26.16	36.38
1.60	4.80	3.81	3.79	3.49	4.31	**2.67**	3.51	6.02
1.90	1.50	1.65	1.61	1.32	1.48	**1.24**	1.33	1.97
$\delta_X = \delta_Y$	$UCL = 12.22$	$UCL = 8.60$	$UCL = 8.73$	$168 = a > c = 164$ $833 = b < d = 837$	$a = 139$ $b = 862$	$a^* = 54$ $b^* = 835$	$169 = a > c = 165$ $832 = b < d = 836$ $\ell = \ell_X = \ell_Y = 3$	$100 = a > c = 70$ $901 = b < d = 931$ $t = t_X = t_Y = 2$
1.00	199.37	217.98	204.28	204.82	205.46	203.39	200.01	197.69
1.10	132.36	165.06	123.51	161.96	152.73	**89.84**	156.98	152.16
1.20	54.54	83.34	47.61	88.38	76.63	**41.70**	84.05	85.69
1.30	20.16	40.16	**18.66**	41.26	37.44	21.39	40.10	43.34
1.40	**8.10**	19.81	8.21	19.64	19.36	11.34	19.17	27.73
1.50	**3.79**	10.54	4.45	9.90	10.72	6.25	9.49	12.10

$O2N2$ have similar performance and they overperform compared to the SN^2 but underperform compared to the SR^2 chart. The OC with asymmetric CLs has the fastest detection capability when small shifts occur, while the χ^2 and the SR^2 CCs are faster in detecting smaller mean shifts.

Scenario II: Shifts in Variability

Here, we study the performance of the classical parametric $|S|$ scheme introduced by Alt (1985) and the nonparametric $QMAX, SMAX, MDMAX, MADMAX$ charts of Osei-Aning et al. (2017), all of which have been proposed for monitoring the process variability. The aforementioned charts are compared in terms of the in-and out-of-control performance to the semiparametric $O2N2$ and $O4$ charts which have been recently suggested by Koutras and Sofikitou (2019) for monitoring simultaneous shifts in mean and variability.

Under this scenario, we also generate data from both bivariate normal and $t_{(5)}$ data, but now we assume that the process mean vector remains constant ($\boldsymbol{\mu} = 1$) and shifts occur only in the variance–covariance matrix of the process. More precisely, the in-control distribution parameters are the ones practised before, i.e.

$$\boldsymbol{\mu}_{in} = \begin{pmatrix} 1 \\ 1 \end{pmatrix}, \quad \boldsymbol{\Sigma}_{in} = \begin{pmatrix} 1.0 & 0.7 \\ 0.7 & 1.0 \end{pmatrix} \quad \text{and} \quad \boldsymbol{\Sigma}_{t(5)} = \begin{pmatrix} 0.60 & 0.42 \\ 0.42 & 0.60 \end{pmatrix} = \frac{3}{5} \boldsymbol{\Sigma}_{in}.$$

The ARL_{out} was evaluated by shifting the process, variance–covariance matrix from $\boldsymbol{\Sigma}_{in}$ to

$$\boldsymbol{\Sigma}_{out} = \begin{pmatrix} \delta_X^2 \sigma_X^2 & \delta_X \delta_Y \sigma_{XY} \\ \delta_X \delta_Y \sigma_{XY} & \delta_Y^2 \sigma_Y^2 \end{pmatrix},$$

where $\sigma_{XY} = \rho \sigma_X \sigma_Y$ and δ_X, δ_Y are used to control the variance shifts per characteristic. For our experimentation, we used values $\sigma_X = \sigma_Y = 1$ and $\rho = 0.7$.

A common sample size ($n = 15$) was used for all charts. For the $O2N2$ and $O4$ charts, the design parameters were calculated by $b = m - a + 1$, $d = m - c + 1$, $r_1 = (n + 1)/2 - t_X$, $s_1 = (n + 1)/2 - t_Y$, $r_2 = (n + 1)/2 + t_X$, $s_2 = (n + 1)/2 - t_Y$ and the reference sample m was determined so that $ARL_{in} \approx 200$.

Tables 3 and 4 illustrate the performance of all charts when the process characteristics are distributed as normal and $t(5)$, respectively. The bold and underlined values indicate the charts with the first and second best performance. When normal data are used and equal shifts occur in variance, the $|S|$ chart has the best performance. However, the $SMAX$ and $MDMAX$ seem to detect shifts more quickly than the other charts. It should be emphasized that this is not the case when larger shifts occur to the X characteristic. In fact, the performance of the $O2N2$ and $O4$ charts, especially using non-normal distribution, is the best among the rest of the charts. This can be attributed to the fact that the CLs of the $O2N2$ and $O4$ charts can be chosen asymmetrically with respect to the characteristics under study, and therefore those charts provide increased flexibility to deal with cases where unequal shifts need to be detected. To be more specific, for $\delta_X > \delta_Y$, one should choose the rest of the parameters so that $\ell_X > \ell_Y$ and $r_1 < r_2$, $s_1 < s_2$.

Table 3 ARL_{out} comparison of the $|S|$, $QMAX$, $SMAX$, $MDMAX$, $MADMAX$, O_2, OC, $O2N2$, $O4$ charts with common $ARL_{in} \approx 200$ for bivariate normal data

St. Dev. Shifts		Parametric chart	Nonparametric charts				Semiparametric charts	
		$|S|$	$QMAX$	$SMAX$	$MDMAX$	$MADMAX$	$O2N2$	$O4$
$\delta_X = \delta_Y$		$UCL = 1.520$	$UCL = 2.020$	$UCL = 1.542$	$UCL = 1.223$	$UCL = 1.908$	$a = c = 52$	$a = c = 15$
							$b = d = 949$	$b = d = 986$
							$\ell = \ell_X = \ell_Y = 10$	$t = t_X = t_Y = 5$
1.00		202.10	200.01	202.51	201.77	200.50	198.01	198.39
1.10		**25.15**	50.57	33.09	37.87	55.81	46.18	50.43
1.20		**6.88**	19.35	9.99	11.84	21.91	16.48	19.03
1.30		**3.07**	9.19	4.45	5.26	10.75	7.50	9.03
1.40		**1.87**	5.32	2.57	3.01	6.32	4.33	5.18
1.50		**1.38**	3.48	1.81	2.02	4.12	2.85	3.33
$\delta_X = \delta_Y$		$UCL = 1.520$	$UCL = 2.020$	$UCL = 1.542$	$UCL = 1.223$	$UCL = 1.908$	$a = c = 13$	$a = c = 19$
							$b = d = 988$	$b = d = 982$
							$13 = \ell_X > \ell_Y = 10$	$5 = t_X > t_Y = 2$
1.00	1.00	199.15	200.34	199.65	201.69	200.32	198.83	204.40
1.00	1.10	65.25	79.62	53.66	61.09	85.78	**46.46**	58.06
1.00	1.20	27.19	32.99	**16.79**	20.19	37.82	16.92	23.17
1.00	1.30	13.86	16.01	**7.21**	8.63	18.77	8.00	11.48
1.00	1.40	8.25	9.04	**3.85**	4.69	11.07	4.57	6.55
1.00	1.50	5.42	5.85	**2.54**	3.00	7.08	3.08	4.27

Table 4 ARL_{out} comparison of the $|S|$, $QMAX$, $SMAX$, $MDMAX$, $MADMAX$, $O2$, OC, $O2N2$, $O4$ charts with common $ARL_{in} \approx 200$ for bivariate $t(5)$ data

St. Dev.	Shifts	Parametric Chart	Nonparametric charts				Semiparametric charts	
		$|S|$	$QMAX$	$SMAX$	$MDMAX$	$MADMAX$	$O2N2$	$O4$
		$UCL = 3.292$	$UCL = 1.864$	$UCL = 2.198$	$UCL = 1.316$	$UCL = 1.729$		
$\delta_X = \delta_Y$							$a = c = 52$	$a = c = 15$
							$b = d = 949$	$b = d = 986$
							$\ell = \ell_X = \ell_Y = 10$	$t = t_X = t_Y = 5$
1.00	1.00	201.46	201.80	199.84	202.91	199.60	204.34	200.21
	1.10	75.80	64.65	111.32	67.94	67.32	64.00	83.49
	1.20	32.10	26.65	61.16	27.31	28.16	25.98	39.09
	1.30	15.78	13.38	35.75	13.06	14.36	12.47	20.85
	1.40	8.40	7.64	21.96	7.16	8.39	7.14	12.42
	1.50	5.06	5.01	13.68	4.47	5.48	4.61	8.08
$\delta_X = \delta_Y$							$a = c = 13$	$a = c = 19$
		$UCL = 3.292$	$UCL = 1.864$	$UCL = 2.198$	$UCL = 1.316$	$UCL = 1.729$	$b = d = 988$	$b = d = 982$
							$13 = \ell_X > \ell_Y = 10$	$5 = t_X > t_Y = 2$
1.00	1.00	203.60	200.49	200.39	202.41	199.41	198.79	204.57
1.10	1.00	124.14	96.63	139.48	99.51	98.23	78.52	89.25
1.20	1.00	79.37	45.41	89.50	44.14	48.02	36.16	42.95
1.30	1.00	53.09	23.09	54.45	21.15	25.04	19.21	23.24
1.40	1.00	36.91	13.21	34.40	11.40	14.87	11.34	14.22
1.50	1.00	26.88	8.43	21.93	6.92	9.61	7.19	9.39

Scenario III: *Simultaneous Shifts in Mean, Variability and Correlation*

Under the last scenario, the Q chart (of Liu (1995)), the $O2N2$ and $O4$ semiparametric schemes (proposed by Koutras and Sofikitou (2019)), as well as the classical parametric, *BV-MAX* chart (introduced by Khoo (2004)) are compared in terms of their ARL performance, when simultaneous shifts occur in the process mean vector and the variance–covariance matrix.

To carry out the comparison, we generate data from a bivariate normal distribution and we assume that the process is in-control when

$$\mu_{in} = \begin{pmatrix} 1 \\ 1 \end{pmatrix}, \quad \Sigma_{in} = \begin{pmatrix} 1 & 0 \\ 0 & 1 \end{pmatrix} \quad \text{and} \quad \rho = 0.$$

These quantities represent the null values of the mean vector, the variance–covariance matrix and the correlation coefficient, respectively. Then, the process means shifts to the out-of-control through equal shifts in each characteristic, namely $\mu = \mu_X = \mu_Y = 1.00(0.30)1.90$, while the matrix Σ shifts to

$$\Sigma = \begin{pmatrix} 1 + \rho^2 & \rho \\ \rho & 1 + \rho^2 \end{pmatrix} \quad \text{and} \quad \rho = 0.1(0.3)0.9.$$

The design parameters of the aforementioned CCs were selected so that an $ARL_{in} = 200$ (approximately) is achieved. In each chart, a test sample of size $n = 5$ was selected. The results of this comparison are reported in Table 5, where the bold-faced figures indicate the CC with the best performance, and the underlined figures indicate the CC with the next better performance. It is obvious that for small shifts in the mean (e.g. when $\mu = 1.00$), the Q chart has always faster detection power than the rest of the schemes. As one should expect, the parametric (*Shewhart*-type) *BV-MAX* chart is capable of detecting quickly only large mean shifts, more specifically $\mu \geq 1.3$.

In the latter case that larger shifts occur in the mean, one may argue that the performance of Q, $O2N2$ and $O4$ charts is very similar. However, a more careful inspection may easily reveal that the $O4$ chart overperforms when larger shifts occur in both μ and ρ, while the $O2N2$ is a better choice when larger shifts occur in μ and relatively small shifts in ρ. It should be stressed that the parameters a, b, c, d and ℓ_X, ℓ_y or t_X, t_y of the semiparametric $O2N2$, $O4$ charts can be appropriately chosen or adjusted to meet the practitioner's needs, like, for example, the need for detecting equal or unequal potential shifts in the process mean and/or variability. All in all and taking also into account the fact that the Q chart requires high computational effort, the $O2N2$ and $O4$ schemes seem to offer a better alternative, especially for non-normal processes.

Table 5 ARL_{out} performance of the Q, BV-MAX, $O2N2$ and $O4$ charts with common $ARL_{in} \approx$ 200 for bivariate normal data

Mean shifts	Parametric BV-MAX chart				
	Design	ρ			
$\mu = \mu_X = \mu_Y$	$ARL_{in} = 199.53$	0.1	0.3	0.6	0.9
1.00	$UCL = 3.023$	193.54	136.37	49.52	15.26
1.30	$n = 5$	**82.48**	**51.98**	**23.90**	**10.80**
1.60		**14.92**	**11.43**	**7.91**	**5.19**
1.90		**3.77**	**3.50**	**3.16**	**2.76**
Mean shifts	Nonparametric Q chart				
	Design	ρ			
$\mu = \mu_X = \mu_Y$	$ARL_{in} = 198.19$	0.1	0.3	0.6	0.9
1.00	$UCL = 0.18$	**186.29**	**119.69**	**39.21**	**13.05**
1.30	$n = 5$	110.76	75.53	30.19	<u>11.42</u>
1.60	$m = 100$	32.21	27.80	15.62	7.93
1.90	$a = 0.005$	9.32	9.19	7.33	5.06
Mean shifts	Semiparametric $O4$ chart				
	Design	ρ			
$\mu = \mu_X = \mu_Y$	$ARL_{in} = 201.60$	0.1	0.3	0.6	0.9
1.00	$m = 255$	189.93	123.68	43.84	14.86
1.30	$n = 5$	106.53	<u>71.59</u>	<u>29.12</u>	<u>11.42</u>
1.60	$a = c = 3$	31.83	24.87	<u>13.08</u>	<u>6.73</u>
1.90	$t = 1$	10.55	8.81	5.85	<u>3.83</u>
Mean shifts	Semiparametric $O2N2$ chart				
	Design	ρ			
$\mu = \mu_X = \mu_Y$	$ARL_{in} = 202.66$	0.1	0.3	0.6	0.9
1.00	$m = 100$	194.64	137.59	55.14	21.37
1.30	$n = 5$	<u>98.11</u>	72.13	34.58	16.08
1.60	$a = c = 5$	<u>25.14</u>	<u>21.63</u>	13.80	8.18
1.90	$\ell_X = \ell_Y = 2$	<u>7.82</u>	<u>7.24</u>	<u>5.62</u>	4.27

5   Conclusion—Future Research

Nonparametric techniques—including, among others, hypothesis testing, decision-making, estimation and prediction procedures— have gained widespread acceptance and have nowadays been used in several practical applications dealing with complex real-world problems. The technology advances in computational power and storage has offered effective tools to deal with the computational aspects related to these problems. As a result, nonparametric methods were quickly appreciated, developed and applied in various research areas, one of which is SPC.

During the past decades, a tremendous increase has been observed in the SPC literature. In the beginning, most charts were proposed for monitoring the location parameter of a single quality characteristic; however very quickly, the CC techniques were spread out in order to cover the need for monitoring the scale or both location and scale of a characteristic. Multivariate surveillance using nonparametric methods attracted the attention of practitioners relatively recently. The development of multivariate nonparametric CCs for mean and CCs for simultaneous monitoring of mean and variability occurred almost at the same time. However, not much work has been done in the detection of potential shifts in scale of either a univariate or a multivariate process. Hopefully, more nonparametric CCs for variance/standard deviation will be studied in the future, since in SPC variability is closely related to the stability of the process.

When it comes to the multivariate SPC, there are two important issues that should be addressed. The first one is the dependence structure of the observations under study and the second is the identification of the out-of-control variables. The vast majority of the multivariate (parametric or nonparametric) CCs fail in detecting correlation shifts. This may have an impact on the determination of the variables which triggered an alarm; variables that are highly correlated to each other have an additive impact in the plotted statistic and may easily shift the process to an out-of-control condition. Future perspectives could be the development of schemes sensitive to correlation shifts (such as the one proposed by Koutras and Sofikitou (2017b)), as well as the refinement/improvement of the techniques that already exist in order to handle the problem of detecting the variable(s) responsible for the out-of-control shift of the monitored process.

Generally speaking, the multivariate CCs currently available in the literature consist of direct extensions from univariate schemes. A majority of CCs are based on multivariate quadratic forms and therefore the test statistics asymptotically follow a χ^2 distribution. Some of these tests require the process distribution to be continuous and symmetric.

In real-life scenarios, symmetry does not often exist, therefore it is important that non-symmetric CLs be used. This is feasible when the CLs pertain percentiles or order statistics from the reference sample which are not symmetrically placed. This appealing property is observed in the median CC and its univariate or multivariate extensions.

The construction of other multivariate nonparametric CCs involves exploiting— in the broader sense—distance measures (such as data, simplicial or spatial depth, Euclidean and Mahalanobis distances, etc.), which are used to estimate how far the future observations are placed from the reference sample or some null historical values. Usually these CCs require extensive computational effort, which makes their applicability difficult and time consuming, especially for large samples and data dimensionality. To overcome this issue, one may try to reduce the number of dimensions using techniques like PCA and then apply the charting techniques only to the most important components. For more information, the interested reader might wish to consult the related articles presented earlier in Section 3.3. However, such techniques may not always provide a small number of components without missing a large proportion of data information. This unveils the need and importance of constructing schemes which remain invariant under transformations and do not require complex computational manipulations.

Another important point that should be mentioned is that currently there is lack of software availability, which makes the implementation of CCs by practitioners quite difficult. A few packages are currently available in *R Software*; for a more detailed description of this topic we refer to Chakraborti and Graham (2019).

It should also be stressed that a lot of nonparametric CCs use a relatively large reference (historical) sample to determine appropriate CLs. This links with the sample size calculations and requirements needed to set up the decision rule. The sample size effect and the impact of Phase I Analysis on Phase II have been studied by a few authors (see Sect. 1); however, this seems to be an open problem for further investigation.

In closing we mention that for monitoring attributes, count or discrete data only parametric CCs have been proposed up to date. Although some discussion can be found on the nonparametric framework of this problem in Qiu and Li (2011) and Qiu et al. (2019), it seems that there is a shortage in the literature on that matter.

References

Abu-Shawiesh, M. O., & Abdullah, M. B. (2001). A new robust bivariate control chart for location. *Communications in Statistics—Simulation and Computation, 30,* 513–529.

Albers, W., & Kallenberg, W. C. M. (2008). Minimum control charts. *Journal of Statistical Planning and Inference, 138,* 539–551.

Albers, W., & Kallenberg, W. C. M. (2009). CUMIN charts. *Metrika, 70,* 111–130.

Alloway, J. A., & Raghavachari, M. (1991). Control chart based on Hodges-Lehmann estimator. *Journal of Quality Technology, 23,* 336–347.

Alt, F. B. (1985). Multivariate quality control. In S. Kotz, N. L. Johnson, & C. R. Read (Eds.), *Encyclopedia of Statistical Sciences.* New York: Wiley.

Amin, R. W., Reynolds, M. R. J., & Bakir, S. T. (1995). Nonparametric quality control charts based on the sign statistic. *Communications in Statistics—Theory and Methods, 24,* 1579–1623.

Amin, R. W., & Searcy, A. J. (1991). A nonparametric exponentially weighted moving average control scheme. *Communications in Statistics—Simulation and Computation, 20,* 1049–1072.

Bakir, S. (2010). A nonparametric test for homogeneity of variances: Application to GPAs of students across academic majors. *American Journal of Business Education, 3,* 47–54.

Bakir, S. T. (2004). A distribution-free shewhart quality control chart based on signed-ranks. *Quality Engineering, 16,* 613–623.

Bakir, S. T. (2008). Distribution-free quality control charts based on signed-rank-like statistics. *Communications in Statistics—Theory and Methods, 35,* 743–757.

Bakir, S. T., & Reynolds, J. M. R. (1979). A nonparametric procedure for process control based on within-group ranking. *Technometrics, 21,* 175–183.

Balakrishnan, N., Paroissin, C., & Turlot, J.-C. (2015). One-sided control charts based on precedence and weighted precedence statistics. *Quality and Reliability Engineering International, 31,* 113–134.

Balakrishnan, N., Triantafyllou, I. S., & Koutras, M. V. (2009). Nonparametric control charts based on runs and Wilcoxon-type rank-sum statistics. *Journal of Statistical Planning and Inference, 139,* 3177–3192.

Balakrishnan, N., Triantafyllou, I. S., & Koutras, M. V. (2010). A distribution-free control chart based on order statistics. *Communications in Statistics—Theory and Methods, 39*, 3652–3677.

Bell, R. C., Jones-Farmer, L. A., & Billor, N. (2014). A distribution-free multivariate phase i location control chart for subgrouped data from elliptical distributions. *Technometrics, 56*, 528–538.

Beltran, L. A. (2006). *Nonparametric multivariate statistical process control using principal component analysis and simplicial depth.* Ph.D. thesis, University of Central Florida.

Bennett, B. M. (1964). Non-parametric test for randomness in a sequence of multinomial trials. *Biometrics, 20*, 182–190.

Bickel, P. J., & Lehmann, E. L. (1976). Descriptive statistics for nonparametric models. III. dispersion. *The Annals of Statistics, 4*, 1139–1158.

Boone, J. M., & Chakraborti, S. (2012). Two simple shewhart-type multivariate nonparametric control charts. *Applied Stochastic Models in Business and Industry, 28*, 130–140.

Capizzi, G., & Masarotto, G. (2017). Phase I distribution-free analysis of multivariate data. *Technometrics, 57*, 484–495.

Chakraborti, S. (2011). Nonparametric (Distribution-Free) quality control charts. In S. Kotz, C. B. Read, N. Balakrishnan, & B. Vidakovic (Eds.), *Encyclopedia of Statistical Sciences.* New Jersey: Wiley.

Chakraborti, S., & Eryilmaz, S. (2007). A nonparametric shewhart-type signed rank control chart based on runs. *Communications in Statistics—Simulation and Computation, 36*, 335–356.

Chakraborti, S., & Graham, M. A. (2008). Control charts, nonparametric. In F. Ruggeri, R. S. Kenett, & F. W. Faltin (Eds.), *Encyclopedia of Statistics in Quality and Reliability.* New York: Wiley.

Chakraborti, S., & Graham, M. A. (2019). Nonparametric control charts: An updated overview and some results. *Quality Engineering, 31*, 523–544.

Chakraborti, S., Human, S. W., & Graham, M. A. (2008). Phase I statistical process control charts: An overview and some results. *Quality Engineering, 21*, 52–62.

Chakraborti, S., Human, S. W., & Graham, M. A. (2011). Nonparametric (Distribution-Free) quality control charts. In N. Balakrishnan (Ed.), *Handbook of Methods and Applications of Statistics: Engineering, Quality Control and Physical Sciences* (pp. 298–329). New York: Wiley.

Chakraborti, S., van der Laan, P., & van de Wiel, M. A. (2004). A class of distribution-free control charts. *Journal of the Royal Statistical Society: Series C (Applied Statistics), 53*, 443–462.

Chakraborti, S., Van Der Laan, P., & Bakir, S. (2001). Nonparametric control charts: An overview and some results. *Journal of Quality Technology, 33*, 304–315.

Chakraborty, N., Chakraborti, S., Human, S. W., & Balakrishnan, N. (2016). A generally weighted moving average signed-rank control chart. *Quality and Reliability Engineering International, 32*, 2835–2845.

Cheng, C.-R., & Shiau, J.-J. (2015). A distribution-free multivariate control chart for phase I applications. *Quality and Reliability Engineering International, 31*, 97–111.

Cheng, S. W., & Thaga, K. (2006). Single variables control charts: An overview. *Quality and Reliability Engineering International, 22*, 811–820.

Choi, K., & Marden, J. (1997). An approach to multivariate rank tests in multivariate analysis of variance. *Journal of the American Statistical Association, 92*, 1581–1590.

Chowdhury, S., Mukherjee, A., & Chakraborti, S. (2014). A new distribution-free control chart for joint monitoring of unknown location and scale parameters of continuous distributions. *Quality and Reliability Engineering International, 30*, 191–204.

Chowdhury, S., Mukherjee, A., & Chakraborti, S. (2015). Distribution-free Phase II CUSUM control chart for joint monitoring of location and scale. *Quality and Reliability Engineering International, 31*, 135–151.

Conover, W. J. (1980). *Practical Nonparametric Statistics.* New York: Wiley.

Crosier, R. B. (1988). Multivariate generalizations for cumulative sum quality-control schemes. *Technometrics, 30*, 291–303.

Cucconi, O. (1968). Un Nuovo Test non Parametrico per il Confronto tra Due Gruppi Campionari (pp. 225–248). XXVII: *Giornale degli Economisti.*

Dai, Y., Zou, C., & Wang, Z. (2004). *Multivariate cusum control chart based on data for preliminary analysis*. Technical Report Department of Statistics, School of Mathematical Sciences, Nankai University.

Das, N. (2008). Nonparametric control chart for controlling variability based on rank test. *Economic Quality Control, 23,* 227–242.

Das, N. (2009). A new multivariate non-parametric control chart based on sign test. *Quality Technology and Quantitative Management, 6,* 155–169.

Das, N., & Bhattacharya, A. (2008). A new nonparametric control chart for controlling variability. *Quality Technology and Quantitative Management, 5,* 351–361.

Gadre, M. P., & Kakade, V. C. (2014). A nonparametric group runs control chart to detect shifts in the process median. *Indian Association of Productivity Quality and Reliability Transactions, 16,* 29–53.

Garde, M. P., & Kakade, V. C. (2019). Some side sensitive group runs based control charts to detect shifts in the process median. *Communications in Statistics-Simulation and Computation,* https://doi.org/10.1080/03610918.2019.1672736.

Ghute, V. B. (2013). Distribution-free control chart for bivariate processes. *Journal of Academia and Industrial Research, 26,* 703–705.

Ghute, V. B., & Shirke, D. T. (2012). A nonparametric signed-rank control chart for bivariate process location. *Quality Technology and Quantitative Management, 9,* 317–328.

Graham, M. A., Chakraborti, S., & Human, S. W. (2011a). A nonparametric EWMA sign chart for location based on individual measurements. *Quality Engineering, 23,* 221–241.

Graham, M. A., Chakraborti, S., & Human, S. W. (2011b). A nonparametric exponentially weighted moving average signed-rank chart for monitoring location. *Computational Statistics and Data Analysis, 55,* 2490–2503.

Graham, M. A., Chakraborti, S., & Human, S. W. (2012). Distribution-free exponentially weighted moving average control charts for monitoring unknown location. *Computational Statistics and Data Analysis, 56,* 2539–2561.

Graham, M. A., Human, S. W., & Chakraborti, S. (2010). A Phase I nonparametric shewhart-type control chart based on the median. *Journal of Applied Statistics, 37,* 1795–1813.

Hackl, P., & Ledolter, J. (1991). A control chart based on ranks. *Journal of Quality Technology, 23,* 117–124.

Hackl, P., & Ledolter, J. (1992). A new nonparametric quality control technique. *Communications in Statistics—Simulation and Computation, 21,* 423–443.

Hamurkaroğlu, C., Mert, M., & Saykan, Y. (2004). Nonparametric control charts based on Mahalanobis depth. *Hacettepe Journal of Mathematics and Statistics, 33,* 57–67.

Haq, A., & Khoo, M. B. C. (2018). A new non-parametric multivariate EWMA sign control chart for monitoring process dispersion. *Communications in Statistics—Theory and Methods, 45,* 3703–3716.

Hawkins, D. M., & Deng, Q. (2010). A nonparametric change-point model. *Journal of Quality Technology, 42,* 165–173.

Hayter, A. J., & Tsui, K.-L. (1994). Identification and quantification in multivariate quality control problems. *Journal of Quality Technology, 26,* 197–208.

Hettmansperger, T. P. (2006). Multivariate location tests. In S. Kotz, C. B. Read, N. Balakrishnan, & B. Vidakovic (Eds.), *Encyclopedia of Statistical Sciences.* New Jersey: Wiley.

Hettmansperger, T. P., & Randles, R. H. (2002). A practical affine equivariant multivariate median. *Biometrika, 89,* 851–860.

Hodges, J. (1955). A bivariate sign test. *The Annals of Mathematical Statistics, 26,* 523–527.

Hodges, J. L., & Lehmann, E. L. (1963). Estimates of location based on rank tests. *The Annals of Mathematical Statistics, 34,* 598–611.

Holland, M. D., & Hawkins, D. M. (2014). A control chart based on a nonparametric change-point model. *Journal of Quality, 46,* 63–77.

Human, S. W., Chakraborti, S., & Smit, C. F. (2010). Nonparametric shewhart-type sign control charts based on runs. *Communications in Statistics—Theory and Methods, 39,* 2046–2062.

Janacek, G. J., & Meikle, S. E. (1997). Control charts based on medians. *Journal of the Royal Statistical Society: Series D (The Statistician)*, *46*, 19–31.

Jensen, W. A., Jones-Farmer, L. A., Champ, C. W., & Woodall, W. H. (2006). Effects of parameter estimation on control chart properties: A literature review. *Journal of Quality Technology*, *38*, 349–364.

Jones-Farmer, L. A., Jordan, V., & Champ, C. W. (2009). Distribution-free control charts for subgroup location. *Journal of Quality Technology*, *41*, 304–316.

Jones-Farmer, L. A., Woodall, W. H., Steiner, S. H., & Champ, C. W. (2014). Effects of parameter estimation on control chart properties: A literature review. *Journal of Quality Technology*, *46*, 265–280.

Kapatou, A. (1996). *Multivariate nonparametric control charts using small samples*. Ph.D. thesis, Virginia Polytechnical Institute and State University.

Kapatou, A., & Reynolds, M. R. J. (1994). Multivariate nonparametric control charts using small samples. In *ASA Proceedings of the Section on Quality and Productivity*, Joint Statistical Meetings (pp 241–246). Alexandria, Virginia: American Statistical Association.

Kapatou, A., & Reynolds, M. R. J. (1998). Multivariate Nonparametric Control Charts For the Case of Uknown Sigma, In *ASA Proceedings of the Section on Quality and Productivity*, Joint Statistical Meetings (pp. 77–82). Alexandria, Virginia: American Statistical Association.

Khoo, M. B. C. (2004). A new bivariate control chart to monitor the multivariate process mean and variance simultaneously. *Quality Engineering*, *17*, 109–118.

Koutras, M. V., & Sofikitou, E. M. (2019) Bivariate semiparametric control charts for simultaneous monitoring of process mean and variance. *Quality and Reliability Engineering International*, *36*, 447–473.

Koutras, M. V., & Sofikitou, E. M. (2017a). A bivariate semiparametric control chart based on order statistics. *Quality and Reliability Engineering International*, *33*, 183–202.

Koutras, M. V., & Sofikitou, E. M. (2017b). A new bivariate semiparametric control chart based on order statistics and concomitants. *Statistics and Probability Letters*, *129*, 340–347.

Lepage, Y. (1971). A combination of Wilcoxon's and Ansari-Bradley's statistics. *Biometrika*, *58*, 213–217.

Li, J., Zhang, X., & Jeske, D. R. (2013). Nonparametric multivariate CUSUM control charts for location and scale changes. *Journal of Nonparametric Statistics*, *25*, 1–20.

Li, S. Y., Tang, L. C., & Ng, S. H. (2010). Nonparametric CUSUM and EWMA control charts for detecting mean shifts. *Journal of Quality Technology*, *42*, 209–226.

Li, Z., Dai, Y., & Wang, Z. (2014). Multivariate change point control chart based on data depth for Phase I analysis. *Communications in Statistics—Simulation and Computation*, *43*, 1490–14507.

Liu, R. Y. (1995). Control charts for multivariate processes. *Journal of the American Statistical Association*, *90*, 1380–1387.

Liu, R. Y., Singh, K., & Teng, J. H. (2004). DDMA-charts: Nonparametric multivariate moving average control charts based on data depth. *Allgemeines Statistiches ARCHIV*, *88*, 235–258.

Lu, S.-L. (2015). An extended nonparametric exponentially weighted moving average sign control chart. *Quality and Reliability Engineering International*, *31*, 3–13.

Malela-Majika, J. C., Chakraborti, S., & Graham, M. A. (2016). Distribution-free Phase II Mann-Whitney control charts with runs-rules. *The International Journal of Advanced Manufacturing Technology*, *86*, 723–735.

McCracken, A. K., & Chakraborti, S. (2013). Control charts for joint monitoring of mean and variance: An overview. *Quality Technology and Quantitative Management*, *10*, 17–36.

McDonald, D. (1990). A CUSUM procedure based on sequential ranks. *Naval Research Logistics*, *37*, 627–646.

Messaoud, A., Weihs, C., & Hering, F. (2004). *A nonparametric multivariate control chart based on data depth*. Technical Report 2004, 61. Universität Dortmund, Sonderforschungsbereich 475 - Komplexitätsreduktion in Multivariaten Datenstrukturen, Dortmund.

Messaoud, A., Weihs, C., & Hering, F. (2008). Detection of chatter vibration in a drilling process using multivariate control charts. *Computational Statistics and Data Analysis*, *52*, 3208–3219.

Mood, A. M. (1954). On the asymptotic efficiency of certain non-parametric two sample test. *The Annals of Mathematical Statistics*, *25*, 514–522.

Mukherjee, A. (2017). Distributionfree Phase-II exponentially weighted moving average schemes for joint monitoring of location and scale based on subgroup samples. *The International Journal of Advanced Manufacturing Technology*, *92*, 101–116.

Mukherjee, A., & Chakraborti, S. (2012). A distribution-free control chart for the joint monitoring of location and scale. *Quality and Reliability Engineering International*, *28*, 335–352.

Mukherjee, A., Graham, M., & Chakraborti, S. (2013). Distribution-free exceedance CUSUM control charts for location. *Communication in Statistics—Simulation and Computation*, *42*, 1153–1187.

Osei-Aning, R., Abbasi, S. A., & Riaz, M. (2017). Bivariate dispersion control charts for monitoring non-normal processes. *Quality and Reliability Engineering International*, *33*, 515–529.

Ou, Y., Wen, D., Wu, Z., & Khoo, M. B. C. (2012). Comparison study on effectiveness and robustness of control charts for monitoring process mean and variance. *Quality and Reliability Engineering International*, *28*, 2–17.

Puri, M. L., & Sen, P. K. (1976). *Nonparametric methods in multivariate analysis*. New York: Wiley.

Qiu, P., & Hawkins, D. (2001). A rank based multivariate CUSUM procedure. *Technometrics*, *43*, 120–132.

Qiu, P., & Hawkins, D. (2003). A nonparametric multivariate cumulative sum procedure for detecting shifts in all directions. *Journal of the Royal Statistical Society: Series D (The Statistician)*, *52*, 151–164.

Qiu, P., He, Z., & Wang, Z. (2019). Nonparametric monitoring of multiple count data. *IISE Transactions*, *51*, 972–984.

Qiu, P., & Li, Z. (2011). On nonparametric statistical process control of univariate processes. *Technometrics*, *53*, 390–405.

Randles, R. H. (2000). A simpler, affine invariant, multivariate, distribution-free sign test. *Journal of the American Statistical Association*, *95*, 1263–1268.

Reynolds, M. R. J., & Stoumbos, Z. G. (2006). Comparisons of some exponentially weighted moving average control charts for monitoring the process mean and variance. *Technometrics*, *48*, 550–567.

Shewhart, W. A. (1926). Quality control charts. *Bell System Technical Journal*, *5*, 593–603.

Shewhart, W. A. (1931). *Economic control of quality of manufactured product*. New York: D. Van Nostrand Company.

Shewhart, W. A. (1939). *Statistical method from the viewpoint of quality control*. Washington: The Graduate School, The Department of Agriculture.

Siegel, S., & Tukey, J. W. (1960). Nonparametric sum of ranks procedure for relative spread in unpaired samples. *Journal of the American Statistical Association*, *55*, 429–445.

Sklar, A. (1959). Fonctions de Répartition à n Dimensions et Leurs Marges. *Publications de l'Institut de Statistique de l'Université de Paris*, *8*, 229–231.

Stoumbos, Z. G., Jones, A., Woodall, W. H., & Reynolds, M. R. (2001). On nonparametric multivariate control charts based on data depth. In H.-J. Lenz & P.-T. Wilrich (Eds.), *Frontiers in Statistical Quality Control* (Vol. 6, pp. 207–227). New York: Physica-Verlag (Springer).

Sukhamte, B. (1954). On certain two-sample nonparametric tests for variances. *The Annals of Mathematical Statistics*, *28*, 188–194.

Triantafyllou, I. S. (2018). Nonparametric control charts based on order statistics: Some advances. *Communications in Statistics—Simulation and Computation*, *47*, 2684–2702.

Zarate, P. B. (2004). *Design of nonparametric control chart for monitoring multivariate processes using principal components analysis and data depth*. Ph.D. thesis University of South Florida.

Zhou, C., Zou, C., Zhang, Y., & Wang, Z. (2009). Nonparametric control chart based on change-point model. *Statistical Papers*, *50*, 13–28.

Zi, X., Zou, C., Zhou, Q., & Wang, J. (2013). A directional multivariate sign EWMA control chart. *Quality Technology and Quantitative Management*, *10*, 115–132.

Zombade, D. M., Ghute, V. B. (2014). Nonparametric control chart for variability using runs rules. *The Experiment*, *24*, 1683–1691.

Zou, C., & Tsung, F. (2011). A multivariate sign EWMA control chart. *Technometrics*, *53*, 84–97.

Zou, C., Wang, Z., & Tsung, F. (2012). A spatial rank-based multivariate EWMA control chart. *Naval Research Logistics*, *59*, 91–110.

Zuo, Y., & Serfling, R. (2000). General notions of statistical depth functions. *The Annals of Statistics*, *28*, 461–482.

The Shewhart Sign Chart with Ties: Performance and Alternatives

Philippe Castagliola, Kim Phuc Tran, Giovanni Celano
and Petros E. Maravelakis

Abstract The Shewhart Sign (SN) control chart is a well-known distribution-free statistical process monitoring tool due to its robustness to the violation of the normality assumption for observations. To the best of our knowledge, there is not yet a thorough understanding of what happens to the statistical properties of the SN control chart in the presence of observations tied to the monitored population quantile, for example, the median: this is an event occurring in practice, in particular when the process runs in-control, because of the measurement device resolution, which inevitably introduces a rounding-off error. In this paper, we tackle the problem and show that when ties occur, the Shewhart SN control chart is no longer distribution-free, even in the presence of a small probability of having ties. To solve the problem, we discuss some procedures to handle the occurrence of ties. The study shows that the best strategy simply consists in implementing a Bernoulli trial approach: in practice, ties are reconsidered by 50% chance as being greater or smaller than the monitored population quantile. We quantitatively show that this approach allows the distribution-free properties of the Shewhart SN to be generally preserved.

Keywords Nonparametric control charts · Measurement error · Device Resolution · ARL · Process location.

P. Castagliola (✉)
Université de Nantes & LS2N UMR CNRS 6004, Nantes, France
e-mail: philippe.castagliola@univ-nantes.fr

K. P. Tran
ENSAIT & GEMTEX, Roubaix, France

G. Celano
Università di Catania, Catania, Italy

P. E. Maravelakis
University of Piraeus, Department of Business Administration, Piraeus, Greece

© Springer Nature Switzerland AG 2020
M. V. Koutras and I. S. Triantafyllou (eds.), *Distribution-Free
Methods for Statistical Process Monitoring and Control*,
https://doi.org/10.1007/978-3-030-25081-2_3

1 Introduction

Control charts have been widely recognized as a primary tool of Statistical Process Monitoring (SPM) that are frequently used for improving process capability and productivity by reducing variability in the process, (see Montgomery 2013). In SPM literature, several control charts for monitoring the process mean and/or the dispersion have been proposed. These control charts are often designed by assuming that the observations are normally distributed. However, when the actual distribution of the observations is not normal or the one anticipated by the quality practitioner, many false alarms can be triggered by the control chart. In this context, the use of a nonparametric (or distribution-free) control chart can be a good solution to overcome the problems related to the distributional assumptions. In the SPM literature, the first work that dealt with a nonparametric control chart was by McGilchrist and Woodyer (1975), who proposed a distribution-free cumulative sum technique applied to the monitoring of rainfall amounts. Then, the properties and design of distribution-free-type control charts have been thoroughly investigated by many authors. For further details see, Bakir and Reynolds (1979), Amin and Searcy (1991), Chakraborti et al. (2001), Li et al. (2010), Graham et al. (2011), Zou and Tsung (2011), Yang et al. (2011), Lu (2015) and Abid et al. (2017).

In recent years, many researchers have paid attention to the development and implementation of new nonparametric control charts. Chakraborty et al. (2016) proposed a distribution-free generally weighted moving average (GWMA) control chart based on the Wilcoxon signed-rank statistic. Capizzi (2015) discussed about the need for a nonparametric approach to Phase I analysis and the use of variable selection-based control charts in multivariate Phase II monitoring. The control charts monitoring a sign statistic have been originally introduced by Amin et al. (1995). EWMA control charts with sign statistics have been investigated by Graham et al. (2011) and Yang et al. (2011). More recently, Celano et al. (2016a) have investigated the statistical performance of a Shewhart Sign (SN) control chart in a process with a finite production horizon and they have shown that it often outperforms the parametric t control chart. Then, Celano et al. (2016b) investigated the statistical performance of the Shewhart SN control chart for finite populations demonstrating the change of the distribution properties of the sign statistic and suggesting a simple rule to select its design parameters. Castagliola et al. (2019) proposed a new Phase II EWMA-type chart for count data, based on the sign statistic and they also provided a methodology to compute the exact run length properties of the proposed chart. Very recently, Qiu (2018) discussed some perspectives on issues related to the robustness of conventional SPM charts and to the strengths and limitations of various nonparametric SPM charts.

A central role in the implementation of a control chart is played by the measurement system. A measuring device inevitably introduces some errors in the observed value, thus making unobservable the actual value of the characteristic under control. Control charts with measurement error have been widely investigated in the literature, in particular with reference to the bias and precision errors, for more details see the recent review by Maleki et al. (2017). However, an important role in the distribution definition of the observed values is played by the measurement system resolution, which introduces a rounding-off error resulting in a discretization of the observed measures. Rounding-off errors also result in "ties", that is, observations having the same observed value, even if their true distribution is continuous. The treatment of ties in a nonparametric approach is of great importance since the choice of techniques for treating them can markedly affect its distribution-free properties, see Putter (1955) and Gibson and Melsa (1976) for more details.

To the best of our knowledge, there is not yet a thorough understanding of what happens to the statistical properties of the Shewhart SN control chart in the presence of observations tied to the monitored population quantile. In this paper, we tackle the problem and show that when ties occur the Shewhart SN control chart is no longer distribution-free, even in the presence of a small probability of ties. To overcome this problem, we discuss some procedures to handle the occurrence of such ties.

The rest of this paper proceeds as follows: in Sect. 2, the Shewhart SN chart "without ties" is briefly introduced; In Sect. 3, the Shewhart SN chart "with ties" is defined; Sect. 4 provides the effect of the measurement system resolution on the Shewhart SN chart. In Sect. 5, we discuss procedures to tackle the occurrence of rounding-off errors. Finally, some concluding remarks and recommendations are made in Sect. 6.

2 The Shewhart SN Chart "Without Ties"

Let X be a quality characteristic following an *unknown* continuous distribution with cumulative distribution function (c.d.f.) $F_X(x|\theta)$ where θ is the location parameter to be monitored. If $\theta = \theta_0$ the process is declared as *in-control* and, if $\theta = \theta_1$, the process is declared as *out-of-control*. In this paper, without loss of generality, we consider $\text{med}(X) = \theta$ as being the median of the distribution $F_X(x|\theta)$ (but other quantiles can also be considered).

Let us suppose that, at time $t = 1, 2, \ldots$, we observe subgroup $\{X_{t,1}, X_{t,2}, \ldots, X_{t,n}\}$ of size $n \geq 1$. Let $S_{t,k} = \text{sign}(X_{t,k} - \theta_0)$, $k = 1, 2, \ldots, n$, where $\text{sign}(x) = -1, 0$ or $+1$ if $x < 0, x = 0$ or $x > 0$, respectively. At this step of the study, following past literature about the Shewhart SN control chart, we assume a perfect calibration of the measurement system, which eliminates *bias* and *linearity* errors; furthermore, the *precision* error is sufficiently small to be neglected. Definitions of these errors can be found in Montgomery (2013). Finally, the gauge resolution is such that the rounding error is eliminated. By definition, the plotting statistic SN_t, at time $t = 1, 2, \ldots$, of the Shewhart SN control chart is

$$\text{SN}_t = \sum_{k=1}^{n} S_{t,k}. \tag{1}$$

Let π_{-1}, π_0, and π_{+1} be the following probabilities:

$$\pi_{-1} = \Pr(S_{t,k} = -1) = \Pr(X_{t,k} < \theta_0) = F_X(\theta_0|\theta),$$
$$\pi_0 = \Pr(S_{t,k} = 0) = \Pr(X_{t,k} = \theta_0),$$
$$\pi_{+1} = \Pr(S_{t,k} = +1) = \Pr(X_{t,k} > \theta_0) = 1 - F_X(\theta_0|\theta).$$

where θ can be either θ_0 or θ_1. As X is assumed to be a continuous random variable, no matter the value of θ (θ_0 or θ_1), we always have $\pi_0 = 0$ (see Fig. 1a, b) and, consequently, each $S_{t,k}$ can only take values -1 or $+1$. Therefore, the random variable SN_t is defined on $\{-n, -n+2, \ldots, n-2, n\}$ and its distribution can be easily obtained by considering the relationship $SN_t = 2D_t - n$, where $D_t = \#\{X_{t,k} > \theta_0, k = 1, \ldots, n\}$, i.e., D_t is the number of observations $\{X_{t,1}, X_{t,2}, \ldots, X_{t,n}\}$ larger than θ_0. As $\Pr(X_{t,k} > \theta_0) = \pi_{+1}$, we have $D_t \sim \text{Bin}(n, \pi_{+1})$, i.e., a binomial random variable of parameters n and π_{+1} and the c.d.f. $F_{SN}(s|n)$ of SN_t is equal to

$$F_{SN}(s|n) = F_{Bin}\left(\frac{n+s}{2}\Big|n, \pi_{+1}\right),$$

where $F_{Bin}(\ldots|n, \pi_{+1})$ is the c.d.f. of the binomial distribution with parameters n and π_{+1}. Moreover, if the process is in-control ($\theta = \theta_0$), we also have $\pi_{-1} = \pi_{+1} = 0.5$ (see Fig. 1a) and the c.d.f. of SN_t reduces to

$$F_{SN}(s|n) = F_{Bin}\left(\frac{n+s}{2}\Big|n, 0.5\right).$$

Because the in-control distribution of SN only depends on n and π_{+1}, which is a constant, (and not on $F_X(x|\theta)$), the Shewhart SN control chart "without ties" is a distribution-free control chart. The control limits (LCL, UCL) and the centerline CL of the Shewhart SN control chart are equal to $LCL = -C$, $CL = 0$, and $UCL = C$ where $C \in \{2, 4, \ldots, n\}$ (if n is an even integer) or $C \in \{1, 3, \ldots, n\}$ (if n is an odd integer) is a constant to be fixed. The process is declared to be in-control if $-C < SN_t < C$ and out-of-control otherwise. The type II probability error β of the Shewhart SN control chart "without ties" is equal to

$$\beta = \Pr(-C < SN_t < C|\theta = \theta_1) = \Pr(-C < SN_t \leq C - 2|\theta = \theta_1),$$

which can be rewritten in terms of the binomial distribution as

$$\beta = F_{Bin}\left(\frac{n+C}{2} - 1\Big|n, \pi_{+1}\right) - F_{Bin}\left(\frac{n-C}{2}\Big|n, \pi_{+1}\right),$$

with $\pi_{+1} = 1 - F_X(\theta_0|\theta_1)$. As the distribution of the Run Length (RL) of the Shewhart SN control chart follows a geometric distribution with parameter $1 - \beta$, its probability mass function (p.m.f.) and c.d.f. are equal to $f_{RL}(\ell) = \Pr(RL = \ell) = (1 - \beta)\beta^{\ell-1}$ and $F_{RL}(\ell) = \Pr(RL \leq \ell) = 1 - \beta^{\ell}$ for $\ell = 1, 2, \ldots$, respectively, its mean $ARL = E(RL) = \frac{1}{1-\beta}$ and its standard-deviation $SDRL = \sigma(RL) = \frac{\sqrt{\beta}}{1-\beta}$.

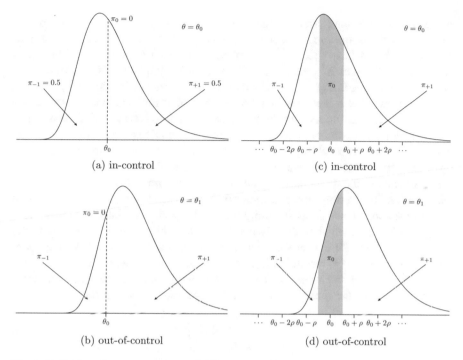

(a) in-control

(c) in-control

(b) out-of-control

(d) out-of-control

Fig. 1 Definition of π_{-1}, π_0, and π_{+1}. Subfigures **a** and **b** correspond to the "without ties" case, while subfigures **c** and **d** correspond to the "with ties" case

3 The Shewhart SN Chart "With Ties"

In practice, due to the imperfection of the measurement system, the real values $X_{t,k}$ of the observations are never observed. Instead, we observe a measured value $X'_{t,k} \neq X_{t,k}$. Many sources of error can be identified in a measurement system. A model to account for three well-known sources of error is

$$X'_{t,k} = \left\lfloor \frac{A + B X_{t,k} + \varepsilon_{t,k}}{\rho} + \frac{1}{2} \right\rfloor \rho, \tag{2}$$

where $\lfloor \ldots \rfloor$ is the floor function, the constants (A, B) account for the *bias-linearity* error, the *precision* error is quantified through the noise $\varepsilon_{t,k}$ and ρ is a parameter quantifying the device *resolution*, which introduces a *rounding-off* error. Equation (2) refers to the linear measurement error model, see Linna and Woodall (2001), who discuss the measurement error effect on the performance of a Shewhart control chart for monitoring the mean without considering the rounding-off error. The resolution of a measuring device is defined as the smallest amount of change in the quality characteristic that the measurement system can faithfully indicate: it is reported as a unit of measure. For example, the resolution of a scale can be equal to 0.01gr; the parameter

ρ is equal to the resolution value: then, for the scale we get $\rho = 0.01$. More specifically, if ρ is the *resolution* of the measurement system then, by definition, we have $X'_{t,k} = x$ if $X_{t,k} \in (x - \frac{\rho}{2}, x + \frac{\rho}{2}]$. For instance, if $\rho = 0.1$ and $\theta_0 = 100$ then possible measured values $X'_{t,k}$ are $\{\ldots, 99.7, 99.8, 99.9, 100, 100.1, 100.2, 100.3, \ldots\}$ and, if the real value is $X_{t,k} = 100.038$, then the measured observation is $X'_{t,k} = 100$, thus generating a *tie*. It is evident that the rounding-off error introduced by the device resolution in the measurement of the true value of a quality characteristic results in a discretization of the observed quality characteristic, if this is a continuous variable, thus increasing the probability of having ties. Here, we are only interested to investigate the effect of the tool resolution on the SN control chart performance: that is, we still maintain the assumption of a perfect tool calibration, $(A, B) = (0, 1)$ and we intentionally overlook the precision error. This last source of error is the sum of the repeatability and reproducibility errors. They are estimated by means of a Gauge R&R study, which is usually based on a dataset of multiple measurements on a sample of $m = 20$–30 parts collected by different appraisers, see Montgomery (2013). Conversely, the resolution is an intrinsic device characteristic and a source of error that cannot be eliminated. Under the assumption of the measurement error only depending on the rounding-off error, the error model above reduces to

$$X'_{t,k} = \left\lfloor \frac{X_{t,k}}{\rho} + \frac{1}{2} \right\rfloor \rho. \tag{3}$$

In the case of rounding-off errors of the measurement system, the statistic $S_{t,k}$ used in (1) must be replaced by $S_{t,k} = \text{sign}(X'_{t,k} - \theta_0)$ and the probabilities π_{-1}, π_0 and π_{+1} must be redefined as

$$\pi_{-1} = \Pr(X'_{t,k} < \theta_0) = \Pr(X_{t,k} \le \theta_0 - \tfrac{\rho}{2}) = F_X(\theta_0 - \tfrac{\rho}{2}|\theta),$$

$$\pi_0 = \Pr(X'_{t,k} = \theta_0) = \Pr(\theta_0 - \tfrac{\rho}{2} < X_{t,k} \le \theta_0 + \tfrac{\rho}{2})$$
$$= F_X(\theta_0 + \tfrac{\rho}{2}|\theta) - F_X(\theta_0 - \tfrac{\rho}{2}|\theta),$$

$$\pi_{+1} = \Pr(X'_{t,k} > \theta_0) = \Pr(X_{t,k} > \theta_0 + \tfrac{\rho}{2}) = 1 - F_X(\theta_0 + \tfrac{\rho}{2}|\theta).$$

In the previous Section, we claimed that because X is a continuous random variable, no matter the value of θ, we always have $\pi_0 = \Pr(X_{t,k} = \theta_0) = 0$. But now, with the new definition of π_0, we actually have $\pi_0 \ne 0$ (see Fig. 1c, d) and, consequently, $S_{t,k}$ will not only take values -1 or $+1$ but it will also take the value 0. As a consequence, the statistical properties of the SN_t statistic change as follows:

- the random variable SN_t is no longer defined on $\{-n, -n + 2, \ldots, n - 2, n\}$ but it is rather defined on $\{-n, -n + 1, \ldots, n - 1, n\}$,
- the p.m.f. $f_{SN}(s|n, \pi_{-1}, \pi_0, \pi_{+1})$ of SN_t is no longer related to the binomial distribution as for the "without ties" case. As far as we know, $f_{SN}(s|n, \pi_{-1}, \pi_0, \pi_{+1})$ does not correspond to a well-known distribution. For this reason, we provide in

Appendix three different ways for evaluating it. From now on, for simplicity, we denote it as $f_{\text{SN}}(s|n)$ instead of $f_{\text{SN}}(s|n, \pi_{-1}, \pi_0, \pi_{+1})$, (but still having in mind that, in the "with ties" case, the in- and out-of-control distribution of SN_t actually depends on π_{-1}, π_0 and π_{+1}). The c.d.f. $F_{\text{SN}}(s|n)$ of SN_t is simply obtained by summing the p.m.f. terms, i.e.,

$$F_{\text{SN}}(s|n) = f_{\text{SN}}(-n|n) + f_{\text{SN}}(-n+1|n) + \cdots + f_{\text{SN}}(s|n).$$

- Because the in-control distribution of SN depends on $F_X(x|\theta)$ (through π_{-1}, π_0 and π_{+1}), the Shewhart SN control chart "with ties" is *no longer* a distribution-free control chart.
- The type II probability error β of the Shewhart SN control chart "with ties" is equal to

$$\beta = \Pr(-C < \text{SN}_t < C|\theta = \theta_1) = \Pr(-C < \text{SN}_t \le C - 1|\theta = \theta_1),$$

which can be rewritten in terms of the SN distribution as

$$\beta = F_{\text{SN}}(C - 1|n) - F_{\text{SN}}(-C|n).$$

The formulas for the p.m.f., c.d.f., ARL, and SDRL of the RL of the Shewhart SN control chart "with ties" are the same as the ones for the "without ties" case.

Now, without loss of generality, we will assume that (i) $F_X(x|\theta)$ belongs to a location-scale family of distributions and it can be rewritten $F_X(x|\theta) = F_Z(\frac{x-\theta}{\sigma})$ where σ is the standard-deviation of X and (ii) $\theta_1 = \theta_0 + \delta\sigma$, where δ is the standardized distribution shift. It is not difficult to prove that

$$\pi_{-1} = F_Z(-\tfrac{\kappa}{2} - \delta),$$
$$\pi_0 = F_Z(\tfrac{\kappa}{2} - \delta) - F_Z(-\tfrac{\kappa}{2} - \delta),$$
$$\pi_{+1} = 1 - F_Z(\tfrac{\kappa}{2} - \delta),$$

where $\kappa = \frac{\varrho}{\sigma}$ is the *standardized resolution*. Replacing $\delta = 0$ in the previous equations allows to obtain the in-control values for π_{-1}, π_0, and π_{+1} as

$$\pi_{-1} = F_Z(-\tfrac{\kappa}{2}),$$
$$\pi_0 = F_Z(\tfrac{\kappa}{2}) - F(-\tfrac{\kappa}{2}),$$
$$\pi_{+1} = 1 - F_Z(\tfrac{\kappa}{2}).$$

As an illustration, Fig. 2 depicts the p.m.f. $f_{\mathrm{SN}}(s|n)$ of SN_t for $n = 20$ (left side), $n = 50$ (right side), and $\kappa \in \{0, 0.05, 0.1, 0.2\}$. These plots assume a normal $(0, 1)$ c.d.f. for $F_z(z)$. As it can be seen, when $\kappa = 0$, SN_t is only defined for $\{-n, -n + 2, \ldots, n - 2, n\}$ (as expected) and, when $\kappa > 0$, SN_t becomes also defined for "intermediate values" $\{-n + 1, -n + 3, \ldots, n - 3, n - 1\}$. The larger is κ the larger are the probabilities associated with these intermediate values and the smaller are the probabilities associated with the initial ones $\{-n, -n + 2, \ldots, n - 2, n\}$.

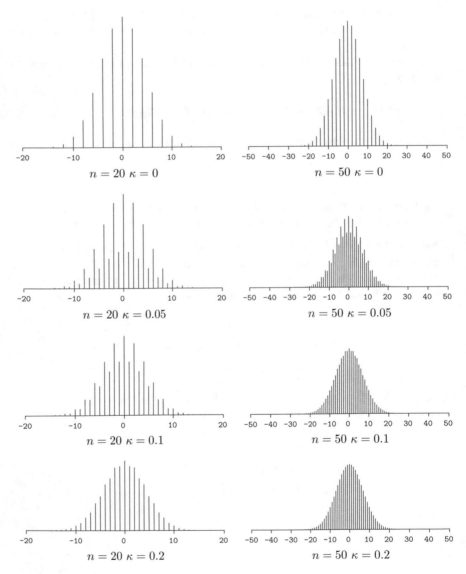

Fig. 2 Examples of p.m.f. $f_{\mathrm{SN}}(s|n)$ of SN_t for $n = 20$ (left side), $n = 50$ (right side), and $\kappa \in \{0, 0.05, 0.1, 0.2\}$

4 Effect of the Measurement System Resolution on the Shewhart SN Chart

In order to evaluate the effect of the measurement system resolution and the related probability to have observations tied to the monitored population quantile θ_0, we have chosen a benchmark of 18 Johnson's type distributions covering a wide range of skewness $\gamma_3 = \frac{\mu_3}{\sigma^3} \in \{0, 2, 5\}$ and kurtosis $\gamma_4 = \frac{\mu_4}{\sigma^4} - 3$, see Fig. 3 (remark: all the plots are on the same scale). A Johnson's distribution depends on four parameters a, $b > 0$, c and $d > 0$ and it is either

- defined on $[c, c + d]$ (bounded, denoted as B in Table 1) and its c.d.f. $F_Z(x)$ is equal to

$$F_Z(x) = \Phi \left(a + b \ln \left(\frac{x - c}{c + d - x} \right) \right),$$

- or defined on $(-\infty, +\infty)$ (unbounded, denoted as U in Table 1) and its c.d.f. $F_Z(x)$ is equal to

$$F_Z(x) = \Phi \left(a + b \sinh^{-1} \left(\frac{x - c}{d} \right) \right),$$

where $\Phi(\dots)$ is the c.d.f. of the normal $(0, 1)$ distribution. The values of parameters a, b, c, and d in Table 1 have been computed in order to fulfill the following constraints: (i) the median $\text{med}(Z) = \theta_0 = 0$, i.e $F_Z(0) = 0.5$, (ii) the standard-deviation $\sigma(Z) = 1$, (iii) the skewness $\gamma_3(Z)$ and kurtosis $\gamma_4(Z)$ coincide with the values γ_3 and γ_4 as in Table 1, respectively. In terms of skewness and kurtosis, cases #1–#6 correspond (without being exactly identical) to some remarkable symmetric distributions: case #1 corresponds to the uniform distribution, case #2 corresponds to the triangular distribution, case #3 corresponds to the normal distribution (here the approximation $b = d = 100$ has been used), and cases #4–#6 correspond to the Student t distribution with 10, 6, and 5 degrees of freedom, respectively.

For the 18 Johnson's distributions listed in Table 1, Table 2 (for n 20) and Table 3 (for $n = 50$) show the ARL values of the Shewhart SN control chart for shifts $\delta \in \{-1, -0.5, -0.2, -0.1, 0, 0.1, 0.2, 0.5, 1\}$ and for standardized resolution $\kappa = 0$ (i.e., "without ties") and $\kappa \in \{0.05, 0.1, 0.2\}$ (i.e., "with ties"). The chart parameter C has been selected assuming a perfect measurement device, (i.e., $\kappa = 0$), and an in-control ARL value *as close as possible* to 370.4 (i.e., the resulting values can be either smaller or larger than 370.4). For the considered sample sizes, we have $(C = 14, \text{ARL}_0 = 388.1)$ when $n = 20$ and $(C = 22, \text{ARL}_0 = 384.3)$ when $n = 50$. Tables 2 and 3 also provide the values $\bar{\pi}_0$, $\bar{\pi}_{+1}$, and $\overline{\text{ARL}}$ corresponding to the average of values π_0, π_{+1} and ARL for the 18 cases under consideration. From Tables 2 and 3, we can draw the following conclusions:

- if $\delta = 0$ and $\kappa = 0$ then all the in-control ARL values are the same, no matter which distribution has been considered among the 18 available ones. For example, we have ARL = 388.1 when $n = 20$ and ARL = 384.3 when $n = 50$, (see values

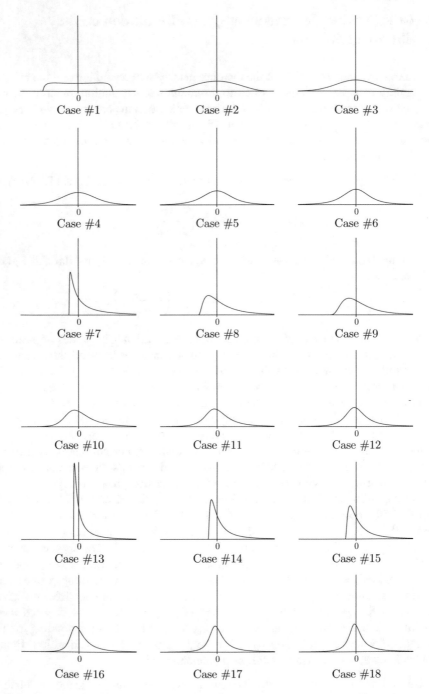

Fig. 3 Benchmark of 18 Johnson's type distributions

Table 1 Benchmark of 18 Johnson's type distributions

Case	γ_3	γ_4	Type	a	b	c	d
1	0	−1.2	B	0.0000e+00	6.4646e−01	−1.8153e+00	3.6306e+00
2	0	−0.6	B	0.0000e+00	1.3983e+00	−3.1097e+00	6.2195e+00
3	0	0.0	U	0.0000e+00	1.0000e+02	0.0000e+00	1.0000e+02
4	0	1.0	U	0.0000e+00	2.3212e+00	0.0000e+00	2.1094e+00
5	0	3.0	U	0.0000e+00	1.6104e+00	0.0000e+00	1.3118e+00
6	0	6.0	U	0.0000e+00	1.3493e+00	0.0000e+00	1.0000e+00
7	2	4.3	B	1.7464e+00	6.9076e−01	−4.8932e−01	6.6213e+00
8	2	6.1	B	3.3279e+00	1.2270e+00	−1.0016e+00	1.6088e+01
9	2	7.9	U	−4.8560e+00	1.8044e+00	−1.4190e+00	1.9332e−01
10	2	10.8	U	−1.0444e+00	1.4320e+00	−6.5538e−01	8.2361e−01
11	2	16.7	U	−5.2977e−01	1.2093e+00	−3.3154e−01	7.3314e−01
12	2	25.5	U	−3.4371e−01	1.0892e+00	−2.0230e−01	6.3054e−01
13	5	39.9	B	3.3715e+00	7.4593e−01	−2.7094e−01	2.5150e+01
14	5	52.6	B	5.2193e+00	9 8134e−01	−4.7316e−01	9.7043e+01
15	5	65.3	U	−4.0187e+00	1.0864c+00	−5.6652e−01	2.8059e−02
16	5	86.4	U	−7.5701e−01	9.8744e−01	−3.2033e−01	3.7954e−01
17	5	128.7	U	−4.3187e−01	9.0797e−01	−1.8538e−01	3.7543e−01
18	5	192.1	U	−2.9868e−01	8.5558e−01	−1.2122e−01	3.4029e−01

in bold in Tables 2 and 3). On the other hand, if $\kappa > 0$, then the in-control ARL values are *no longer* equal, (they actually depend on the specific distribution), and they tend to become larger as γ_4 increase. For instance, in Table 2, for $\kappa = 0.2$ and $\delta = 0$, we have ARL = 722.2 for case #1 for which $\gamma_4 = −1.2$ while we have ARL = 2274.8 for case #13 for which $\gamma_4 = 39.9$.

- if $\gamma_3 = 0$ (cases #1–#6) then, no matter the value of κ, the ARL values are symmetric, i.e., they are the same for shifts δ and $−\delta$. On the other hand, if $\gamma_3 \neq 0$ (cases #7–#18) then the ARL values are asymmetric. More specifically, negative shifts $−\delta$ give larger ARL values than positive ones. For instance, for $n = 20$, $\kappa = 0.1$, case #14, we have ARL = 197.3 when $\delta = −0.1$ while we have ARL = 137.0 when $\delta = 0.1$. This asymmetry also holds for the $\overline{\text{ARL}}$ values.
- for each distribution and shift size δ, the larger is κ the larger are the ARL values. For instance, in Table 2, for case #9 and shift $\delta = −0.2$, we have ARL = 72.5 for $\kappa = 0$, ARL = 89.5 for $\kappa = 0.05$, ARL = 105.8 for $\kappa = 0.1$ and ARL = 138.9 for $\kappa = 0.2$. The same happens to the $\overline{\text{ARL}}$ values. This clearly highlights the fact that an increasing value of κ not only affects the distribution-free property of the Shewhart SN control chart when the process runs in-control, but also deteriorates its detection efficiency.

Table 2 ARL values of the Shewhart SN control chart for shifts $\delta \in \{-1, -0.5, -0.2, -0.1, 0, 0.1, 0.2, 0.5, 1\}$, for standardized resolution $\kappa = 0$ (i.e., "without ties") and $\kappa \in \{0.05, 0.1, 0.2\}$ (i.e., "with ties"), when $n = 20$

Case	$\kappa = 0$									$\kappa = 0.05$								
	δ									δ								
	−1	−0.5	−0.2	−0.1	0	0.1	0.2	0.5	1	−1	−0.5	−0.2	−0.1	0	0.1	0.2	0.5	1
1	2.8	25.9	164.7	296.1	**388.1**	296.1	164.7	25.9	2.8	3.0	29.4	193.8	354.3	468.8	354.3	193.8	29.4	3.0
2	1.9	14.3	118.2	258.3	**388.1**	258.3	118.1	14.3	1.9	2.0	16.2	142.9	320.5	490.4	320.5	142.8	16.2	2.0
3	1.7	10.7	99.0	238.4	**388.1**	238.4	99.0	10.7	1.7	1.7	12.2	121.2	301.3	502.1	301.3	121.2	12.2	1.7
4	1.5	8.3	83.3	219.5	**388.1**	219.5	83.3	8.3	1.5	1.5	9.4	103.0	282.0	513.8	282.0	103.0	9.4	1.5
5	1.3	6.3	67.3	197.0	**388.1**	197.0	67.3	6.3	1.3	1.4	7.1	84.2	258.2	528.6	258.2	84.2	7.1	1.4
6	1.3	5.0	55.2	177.3	**388.1**	177.3	55.2	5.0	1.3	1.3	5.6	69.8	236.3	542.9	236.3	69.8	5.6	1.3
7	2.1	8.3	59.3	169.5	**388.1**	130.4	22.6	1.0	1.0	2.2	9.2	73.0	224.4	563.5	184.7	30.5	1.0	1.0
8	1.9	8.6	70.1	193.1	**388.1**	173.5	46.4	2.1	1.0	2.0	9.6	86.5	252.1	537.9	234.2	60.3	2.4	1.0
9	1.8	8.5	72.5	198.7	**388.1**	183.6	53.8	3.0	1.0	1.9	9.5	89.5	258.7	532.4	244.9	69.0	3.4	1.0
10	1.6	6.7	60.9	181.4	**388.1**	169.0	46.9	3.0	1.0	1.6	7.4	75.9	239.9	544.2	228.2	60.3	3.4	1.0
11	1.4	5.0	48.5	160.0	**388.1**	149.4	38.2	2.7	1.0	1.4	5.6	61.1	215.7	560.9	205.1	49.4	3.0	1.0
12	1.3	4.0	39.1	141.1	**388.1**	132.0	31.4	2.4	1.0	1.3	4.4	49.7	193.6	577.6	184.0	40.7	2.7	1.0
13	1.3	2.8	18.0	71.6	**388.1**	29.8	1.6	1.0	1.0	1.3	3.0	22.5	102.9	714.2	46.5	1.8	1.0	1.0
14	1.3	3.8	29.5	109.0	**388.1**	73.6	8.5	1.0	1.0	1.4	4.1	37.2	152.4	629.7	110.3	11.3	1.0	1.0
15	1.4	4.2	34.3	122.3	**388.1**	90.0	12.7	1.0	1.0	1.4	4.6	43.2	169.2	609.6	132.4	17.0	1.0	1.0
16	1.3	3.4	29.1	111.8	**388.1**	89.2	14.7	1.2	1.0	1.3	3.8	36.9	156.6	618.2	130.6	19.5	1.3	1.0
17	1.2	2.7	22.5	94.9	**388.1**	78.1	13.1	1.4	1.0	1.2	2.9	28.7	135.4	639.6	115.4	17.2	1.4	1.0
18	1.1	0.3	17.6	80.0	**388.1**	66.8	11.1	1.4	1.0	1.1	2.4	22.5	116.0	663.4	99.7	14.4	1.4	1.0
$\overline{\text{ARL}}$	1.6	7.3	60.5	167.8	388.1	152.9	49.4	5.1	1.2	1.6	8.1	74.5	220.5	568.8	203.8	61.5	5.7	1.3
$\bar{\pi}_0$	0.0000	0.0000	0.0000	0.0000	0.0000	0.0000	0.0000	0.0000	0.0000	0.0091	0.0167	0.0244	0.0277	0.0312	0.0348	0.0384	0.0182	0.0063
$\bar{\pi}_{+1}$	0.1426	0.2671	0.3889	0.4411	0.5000	0.5661	0.6395	0.8200	0.9316	0.1381	0.2589	0.3769	0.4274	0.4846	0.5489	0.6204	0.8105	0.9284

(continued)

Table 2 (continued)

Case	κ = 0.1									κ = 0.2								
	δ									δ								
	−1	−0.5	−0.2	−0.1	0	0.1	0.2	0.5	1	−1	−0.5	−0.2	−0.1	0	0.1	0.2	0.5	1
1	3.1	32.5	222.4	413.0	551.7	413.0	222.4	32.5	3.1	3.3	37.7	277.1	530.5	722.2	530.5	277.1	37.7	3.3
2	2.1	17.9	166.6	382.7	595.7	382.7	166.4	17.9	2.1	2.2	20.6	211.6	507.6	815.8	507.6	211.4	20.6	2.2
3	1.8	13.4	142.1	363.6	619.6	363.6	142.1	13.4	1.8	1.8	15.4	182.0	489.7	868.5	489.7	182.0	15.4	1.8
4	1.6	10.3	121.4	343.7	643.5	343.7	121.4	10.3	1.6	1.6	11.8	156.6	469.5	923.1	469.5	156.6	11.8	1.6
5	1.4	7.7	99.7	318.0	673.9	318.0	99.7	7.7	1.4	1.4	8.7	129.8	442.2	995.9	442.2	129.8	8.7	1.4
6	1.3	6.0	82.9	293.7	703.3	293.7	82.9	6.0	1.3	1.3	6.7	108.8	415.6	1070.0	415.6	108.8	6.7	1.3
7	2.3	10.0	86.4	281.6	747.0	227.9	35.0	1.0	1.0	2.4	11.4	115.7	425.3	1183.5	294.3	40.2	1.0	1.0
8	2.0	10.5	102.2	311.9	693.2	289.4	71.4	2.5	1.0	2.1	12.0	134.7	447.5	1044.1	391.7	89.8	2.6	1.0
9	1.9	10.4	105.8	319.1	681.8	302.1	81.8	3.6	1.0	2.0	11.9	138.9	452.5	1015.7	411.5	104.1	3.9	1.0
10	1.6	8.1	90.1	298.2	706.2	283.4	71.6	3.6	1.0	1.7	9.2	119.0	428.8	1077.5	392.9	92.4	3.9	1.0
11	1.4	6.0	72.7	270.7	741.0	256.9	58.8	3.2	1.0	1.5	6.7	96.9	397.5	1171.3	364.7	77.0	3.5	1.0
12	1.3	4.7	59.2	245.0	776.2	232.2	48.5	2.8	1.1	1.3	5.2	79.7	367.8	1273.1	337.8	64.4	3.1	1.1
13	1.3	3.2	26.6	136.9	1095.1	54.4	1.6	1.0	1.0	1.3	3.5	37.4	266.1	2274.8	60.0	2.7	1.0	1.0
14	1.4	4.4	44.3	197.3	891.4	137.0	12.6	1.0	1.0	1.4	4.9	60.9	329.6	1638.6	184.7	14.7	1.0	1.0
15	1.4	5.0	51.4	217.3	846.2	165.3	19.4	1.0	1.0	1.5	5.5	70.2	350.2	1489.2	225.9	23.5	1.0	1.0
16	1.3	4.0	44.0	201.9	865.0	165.5	22.9	1.3	1.0	1.3	4.4	60.4	327.5	1558.8	242.6	30.3	1.3	1.0
17	1.2	3.1	34.2	175.9	913.3	148.0	20.3	1.4	1.0	1.2	3.4	47.5	294.2	1739.2	229.6	27.9	1.5	1.0
18	1.1	2.5	26.8	152.0	969.1	129.2	17.0	1.5	1.0	1.1	2.7	37.5	263.9	1967.5	211.4	23.9	1.5	1.0
ARL	1.6	8.9	87.7	273.5	761.8	250.3	72.0	6.2	1.3	1.7	10.1	114.7	400.3	1268.3	344.6	92.0	7.0	1.3
$\bar{\pi}_0$	0.0182	0.0334	0.0489	0.0555	0.0625	0.0697	0.0763	0.0377	0.0126	0.0364	0.0669	0.0980	0.1111	0.1250	0.1394	0.1451	0.0796	0.0254
$\bar{\pi}_{+1}$	0.1338	0.2510	0.3653	0.4142	0.4697	0.5321	0.6019	0.7989	0.9250	0.1255	0.2358	0.3431	0.3889	0.4411	0.5000	0.5661	0.7720	0.9177

Table 3 ARL values of the Shewhart SN control chart for shifts $\delta \in \{-1, -0.5, -0.2, -0.1, 0, 0.1, 0.2, 0.5, 1\}$, for standardized resolution $\kappa = 0$ (i.e., "without ties") and $\kappa \in \{0.05, 0.1, 0.2\}$ (i.e., "with ties"), when $n = 50$

Case	$\kappa = 0$									$\kappa = 0.05$								
	δ									δ								
	−1	−0.5	−0.2	−0.1	0	0.1	0.2	0.5	1	−1	−0.5	−0.2	−0.1	0	0.1	0.2	0.5	1
1	1.1	6.2	74.6	208.3	**384.3**	208.3	74.6	6.2	1.1	1.1	6.9	88.6	255.6	484.0	255.6	88.6	6.9	1.1
2	1.0	3.3	45.1	159.9	**384.3**	159.9	45.0	3.3	1.0	1.0	3.6	54.1	200.7	503.2	200.7	54.0	3.6	1.0
3	1.0	2.5	34.9	138.3	**384.3**	138.3	34.9	2.5	1.0	1.0	2.7	41.9	175.1	512.7	175.1	41.9	2.7	1.0
4	1.0	2.0	27.4	119.6	**384.3**	119.6	27.4	2.0	1.0	1.0	2.2	32.9	152.5	521.7	152.5	32.9	2.2	1.0
5	1.0	1.6	20.5	99.6	**384.3**	99.6	20.5	1.6	1.0	1.0	1.7	24.5	127.9	532.6	127.9	24.5	1.7	1.0
6	1.0	1.4	15.8	83.8	**384.3**	83.8	15.8	1.4	1.0	1.0	1.5	18.8	108.2	542.6	108.2	18.8	1.5	1.0
7	1.0	2.0	17.3	78.0	**384.3**	52.1	5.3	1.0	1.0	1.0	2.1	20.5	100.7	556.5	67.4	6.1	1.0	1.0
8	1.0	2.1	21.6	96.3	**384.3**	81.0	12.6	1.0	1.0	1.0	2.2	25.8	123.7	539.2	104.6	15.0	1.1	1.0
9	1.0	2.1	22.6	101.0	**384.3**	88.7	15.2	1.1	1.0	1.0	2.2	27.0	129.6	535.3	114.4	18.2	1.1	1.0
10	1.0	1.7	17.9	87.0	**384.3**	77.7	12.8	1.1	1.0	1.0	1.8	21.4	112.2	543.5	100.5	15.2	1.1	1.0
11	1.0	1.4	13.3	71.2	**384.3**	64.1	9.9	1.1	1.0	1.0	1.5	15.8	92.3	554.8	83.1	11.7	1.1	1.0
12	1.0	1.3	10.2	58.7	**384.3**	53.1	7.8	1.1	1.0	1.0	1.3	12.0	76.3	565.7	69.0	9.1	1.1	1.0
13	1.0	1.1	4.2	22.2	**384.3**	7.3	1.0	1.0	1.0	1.0	1.1	4.7	28.7	650.8	8.6	1.0	1.0	1.0
14	1.0	1.2	7.2	40.1	**384.3**	23.1	2.1	1.0	1.0	1.0	1.3	8.4	52.1	598.4	29.4	2.2	1.0	1.0
15	1.0	1.3	8.7	47.4	**384.3**	30.5	2.9	1.0	1.0	1.0	1.3	10.1	61.6	585.9	39.2	3.2	1.0	1.0
16	1.0	1.2	7.1	41.5	**384.3**	30.1	3.4	1.0	1.0	1.0	1.2	8.3	54.1	591.3	38.9	3.8	1.0	1.0
17	1.0	1.1	5.3	32.9	**384.3**	25.1	3.0	1.0	1.0	1.0	1.1	6.1	42.7	604.6	32.3	3.4	1.0	1.0
18	1.0	1.0	4.1	25.9	**384.3**	20.3	2.6	1.0	1.0	1.0	1.1	4.6	33.6	619.5	26.1	2.9	1.0	1.0
$\overline{\text{ARL}}$	1.0	1.9	19.9	84.0	384.3	75.7	16.5	1.6	1.0	1.0	2.0	23.6	107.1	557.9	96.3	19.6	1.7	1.0
$\bar{\pi}_0$	0.0000	0.0000	0.0000	0.0000	0.0000	0.0000	0.0000	0.0000	0.0000	0.0091	0.0167	0.0244	0.0277	0.0312	0.0348	0.0384	0.0182	0.0063
$\bar{\pi}_{+1}$	0.1426	0.2671	0.3889	0.4411	0.5000	0.5661	0.6395	0.8200	0.9316	0.1381	0.2589	0.3769	0.4274	0.4846	0.5489	0.6204	0.8105	0.9284

(continued)

Table 3 (continued)

| Case | κ = 0.1 | | | | | | | | | κ = 0.2 | | | | | | | | |
| | δ | | | | | | | | | δ | | | | | | | | |
	−1	−0.5	−0.2	−0.1	0	0.1	0.2	0.5	1	−1	−0.5	−0.2	−0.1	0	0.1	0.2	0.5	1
1	1.1	7.1	96.4	284.4	548.7	284.4	96.4	7.1	1.1	1.1	7.4	107.6	329.3	656.7	329.3	107.6	7.4	1.1
2	1.0	3.7	58.6	223.7	577.3	223.7	58.5	3.7	1.0	1.0	3.8	65.9	264.4	718.6	264.4	65.8	3.8	1.0
3	1.0	2.8	45.3	195.3	592.2	195.3	45.3	2.8	1.0	1.0	2.8	51.2	233.8	754.9	233.8	51.2	2.8	1.0
4	1.0	2.2	35.4	170.3	607.1	170.3	35.4	2.2	1.0	1.0	2.2	40.2	206.5	793.4	206.5	40.2	2.2	1.0
5	1.0	1.7	26.3	143.1	626.1	143.1	26.3	1.7	1.0	1.0	1.8	30.0	176.5	845.8	176.5	30.0	1.8	1.0
6	1.0	1.5	20.1	121.4	644.7	121.4	20.1	1.5	1.0	1.0	1.5	23.1	152.1	900.1	152.1	23.1	1.5	1.0
7	1.0	2.2	22.2	115.4	671.6	72.1	6.1	1.0	1.0	1.0	2.2	26.5	160.5	957.7	74.4	5.5	1.0	1.0
8	1.0	2.3	27.9	140.0	638.1	115.1	15.7	1.0	1.0	1.0	2.3	32.6	181.8	876.9	133.1	16.8	1.0	1.0
9	1.0	2.2	29.2	146.2	631.0	126.5	19.2	1.1	1.0	1.0	2.3	34.0	186.8	858.0	148.8	21.0	1.1	1.0
10	1.0	1.8	23.0	126.7	646.4	111.5	16.1	1.1	1.0	1.0	1.9	26.8	163.7	903.8	134.1	17.8	1.1	1.0
11	1.0	1.5	16.9	104.5	668.8	92.7	12.4	1.1	1.0	1.0	1.5	19.8	137.6	973.4	114.3	13.9	1.1	1.0
12	1.0	1.3	12.8	86.6	691.9	77.3	9.6	1.1	1.0	1.0	1.3	15.1	116.4	1049.5	97.7	11.0	1.1	1.0
13	1.0	1.1	5.0	34.2	899.4	8.1	1.0	1.0	1.0	1.0	1.1	5.0	61.4	1276.6	5.0	1.0	1.0	1.0
14	1.0	1.3	9.0	60.6	767.0	31.2	2.2	1.0	1.0	1.0	1.3	0.8	92.9	1240.1	31.5	1.9	1.0	1.0
15	1.0	1.3	10.9	71.2	737.6	42.3	3.2	1.0	1.0	1.0	1.4	13.0	105.0	1166.7	45.3	3.1	1.0	1.0
16	1.0	1.2	8.8	62.4	751.4	43.0	3.9	1.0	1.0	1.0	1.2	13.6	91.8	1238.1	51.6	4.2	1.0	1.0
17	1.0	1.1	6.5	49.5	785.6	36.2	3.5	1.0	1.0	1.0	1.1	7.8	74.4	1375.6	46.7	4.0	1.0	1.0
18	1.0	1.1	4.9	39.1	825.6	29.5	3.0	1.0	1.0	1.0	1.1	5.9	60.5	1545.6	40.1	3.4	1.0	1.0
$\overline{ARL}$	1.0	2.1	25.5	120.8	684.0	106.9	21.0	1.7	1.0	1.0	2.1	29.3	155.3	1007.4	127.0	23.4	1.8	1.0
$\bar{\pi}_0$	0.0182	0.0334	0.0489	0.0555	0.0625	0.0697	0.0763	0.0377	0.0126	0.0364	0.0669	0.0980	0.1111	0.1250	0.1394	0.1451	0.0796	0.0254
$\bar{\pi}_{+1}$	0.1338	0.2510	0.3653	0.4142	0.4697	0.5321	0.6019	0.7989	0.9250	0.1255	0.2358	0.3431	0.3889	0.4411	0.5000	0.5661	0.7720	0.9177

As the SN control chart is a Shewhart-type one, we simply have SDRL = $\sqrt{\text{ARL}(\text{ARL} - 1)} \simeq$ ARL. This means that the SDRL of the SN control chart (with or without ties) varies the same way as its ARL does, i.e., when the ARL increases (decreases) the SDRL increases (decreases). Therefore, all the conclusions stated above concerning the ARL also holds for the SDRL.

We only presented the results for $n = 20$ and $n = 50$, but all the conclusions above remain nevertheless valid for any value of n. Now a question arises: are there suitable procedures to handle the occurrence of tied observations to θ_0 and to restore the statistical properties of the SN control chart? Throughout the lines of the next section we tackle this problem.

5 Procedures to Tackle the Occurrence of Rounding-Off Errors

The selection of a suitable procedure to reduce or, possibly, eliminate the effect of the rounding-off error is essential to avoid the limitation of the SN statistic implementation for statistical process monitoring purposes. Some reference to how it is possible to handle the occurrence of ties can be found in the pioneer paper about the SN control charts by Amin et al. (1995), who suggested to maintain the values $S_{t,k} = 0$ in the computation of the SN_t statistic and to use the control limits (LCL, UCL) = $(-C, C)$: this approach totally overlooks the effect of the rounding-off error, which changes the distributional properties of the SN_t statistic, and must be avoided. In their survey paper about nonparametric control charts Chakraborti et al. (2001) suggested to remove ties from the sample and to update the sample size n, if their probability of occurrence is small. This approach makes harder to define *a priori* the statistical performance of the SN control chart because varying n requires to redefine the parameter C; in fact, a control chart with randomly variable control limits should be set up. Furthermore, the same authors state that when the probability of occurrence of values $S_{t,k} = 0$ is high, a more sophisticated analysis is needed. In the following, we suggest and test two different procedures to tackle the problem.

5.1 Procedure 1: Computing New Chart Control Limits C'

In order to make the Shewhart SN control chart less sensitive to departures from the $\kappa = 0$ case, we suggest defining dedicated chart parameters C' depending on the actual value of the standardized resolution κ and offering an in-control ARL value *as close as possible* to its nominal value. The selection of these values requires to have reliable estimates about the type of distribution of the observed measures and its parameters. For the 18 distributions displayed in Table 1, Table 4 (for $n = 20$) and Table 5 (for $n = 50$) provide these alternative chart parameters C' as well as the

Table 4 Alternative chart parameters C' and corresponding ARL values of the Shewhart SN control chart for shifts $\delta \in \{-1, -0.5, -0.2, -0.1, 0, 0.1, 0.2, 0.5, 1\}$, for $\kappa \in \{0, 0.05, 0.1, 0.2\}$, when $n = 20$

Case		$\kappa = 0$										$\kappa = 0.05$									
		δ											δ									
	C	−1	−0.5	−0.2	−0.1	0	0.1	0.2	0.5	1	C'	−1	−0.5	−0.2	−0.1	0	0.1	0.2	0.5	1
1	14	2.8	25.9	164.7	296.1	388.1	296.1	164.7	25.9	2.8	13	2.6	23.5	146.2	260.8	340.3	260.8	146.2	23.5	2.6
2	14	1.9	14.3	118.2	258.3	388.1	258.3	118.2	14.3	1.9	13	1.8	12.9	103.7	224.2	334.5	224.2	103.7	12.9	1.8
3	14	1.7	10.7	99.0	238.4	388.1	233.4	99.0	10.7	1.7	13	1.6	9.7	86.5	205.8	332.2	205.8	86.5	9.7	1.6
4	14	1.5	8.3	83.3	219.5	388.1	219.5	83.3	8.3	1.5	13	1.4	7.6	72.5	188.6	330.3	188.6	72.5	7.6	1.4
5	14	1.3	6.3	67.3	197.0	388.1	197.0	67.3	6.3	1.3	13	1.3	5.7	58.4	168.5	328.4	168.5	58.4	5.7	1.3
6	14	1.3	5.0	55.2	177.3	388.1	177.3	55.2	5.0	1.3	13	1.2	4.5	47.9	151.2	327.2	151.2	47.9	4.5	1.2
7	14	2.1	8.3	59.3	169.5	388.1	130.4	22.6	1.0	1.0	13	2.0	7.7	52.0	145.6	326.3	109.2	19.1	1.0	1.0
8	14	1.9	8.6	70.1	193.1	388.1	173.5	46.4	2.1	1.0	13	1.8	7.9	61.3	165.8	327.6	146.9	39.7	2.0	1.0
9	14	1.8	8.5	72.5	198.7	388.1	183.6	53.8	3.0	1.0	13	1.7	7.8	63.3	170.6	328.1	156.0	46.2	2.8	1.0
10	14	1.6	6.7	60.9	181.4	388.1	169.0	46.9	3.0	1.0	13	1.5	6.1	53.0	155.3	327.1	143.5	40.3	2.8	1.0
11	14	1.4	5.0	48.5	160.0	388.1	149.4	38.2	2.7	1.0	13	1.3	4.5	42.1	136.5	326.4	126.7	32.9	2.5	1.0
12	14	1.3	4.0	39.1	141.1	388.1	132.0	31.4	2.4	1.0	13	1.2	3.7	34.0	120.3	326.2	111.9	27.1	2.3	1.0
13	14	1.3	2.8	18.0	71.6	388.1	29.8	1.6	1.0	1.0	13	1.2	2.6	15.8	61.9	339.6	25.5	1.4	1.0	1.0
14	14	1.3	3.8	29.5	109.0	388.1	73.6	8.5	1.0	1.0	13	1.3	3.5	25.8	93.5	328.6	62.1	7.3	1.0	1.0
15	14	1.4	4.2	34.3	122.3	388.1	90.0	12.7	1.0	1.0	13	1.4	3.9	29.9	104.7	327.2	75.8	10.9	1.0	1.0
16	14	1.3	3.4	29.1	111.8	388.1	89.2	14.7	1.2	1.0	13	1.2	3.2	25.4	95.6	327.7	75.5	12.7	1.2	1.0
17	14	1.2	2.7	22.5	94.9	388.1	78.1	13.1	1.4	1.0	13	1.2	2.5	19.7	81.2	329.5	66.5	11.4	1.3	1.0
18	14	1.1	2.3	17.6	80.0	388.1	66.8	11.1	1.4	1.0	13	1.1	2.1	15.4	68.7	332.1	57.2	9.7	1.3	1.0
ARL		1.6	7.3	60.5	167.8	388.1	152.9	49.4	5.1	1.2	ARL	1.5	6.7	52.9	144.4	330.0	130.9	43.0	4.7	1.2

(continued)

Table 4 (continued)

Case		κ = 0.1										κ = 0.2									
						δ											δ				
	C	−1	−0.5	−0.2	−0.1	0	0.1	0.2	0.5	1	C′	−1	−0.5	−0.2	−0.1	0	0.1	0.2	0.5	1
1	13	2.5	22.5	140.0	250.1	326.7	250.1	140.0	22.5	2.5	13	2.5	22.3	142.7	258.8	341.2	258.8	142.7	22.3	2.5
2	13	1.8	12.4	100.3	217.9	326.5	217.9	100.3	12.4	1.8	13	1.7	12.5	106.0	236.4	361.0	236.4	106.0	12.5	1.7
3	13	1.6	9.4	84.2	201.8	327.8	201.8	84.2	9.4	1.6	13	1.5	9.5	90.7	224.5	374.2	224.5	90.7	9.5	1.5
4	13	1.4	7.3	71.0	186.6	329.9	186.6	71.0	7.3	1.4	13	1.4	7.4	77.8	212.8	388.9	212.8	77.8	7.4	1.4
5	13	1.3	5.5	57.6	168.8	333.5	168.8	57.6	5.5	1.3	13	1.3	5.6	64.5	198.4	409.8	198.4	64.5	5.6	1.3
6	13	1.2	4.4	47.6	153.2	337.9	153.2	47.6	4.4	1.2	13	1.2	4.5	54.3	185.3	431.9	185.3	54.3	4.5	1.2
7	13	2.0	7.4	51.4	148.9	345.7	112.6	19.4	1.0	1.0	13	2.0	7.4	58.7	190.0	466.3	130.1	20.6	1.0	1.0
8	13	1.8	7.7	60.3	166.8	336.3	148.5	39.8	1.9	1.0	13	1.8	7.7	67.6	201.1	423.9	174.0	44.6	1.9	1.0
9	13	1.7	7.5	62.3	171.0	334.6	156.9	46.0	2.7	1.0	13	1.7	7.6	69.5	203.6	415.5	183.4	51.5	2.8	1.0
10	13	1.5	5.9	52.5	157.2	338.4	145.8	40.4	2.7	1.0	13	1.5	6.0	59.6	191.7	434.1	174.6	46.0	2.8	1.0
11	13	1.3	4.5	42.1	140.2	344.7	130.5	33.3	2.5	1.0	13	1.3	4.5	48.8	176.7	463.0	161.6	38.8	2.5	1.0
12	13	1.2	3.6	34.2	125.2	351.9	116.9	27.5	2.2	1.0	13	1.2	3.6	40.4	163.0	494.9	149.4	32.7	2.3	1.0
13	13	1.2	2.6	16.1	69.4	439.0	27.4	1.3	1.0	1.0	12	1.1	2.0	11.6	56.9	310.3	13.4	1.4	1.0	1.0
14	13	1.3	3.4	26.1	100.4	380.0	67.0	7.6	1.0	1.0	12	1.2	2.5	17.5	69.9	248.2	40.0	5.0	1.0	1.0
15	13	1.3	3.8	30.2	111.0	368.3	81.0	11.3	1.0	1.0	12	1.2	2.7	19.8	74.0	232.1	48.6	7.4	1.0	1.0
16	13	1.2	3.1	25.7	102.3	373.2	81.6	13.2	1.2	1.0	12	1.1	2.3	17.3	69.5	239.8	52.1	9.4	1.1	1.0
17	13	1.1	2.5	20.1	88.5	386.1	73.1	11.9	1.3	1.0	12	1.1	1.9	14.1	62.7	259.3	49.4	8.8	1.2	1.0
18	13	1.1	2.1	15.9	76.1	401.9	64.0	10.2	1.3	1.0	12	1.0	1.7	11.5	56.5	283.1	45.6	7.8	1.2	1.0
ARL		1.5	6.4	52.1	146.4	354.6	132.4	42.4	4.5	1.2		1.4	6.2	54.0	157.3	365.4	141.0	45.0	4.5	1.2

Table 5 Alternative chart parameters C' and corresponding ARL values of the Shewhart SN control chart for shifts $\delta \in \{-1, -0.5, -0.2, -0.1, 0, 0.1, 0.2, 0.5, 1\}$, for $\kappa \in \{0, 0.05, 0.1, 0.2\}$, when $n = 50$

Case		κ = 0										κ = 0.05								
						δ											δ				
	C	−1	−0.5	−0.2	−0.1	0	0.1	0.2	0.5	1	C′	−1	−0.5	−0.2	−0.1	0	0.1	0.2	0.5	1
1	22	1.1	6.2	74.6	208.3	384.3	208.3	74.6	6.2	1.1	21	1.1	5.8	66.5	184.0	336.7	184.0	66.5	5.8	1.1
2	22	1.0	3.3	45.1	159.9	384.3	159.9	45.1	3.3	1.0	21	1.0	3.1	40.3	141.2	336.3	141.2	40.3	3.1	1.0
3	22	1.0	2.5	34.9	138.3	384.3	138.3	34.9	2.5	1.0	21	1.0	2.4	31.3	122.3	337.0	122.3	31.3	2.4	1.0
4	22	1.0	2.0	27.4	119.6	384.3	119.6	27.4	2.0	1.0	21	1.0	1.9	24.6	106.0	338.2	106.0	24.6	1.9	1.0
5	22	1.0	1.6	20.5	99.6	384.3	99.5	20.5	1.6	1.0	21	1.0	1.6	18.4	88.7	340.3	88.7	18.4	1.6	1.0
6	22	1.0	1.4	15.8	83.8	384.3	83.8	15.8	1.4	1.0	21	1.0	1.4	14.3	75.0	342.8	75.0	14.3	1.4	1.0
7	22	1.0	2.0	17.3	78.0	384.3	52.1	5.3	1.0	1.0	21	1.0	1.9	15.7	70.3	347.0	46.8	4.9	1.0	1.0
8	22	1.0	2.1	21.6	96.3	384.3	81.0	12.6	1.0	1.0	21	1.0	2.0	19.5	86.1	341.8	72.2	11.4	1.0	1.0
9	22	1.0	2.1	22.6	101.0	384.3	88.7	15.2	1.1	1.0	21	1.0	2.0	20.4	90.1	340.9	79.0	13.7	1.1	1.0
10	22	1.0	1.7	17.9	87.0	384.3	77.7	12.8	1.1	1.0	21	1.0	1.6	16.2	77.9	343.0	69.5	11.6	1.1	1.0
11	22	1.0	1.4	13.3	71.2	384.3	64.1	9.9	1.1	1.0	21	1.0	1.4	12.1	64.1	346.5	57.6	9.0	1.1	1.0
12	22	1.0	1.3	10.2	58.7	384.3	53.1	7.8	1.1	1.0	21	1.0	1.2	9.3	53.2	350.3	48.1	7.1	1.1	1.0
13	22	1.0	1.1	4.2	22.2	384.3	7.3	1.0	1.0	1.0	21	1.0	1.1	3.9	20.8	390.6	6.7	1.0	1.0	1.0
14	22	1.0	1.2	7.2	40.1	384.3	23.1	2.1	1.0	1.0	21	1.0	1.2	6.7	36.8	364.3	21.1	1.9	1.0	1.0
15	22	1.0	1.3	8.7	47.4	384.3	30.5	2.9	1.0	1.0	21	1.0	1.3	7.9	43.3	358.6	27.8	2.7	1.0	1.0
16	22	1.0	1.2	7.1	41.5	384.3	30.1	3.4	1.0	1.0	21	1.0	1.2	6.6	38.1	361.0	27.6	3.2	1.0	1.0
17	22	1.0	1.1	5.3	32.9	384.3	25.1	3.0	1.0	1.0	21	1.0	1.1	4.9	30.3	367.3	23.2	2.8	1.0	1.0
18	22	1.0	1.0	4.1	25.9	384.3	20.3	2.6	1.0	1.0	21	1.0	1.0	3.8	24.1	374.6	18.9	2.4	1.0	1.0
ARL		1.0	1.9	19.9	84.0	384.3	75.7	16.5	1.6	1.0	ARL	1.0	1.8	17.9	75.1	351.0	67.5	14.8	1.6	1.0

(continued)

Table 5 (continued)

Case		κ = 0.1										κ = 0.2									
		δ										δ									
	C	−1	−0.5	−0.2	−0.1	0	0.1	0.2	0.5	1	C′	−1	−0.5	−0.2	−0.1	0	0.1	0.2	0.5	1
1	21	1.1	5.8	66.5	184.0	336.7	184.0	66.5	5.8	1.1	21	1.1	5.7	66.8	186.5	344.5	186.5	66.8	5.7	1.1
2	21	1.0	3.1	40.3	141.2	336.3	141.2	40.3	3.1	1.0	21	1.0	3.1	41.0	146.3	354.9	146.3	41.0	3.1	1.0
3	21	1.0	2.4	31.3	122.3	337.0	122.3	31.3	2.4	1.0	21	1.0	2.4	32.0	128.0	361.5	128.0	32.0	2.4	1.0
4	21	1.0	1.9	24.6	106.0	338.2	106.0	24.6	1.9	1.0	21	1.0	1.9	25.3	112.0	368.5	112.0	25.3	1.9	1.0
5	21	1.0	1.6	18.4	88.7	340.3	88.7	18.4	1.6	1.0	21	1.0	1.6	19.1	94.7	377.9	94.7	19.1	1.6	1.0
6	21	1.0	1.4	14.3	75.0	342.8	75.0	14.3	1.4	1.0	21	1.0	1.4	14.8	80.7	387.5	80.7	14.8	1.4	1.0
7	21	1.0	1.9	15.7	70.3	347.0	46.8	4.9	1.0	1.0	21	1.0	1.9	16.3	77.1	401.7	49.0	4.8	1.0	1.0
8	21	1.0	2.0	19.5	86.1	341.8	72.2	11.4	1.0	1.0	21	1.0	2.0	20.2	92.8	384.0	76.7	11.7	1.0	1.0
9	21	1.0	2.0	20.4	90.1	340.9	79.0	13.7	1.1	1.0	21	1.0	2.0	21.1	96.7	380.4	83.9	14.2	1.1	1.0
10	21	1.0	1.6	16.2	77.9	343.0	69.5	11.6	1.1	1.0	21	1.0	1.6	16.8	84.2	388.4	74.4	11.9	1.1	1.0
11	21	1.0	1.4	12.1	64.1	346.5	57.6	9.0	1.1	1.0	21	1.0	1.4	12.6	70.0	400.2	62.3	9.3	1.1	1.0
12	21	1.0	1.2	9.3	53.2	350.3	48.1	7.1	1.1	1.0	21	1.0	1.2	9.7	58.5	412.5	52.4	7.4	1.1	1.0
13	21	1.0	1.1	3.9	20.8	390.6	6.7	1.0	1.0	1.0	20	1.0	1.1	3.3	17.5	312.4	4.8	1.0	1.0	1.0
14	21	1.0	1.2	6.7	36.8	364.3	21.1	1.9	1.0	1.0	21	1.0	1.2	7.0	41.6	453.3	22.1	1.9	1.0	1.0
15	21	1.0	1.3	7.9	43.3	358.6	27.8	2.7	1.0	1.0	21	1.0	1.3	8.3	48.5	437.0	29.4	2.7	1.0	1.0
16	21	1.0	1.2	6.6	38.1	361.0	27.6	3.2	1.0	1.0	21	1.0	1.2	6.8	42.8	444.5	29.9	3.2	1.0	1.0
17	21	1.0	1.1	4.9	30.3	367.3	23.2	2.8	1.0	1.0	20	1.0	1.1	4.2	24.3	279.4	18.3	2.5	1.0	1.0
18	21	1.0	1.0	3.8	24.1	374.6	18.9	2.4	1.0	1.0	20	1.0	1.0	3.3	19.7	291.2	15.3	2.2	1.0	1.0
$\overline{ARL}$		1.0	1.8	17.9	75.1	351.0	67.5	14.8	1.6	1.0		1.0	1.8	18.3	79.0	376.6	70.4	15.1	1.6	1.0

corresponding ARL values for shifts $\delta \in \{-1, -0.5, -0.2, -0.1, 0, 0.1, 0.2, 0.5, 1\}$ and for standardized resolutions $\kappa \in \{0.05, 0.1, 0.2\}$. Results for $\kappa = 0$ are also provided as information, (in this case, we have $C' = C$). As for Tables 2 and 3, the average value $\overline{ARL}$ is also given. From the results displayed in Tables 4 and 5, we can conclude that with these alternative chart parameters C', no matter the value of κ or the considered distributions #1–#18:

- the in-control ARL values are much closer to 370.4 than the ones obtained with the same chart parameter C. For example, if $n = 20$ and $\kappa = 0.2$, the in-control $(\delta = 0)$ ARL values obtained with the same chart parameter $C = 14$ range from 722.2 to 2274.8 with an average value $\overline{ARL} = 1268.3$ (i.e., very far from 370.4), see Table 2. On the other hand, using the alternative chart parameters C' in Table 4 (they are all equal to either 13 or 12) allows obtaining in-control ARL values ranging from 232.1 to 494.9 with an average value $\overline{ARL} = 365.4$ (i.e., much closer to 370.4).
- the out-of-control ARL values are smaller than the ones obtained with the same chart parameter C. This can be verified by comparing the individual ARL values (for each case) as well as the overall $\overline{ARL}$ values in Tables 2 and 3 to those shown in Tables 4 and 5. For example, if $n = 50$ and $\kappa = 0.1$, then the average ARL values are equal to 1.0, 2.1, 25.5, 120.8, 106.9, 21.0, 1.7, 1.0 with the same chart parameter $C = 22$ when $\delta \in \{-1, -0.5, -0.2, -0.1, 0.1, 0.2, 0.5, 1\}$ (see Table 3) while, if we use the alternative chart parameters C' in Table 5 (they are all equal to 21), we obtain 1.0, 1.8, 17.9, 75.1, 67.5, 14.8, 1.6, 1.0.

5.2 Procedure 2: Bernoulli Trial Approach

A second possible strategy consists in *randomly reassigning* to each value $S_{t,k} = 0$ either the value $S_{t,k} = -1$ or $S_{t,k} = +1$ using a simple "flip-a-coin" scheme, (i.e., if $S_{t,k} = 0$ at a first place then transform it into $S_{t,k} = 2D_{t,k} - 1$ where $D_{t,k} \sim \mathrm{Ber}(0.5)$ is a Bernoulli random variable with probability $p = 0.5$. Applying this strategy is equivalent to consider the Shewhart SN control chart in the "without ties" case with probabilities

$$\pi'_{-1} = \pi_{-1} + \frac{\pi_0}{2}$$
$$\pi'_0 = 0$$
$$\pi'_{+1} = \pi_{+1} + \frac{\pi_0}{2}$$

i.e., the probability π_0 is equally allocated on both sides for values $S_{t,k} = -1$ and $S_{t,k} = +1$. We suggest using this "flip-a-coin" strategy whether the underlying unknown distribution is symmetric or not. In order to evaluate the effect of this strategy on the statistical performance of the Shewhart SN control chart, we have first computed in Table 6 the in-control $(\delta = 0)$ values of π_0, π_{+1}, and π'_{+1} for the 18 distributions in Table 1 and for $\kappa \in \{0, 0.05, 0.1, 0.2\}$. From the results in Table 6, we can conclude that

- when $\kappa = 0$, we have $\pi_0 = 0$, $\pi_{+1} = \pi'_{+1} = 0.5$, no matter the considered distribution. This is, of course, an expected result.
- For the *symmetric* cases #1–#6, the values of π_0 and π_{+1} vary from one case to another but we always have $\pi'_{+1} = 0.5$. This is also an expected result.
- For the *asymmetric* cases #7–#18, all the values of π_0, π_{+1}, and π'_{+1} vary but, quite surprisingly, the value of π'_{+1} remains very close to 0.5 (and so is $\pi'_{-1} = 1 - \pi'_{+1}$). In the most extreme case (distribution #13 and $\kappa = 0.2$), we have $\pi'_{+1} = 0.5209$ but, in all the other cases we always have $|\pi'_{+1} - 0.5| < 0.01$.

This allows us to conclude that, in the "with ties" case, when the underlying distribution is symmetric implementing the "flip-a-coin" strategy allows the Shewhart SN control chart to be a distribution-free control chart, as in the "without ties" case. In general, unless the skewness of the underlying distribution or the value of κ are very large, it allows the Shewhart SN control chart to be *approximately* distribution-free.

For the 18 distributions in Table 1, Table 7 (for $n = 20$), and Table 8 (for $n = 50$) show the ARL values of the Shewhart SN control chart using the "flip-a-coin" strategy for shifts $\delta \in \{-1.0, -0.5, -0.2, -0.1, 0, 0.1, 0.2, 0.5, 1.0\}$ and for standardized resolution $\kappa = 0$ (i.e., "without ties") and $\kappa \in \{0.05, 0.1, 0.2\}$ (i.e., "with ties"). The chart parameter C is the one corresponding to the Shewhart SN control chart in the "without ties" case, i.e., $C = 14$ when $n = 20$ and $C = 22$ when $n = 50$. From Tables 7 and 8, we can draw the following conclusions:

- Concerning the in-control situation, except for the extreme case #13, all the ARL values are now similar, no matter the value of κ, (the situation was totally different in the "with ties" case in Tables 2 and 3). The in-control ARL values in terms of κ lead to the conclusion that this Procedure gives much better in-control results than Procedure 1 since the effect of the rounding error is almost eliminated. Consequently, we may state that using this Procedure the in-control distribution-free property of the SN control chart is preserved.
- Concerning the out-of-control situation, still excluding the extreme case #13, for a specific case and shift value δ, the ARL values are very similar, no matter the value of κ. For instance, in Table 7, for case #9 and shift $\delta = -0.2$, we have ARL = 72.5 for $\kappa = 0$, ARL = 72.6 for $\kappa = 0.05$, ARL = 73.2 for $\kappa = 0.1$ and ARL = 75.5 for $\kappa = 0.2$ (these values were ARL = 89.5 for $\kappa = 0.05$, ARL = 105.8 for $\kappa = 0.1$ and ARL = 138.9 for $\kappa = 0.2$ in Table 2 without the use of the "flip-a-coin" strategy).

Based on these findings, we suggest implementing this second strategy when facing the problem of ties during the implementation of the Shewhart SN control chart. Finally, it is worth noting that this conclusion can be extended to any control chart implementing the SN statistic.

Table 6 In-control values of π_0, π_{+1} and π'_{+1} for the 18 cases in Table 1 and for $\kappa \in \{0, 0.05, 0.1, 0.2\}$

Case	$\kappa = 0$	$\kappa = 0.05$	$\kappa = 0.1$	$\kappa = 0.2$
1	0.0000, 0.5000, 0.5000	0.0142, 0.4929, 0.5000	0.0284, 0.4858, 0.5000	0.0568, 0.4716, 0.5000
2	0.0000, 0.5000, 0.5000	0.0179, 0.4910, 0.5000	0.0359, 0.4821, 0.5000	0.0717, 0.4642, 0.5000
3	0.0000, 0.5000, 0.5000	0.0199, 0.4900, 0.5000	0.0399, 0.4801, 0.5000	0.0797, 0.4602, 0.5000
4	0.0000, 0.5000, 0.5000	0.0219, 0.4890, 0.5000	0.0439, 0.4781, 0.5000	0.0876, 0.4562, 0.5000
5	0.0000, 0.5000, 0.5000	0.0245, 0.4878, 0.5000	0.0489, 0.4755, 0.5000	0.0976, 0.4512, 0.5000
6	0.0000, 0.5000, 0.5000	0.0269, 0.4865, 0.5000	0.0538, 0.4731, 0.5000	0.1072, 0.4464, 0.5000
7	0.0000, 0.5000, 0.5000	0.0304, 0.4852, 0.5004	0.0609, 0.4710, 0.5014	0.1227, 0.4444, 0.5058
8	0.0000, 0.5000, 0.5000	0.0261, 0.4871, 0.5001	0.0521, 0.4745, 0.5006	0.1043, 0.4503, 0.5024
9	0.0000, 0.5000, 0.5000	0.0251, 0.4876, 0.5002	0.0503, 0.4754, 0.5005	0.1004, 0.4516, 0.5018
10	0.0000, 0.5000, 0.5000	0.0271, 0.4865, 0.5001	0.0542, 0.4733, 0.5004	0.1082, 0.4475, 0.5016
11	0.0000, 0.5000, 0.5000	0.0300, 0.4851, 0.5001	0.0599, 0.4704, 0.5004	0.1193, 0.4419, 0.5015
12	0.0000, 0.5000, 0.5000	0.0328, 0.4837, 0.5001	0.0655, 0.4676, 0.5004	0.1303, 0.4363, 0.5015
13	0.0000, 0.5000, 0.5000	0.0556, 0.4735, 0.5013	0.1119, 0.4491, 0.5051	0.2293, 0.4062, 0.5209
14	0.0000, 0.5000, 0.5000	0.0416, 0.4798, 0.5006	0.0833, 0.4606, 0.5022	0.1675, 0.4250, 0.5088
15	0.0000, 0.5001, 0.5001	0.0382, 0.4814, 0.5005	0.0765, 0.4635, 0.5017	0.1534, 0.4300, 0.5068
16	0.0000, 0.5000, 0.5000	0.0396, 0.4805, 0.5003	0.0792, 0.4517, 0.5013	0.1578, 0.4261, 0.5050
17	0.0000, 0.5000, 0.5000	0.0432, 0.4787, 0.5003	0.0863, 0.4580, 0.5011	0.1710, 0.4189, 0.5043
18	0.0000, 0.5000, 0.5000	0.0472, 0.4767, 0.5003	0.0941, 0.4540, 0.5011	0.1857, 0.4112, 0.5041

Table 7 ARL values of the Shewhart SN control chart using the "flip-a-coin" strategy for shifts $\delta \in \{-1, -0.5, -0.2, -0.1, 0, 0.1, 0.2, 0.5, 1\}$, for standardized resolution $\kappa = 0$ (i.e., "without ties") and $\kappa \in \{0.05, 0.1, 0.2\}$ (i.e., "with ties"), when $n = 20$

Case	$\kappa = 0$									$\kappa = 0.05$								
	δ									δ								
	-1	-0.5	-0.2	-0.1	0	0.1	0.2	0.5	1	-1	-0.5	-0.2	-0.1	0	0.1	0.2	0.5	1
1	2.8	25.9	164.7	296.1	388.1	296.1	164.7	25.9	2.8	2.8	25.9	164.7	296.1	388.1	296.1	164.7	25.9	2.8
2	1.9	14.3	118.2	258.3	388.1	258.3	118.2	14.3	1.9	1.9	14.3	118.2	258.3	388.1	258.3	118.2	14.3	1.9
3	1.7	10.7	99.0	238.4	388.1	238.4	99.0	10.7	1.7	1.7	10.8	99.1	238.5	388.1	238.5	99.1	10.8	1.7
4	1.5	8.3	83.3	219.5	388.1	219.5	83.3	8.3	1.5	1.5	8.3	83.3	219.6	388.1	219.6	83.3	8.3	1.5
5	1.3	6.3	67.3	197.0	388.1	197.0	67.3	6.3	1.3	1.3	6.3	67.4	197.2	388.1	197.2	67.4	6.3	1.3
6	1.3	5.0	55.2	177.3	388.1	177.3	55.2	5.0	1.3	1.3	5.0	55.4	177.5	388.1	177.5	55.4	5.0	1.3
7	2.1	8.3	59.3	169.5	388.1	130.4	22.6	1.0	1.0	2.1	8.3	59.6	170.5	388.1	129.0	22.3	1.0	1.0
8	1.9	8.6	70.1	193.1	388.1	173.5	46.4	2.1	1.0	1.9	8.7	70.3	193.7	388.1	172.9	46.3	2.1	1.0
9	1.8	8.5	72.5	198.7	388.1	183.6	53.8	3.0	1.0	1.8	8.5	72.6	199.2	388.1	183.2	53.7	3.0	1.0
10	1.6	6.7	60.9	181.4	388.1	169.0	46.9	3.0	1.0	1.6	6.7	61.1	181.9	388.1	168.8	46.9	3.0	1.0
11	1.4	5.0	48.5	160.0	388.1	149.4	38.2	2.7	1.0	1.4	5.0	48.6	160.5	388.1	149.3	38.3	2.7	1.0
12	1.3	4.0	39.1	141.1	388.1	132.0	31.4	2.4	1.0	1.3	4.0	39.3	141.6	388.1	132.0	31.5	2.4	1.0
13	1.3	2.8	18.0	71.6	388.1	29.8	1.6	1.0	1.0	1.3	2.8	18.1	72.8	387.8	28.6	1.6	1.0	1.0
14	1.3	3.8	29.5	109.0	388.1	73.6	8.5	1.0	1.0	1.3	3.8	29.7	110.1	388.0	72.5	8.4	1.0	1.0
15	1.4	4.2	34.3	122.3	388.1	90.0	12.7	1.0	1.0	1.4	4.2	34.5	123.3	388.0	89.1	12.6	1.0	1.0
16	1.3	3.4	29.1	111.8	388.1	89.2	14.7	1.2	1.0	1.3	3.5	29.3	112.7	388.1	88.8	14.7	1.2	1.0
17	1.2	2.7	22.5	94.9	388.1	78.1	13.1	1.4	1.0	1.2	2.7	22.7	95.7	388.1	78.1	13.2	1.4	1.0
18	1.1	2.3	17.6	80.0	388.1	66.8	11.1	1.4	1.0	1.1	2.3	17.8	80.8	388.1	67.0	11.2	1.4	1.0

(continued)

Table 7 (continued)

$\kappa = 0.1$

Case	$\delta=-1$	-0.5	-0.2	-0.1	0	0.1	0.2	0.5	1
1	2.8	25.9	164.6	296.1	388.1	296.1	164.6	25.9	2.8
2	1.9	14.3	118.3	258.4	388.1	258.4	118.3	14.3	1.9
3	1.7	10.8	99.2	238.7	388.1	238.7	99.2	10.8	1.7
4	1.5	8.4	83.5	219.9	388.1	219.9	83.5	8.4	1.5
5	1.3	6.3	67.6	197.6	388.1	197.6	67.5	6.3	1.3
6	1.3	5.0	55.7	178.0	388.1	178.0	55.7	5.0	1.3
7	2.1	8.4	60.4	173.7	387.8	124.8	21.4	1.0	1.0
8	1.9	8.7	70.9	195.6	388.0	171.2	45.9	2.1	1.0
9	1.8	8.6	73.2	200.7	388.0	182.1	53.5	3.0	1.0
10	1.6	6.7	61.6	183.4	388.1	168.1	46.8	3.0	1.0
11	1.4	5.1	49.1	162.0	388.1	149.0	38.4	2.7	1.0
12	1.3	4.0	39.8	143.3	388.1	132.1	31.7	2.5	1.0
13	1.3	2.8	18.6	76.7	384.3	25.1	1.5	1.0	1.0
14	1.4	3.8	30.3	113.4	387.4	69.5	8.2	1.0	1.0
15	1.4	4.3	35.1	126.3	387.6	86.3	12.3	1.0	1.0
16	1.3	3.5	29.9	115.4	387.8	87.7	14.8	1.2	1.0
17	1.2	2.7	23.2	98.3	387.9	78.0	13.4	1.4	1.0
18	1.1	2.3	18.3	83.4	387.9	67.6	11.4	1.4	1.0

$\kappa = 0.2$

Case	$\delta=-1$	-0.5	-0.2	-0.1	0	0.1	0.2	0.5	1
1	2.8	25.9	164.6	296.1	388.1	296.1	164.6	25.9	2.8
2	1.9	14.4	118.8	258.8	388.1	258.8	118.8	14.4	1.9
3	1.7	10.9	99.9	239.4	388.1	239.4	99.9	10.9	1.7
4	1.5	8.5	84.4	221.0	388.1	221.0	84.4	8.5	1.5
5	1.3	6.4	68.7	199.2	388.1	199.2	68.7	6.4	1.3
6	1.3	5.1	56.9	180.2	388.1	180.2	56.9	5.1	1.3
7	2.1	8.6	63.8	186.9	383.3	109.1	18.4	1.0	1.0
8	1.9	8.9	73.5	203.2	387.2	164.4	44.4	2.2	1.0
9	1.8	8.8	75.5	206.8	387.6	177.6	52.5	3.1	1.0
10	1.6	6.9	63.8	189.4	387.7	165.2	46.7	3.1	1.0
11	1.4	5.2	51.2	168.2	387.7	147.9	38.9	2.8	1.0
12	1.3	4.1	41.8	149.8	387.8	132.4	32.6	2.6	1.0
13	1.3	2.9	20.9	95.0	333.0	15.2	2.0	1.0	1.0
14	1.4	3.9	32.9	127.8	377.2	58.6	7.3	1.0	1.0
15	1.4	4.4	37.8	139.3	381.6	76.4	11.3	1.0	1.0
16	1.3	3.6	32.2	126.7	384.5	83.9	15.1	1.3	1.0
17	1.2	2.8	25.3	109.1	385.4	147.2	38.9	1.4	1.0
18	1.1	2.3	20.2	94.0	385.7	69.9	12.6	1.4	1.0

Table 8 ARL values of the Shewhart SN control chart using the "flip-a-coin" strategy for shifts $\delta \in \{-1, -0.5, -0.2, -0.1, 0, 0.1, 0.2, 0.5, 1\}$, for standardized resolution $\kappa = 0$ (i.e., "without ties") and $\kappa \in \{0.05, 0.1, 0.2\}$ (i.e., "with ties"), when $n = 50$

| Case | $\kappa = 0$ | | | | | | | | | $\kappa = 0.05$ | | | | | | | | |
| | δ | | | | | | | | | δ | | | | | | | | |
	-1	-0.5	-0.2	-0.1	0	0.1	0.2	0.5	1	-1	-0.5	-0.2	-0.1	0	0.1	0.2	0.5	1
1	1.1	6.2	74.6	208.3	384.3	208.3	74.6	6.2	1.1	1.1	6.2	74.5	208.3	384.3	208.3	74.5	6.2	1.1
2	1.0	3.3	45.1	159.9	384.3	159.9	45.1	3.3	1.0	1.0	3.3	45.1	160.0	384.3	160.0	45.1	3.3	1.0
3	1.0	2.5	34.9	138.3	384.3	138.3	34.9	2.5	1.0	1.0	2.5	34.9	138.3	384.3	138.3	34.9	2.5	1.0
4	1.0	2.0	27.4	119.6	384.3	119.6	27.4	2.0	1.0	1.0	2.0	27.4	119.7	384.3	119.7	27.4	2.0	1.0
5	1.0	1.6	20.5	99.6	384.3	99.6	20.5	1.6	1.0	1.0	1.7	20.5	99.7	384.3	99.7	20.5	1.7	1.0
6	1.0	1.4	15.8	83.8	384.3	83.8	15.8	1.4	1.0	1.0	1.4	15.8	84.0	384.3	84.0	15.8	1.4	1.0
7	1.0	2.0	17.3	78.0	384.3	52.1	5.3	1.0	1.0	1.0	2.0	17.4	78.8	384.2	51.3	5.3	1.0	1.0
8	1.0	2.1	21.6	96.3	384.3	81.0	12.6	1.0	1.0	1.0	2.1	21.7	96.8	384.3	80.6	12.6	1.0	1.0
9	1.0	2.1	22.6	101.0	384.3	88.7	15.2	1.1	1.0	1.0	2.1	22.7	101.4	384.3	88.4	15.2	1.1	1.0
10	1.0	1.7	17.9	87.0	384.3	77.7	12.8	1.1	1.0	1.0	1.7	18.0	87.4	384.3	77.5	12.8	1.1	1.0
11	1.0	1.4	13.3	71.2	384.3	64.1	9.9	1.1	1.0	1.0	1.4	13.4	71.6	384.3	64.0	9.9	1.1	1.0
12	1.0	1.3	10.2	58.7	384.3	53.1	7.8	1.1	1.0	1.0	1.3	10.2	59.1	384.3	53.1	7.8	1.1	1.0
13	1.0	1.1	4.2	22.2	384.3	7.3	1.0	1.0	1.0	1.0	1.1	4.2	22.8	383.7	7.0	1.0	1.0	1.0
14	1.0	1.2	7.2	40.1	384.3	23.1	2.1	1.0	1.0	1.0	1.2	7.3	40.7	384.2	22.6	2.1	1.0	1.0
15	1.0	1.3	8.7	47.4	384.3	30.5	2.9	1.0	1.0	1.0	1.3	8.7	48.0	384.2	30.1	2.9	1.0	1.0
16	1.0	1.2	7.1	41.5	384.3	30.1	3.4	1.0	1.0	1.0	1.2	7.2	42.0	384.3	29.9	3.4	1.0	1.0
17	1.0	1.1	5.3	32.9	384.3	25.1	3.0	1.0	1.0	1.0	1.1	5.4	33.3	384.3	25.0	3.1	1.0	1.0
18	1.0	1.0	4.1	25.9	384.3	20.3	2.6	1.0	1.0	1.0	1.0	4.1	26.3	384.3	20.4	2.6	1.0	1.0

(continued)

Table 8 (continued)

Case	$\kappa = 0.1$									$\kappa = 0.2$								
	δ									δ								
	-1	-0.5	-0.2	-0.1	0	0.1	0.2	0.5	1	-1	-0.5	-0.2	-0.1	0	0.1	0.2	0.5	1
1	1.1	6.2	74.5	208.3	384.3	208.3	74.5	6.2	1.1	1.1	6.2	74.5	208.2	384.3	208.2	74.5	6.2	1.1
2	1.0	3.3	45.1	160.1	384.3	160.1	45.1	3.3	1.0	1.0	3.3	45.4	160.6	384.3	160.6	45.4	3.3	1.0
3	1.0	2.5	35.0	138.5	384.3	138.5	35.0	2.5	1.0	1.0	2.6	35.3	139.3	384.3	139.3	35.3	2.6	1.0
4	1.0	2.0	27.5	119.9	384.3	119.9	27.5	2.0	1.0	1.0	2.1	27.9	120.9	384.3	120.9	27.9	2.1	1.0
5	1.0	1.7	20.6	100.1	384.3	100.1	20.6	1.7	1.0	1.0	1.7	21.0	101.4	384.3	101.4	21.0	1.7	1.0
6	1.0	1.4	15.9	84.4	384.3	84.4	15.9	1.4	1.0	1.0	1.4	16.4	86.1	384.3	86.1	16.4	1.4	1.0
7	1.0	2.0	17.7	81.1	383.5	48.8	5.0	1.0	1.0	1.0	2.1	19.1	91.3	372.2	40.1	4.3	1.0	1.0
8	1.0	2.1	22.0	98.4	384.2	79.3	12.4	1.0	1.0	1.0	2.1	23.1	104.9	382.2	74.4	11.9	1.0	1.0
9	1.0	2.1	22.9	102.7	384.2	87.5	15.1	1.1	1.0	1.0	2.1	23.9	108.0	383.1	84.1	14.8	1.1	1.0
10	1.0	1.7	18.2	88.6	384.2	77.0	12.7	1.1	1.0	1.0	1.8	19.0	93.3	383.4	75.0	12.7	1.1	1.0
11	1.0	1.4	13.5	72.7	384.2	63.8	9.9	1.1	1.0	1.0	1.5	14.3	77.1	383.4	63.1	10.1	1.1	1.0
12	1.0	1.3	10.4	60.1	384.2	53.1	7.9	1.1	1.0	1.0	1.3	11.0	64.3	383.5	53.3	8.1	1.1	1.0
13	1.0	1.1	4.3	24.4	374.8	6.0	1.0	1.0	1.0	1.0	1.1	4.9	32.9	266.9	3.5	1.0	1.0	1.0
14	1.0	1.2	7.5	42.5	382.5	21.4	2.0	1.0	1.0	1.0	1.3	8.3	50.6	357.5	17.0	1.8	1.0	1.0
15	1.0	1.3	8.9	49.7	383.2	28.8	2.9	1.0	1.0	1.0	1.3	9.7	57.6	367.9	24.3	2.7	1.0	1.0
16	1.0	1.2	7.3	43.5	383.7	29.4	3.4	1.0	1.0	1.0	1.2	8.0	50.0	375.2	27.7	3.5	1.0	1.0
17	1.0	1.1	5.5	34.5	383.8	25.0	3.1	1.0	1.0	1.0	1.1	6.1	40.1	377.4	25.0	3.3	1.0	1.0
18	1.0	1.1	4.2	27.4	383.9	20.6	2.7	1.0	1.0	1.0	1.1	4.7	32.4	378.2	21.5	2.9	1.0	1.0

6 Conclusions

Statistical process monitoring tools work by measuring at scheduled sampling instants observations of the quality characteristic to be controlled. The role of the measurement device and the sources of error associated with the measure collection are of primary importance. Too often, the bad effects related to the measurement system imprecision are overlooked and the monitoring process is not reliable. In this paper, we have shown the delicate role played by the device resolution when measures of a quality characteristic are collected to run a Shewhart SN control chart. In particular, the rounding-off error changes the statistical distribution of the SN statistic and any control chart based on the SN statistic loses the distribution-free property. We have quantified for a comprehensive set of distributions how the control chart's performance changes for both the in- and out-of-control conditions. To propose a solution to this problem, we have discussed two procedures to cope with ties. We have found that one of these procedures, based on a Bernoulli trial approach, is very efficient to maintain the distribution-free property for the SN control charts and it is immediately applicable to any situation.

Finally, we want to highlight the need to study the effect of measurement error to the already established nonparametric control charts in the literature. Such studies will help practitioners to implement on-line nonparametric control charts while improving the reliability of the results at the same time.

Appendix

In this Appendix, we provide three different ways for evaluating $f_{SN}(s|n)$ in the "with ties" case.

The first way to compute $f_{SN}(s|n)$ is a recursive one. The probability to have $SN_t = S_{t,1} + \cdots + S_{t,n} = s$ is equal to the probability to have either

- $(S_{t,1} + \cdots + S_{t,n-1} = s + 1) \cap (S_{t,n} = -1)$, or
- $(S_{t,1} + \cdots + S_{t,n-1} = s) \cap (S_{t,n} = 0)$, or
- $(S_{t,1} + \cdots + S_{t,n-1} = s - 1) \cap (S_{t,n} = +1)$.

Therefore, we have

$$f_{SN}(s|n) = \pi_{-1} f_{SN}(s + 1|n - 1) + \pi_0 f_{SN}(s|n - 1) + \pi_{+1} f_{SN}(s - 1|n - 1),$$

if $s \in \{-n, -n + 1, \ldots, n\}$ and $f_{SN}(s|n) = 0$, otherwise, with $f_{SN}(s|1) = \pi_s$ for $s \in \{-1, 0, +1\}$.

The second approach consists in considering the possible values of SN_t as the states $\{-n, -n + 1, \ldots, n\}$ of a discrete-time Markov chain with transition probability matrix

$$
\mathbf{P} = \left(
\begin{array}{c|ccccccccc}
\text{States} & -n & -n+1 & -n+2 & \cdots & n-2 & n-1 & n & * \\
\hline
-n & \pi_0 & \pi_{+1} & 0 & \cdots & \cdots & & \cdots & 0 & \pi_{-1} \\
-n+1 & \pi_{-1} & \pi_0 & \pi_{+1} & \ddots & & & & 0 \\
-n+2 & 0 & \pi_{-1} & \pi_0 & \pi_{+1} & \ddots & & & \vdots \\
\vdots & \vdots & & \ddots & \ddots & \ddots & \ddots & \ddots & \vdots \\
n-2 & \vdots & & & \ddots & \pi_{-1} & \pi_0 & \pi_{+1} & \ddots & \vdots \\
n-1 & \vdots & & & & \ddots & \pi_{-1} & \pi_0 & \pi_{+1} & 0 \\
n & \vdots & & & & & \ddots & \pi_{-1} & \pi_0 & \pi_{+1} \\
* & 0 & \cdots & & \cdots & & \cdots & & 0 & 0 & 1
\end{array}
\right).
$$

The state "$*$" is an extra absorbing state. If we start with the initial probability vector $\boldsymbol{\pi} = (0, \ldots, 0, 1, 0, \ldots, 0)^\mathsf{T}$, then the p.m.f. vector

$$
\mathbf{f} = (f_{\mathrm{SN}}(-n|n),\ f_{\mathrm{SN}}(-n+1|n),\ \ldots,\ f_{\mathrm{SN}}(n|n))^\mathsf{T}
$$

containing all the values $f_{\mathrm{SN}}(s|n)$ for $s \in \{-n, -n+1, \ldots, n\}$ can simply be obtained using the classical formula $\mathbf{f} = \boldsymbol{\pi}\mathbf{P}^n$.

The third way to compute $f_{\mathrm{SN}}(s|n)$ consists in summing all the possible combinations, like for the binomial distribution. If n_{-1} is the number of occurrences of $S_{t,k} = -1$ in $\{S_{t,1}, S_{t,2}, \ldots, S_{t,n}\}$ then SN_t is necessarily within $-n_{-1}$ and $-n_{-1} + n - n_{-1} = n - 2n_{-1}$. Consequently, if $\mathrm{SN}_t = s$ we have $-s \le n_{-1} \le \frac{n-s}{2}$ and, as n_{-1} is a positive integer, we have $\max(0, -s) \le n_{-1} \le \lfloor \frac{n-s}{2} \rfloor$. Therefore, for each integer value $i \in \{\max(0, -s), \ldots, \lfloor \frac{n-s}{2} \rfloor\}$ we have to count all the combinations of

- i occurrences of $S_{t,k} = -1$,
- $n - s - 2i$ occurrences of $S_{t,k} = 0$,
- $s + i$ occurrences of $S_{t,k} = 1$,

leading to the following formula:

$$
f_{\mathrm{SN}}(s|n) = \sum_{i=\max(0,-s)}^{\lfloor \frac{n-s}{2} \rfloor} \binom{n}{i}\binom{n-i}{s+i} \pi_{-1}^i \pi_0^{n-s-2i} \pi_{+1}^{s+i}.
$$

The three methods presented above have been tested and, no matter the values of n, π_{-1}, π_0 and π_{+1}, they always give the same results.

References

Abid, M., Nazira, H. Z., Riaz, M., & Lin, Z. (2017). An efficient nonparametric EWMA Wilcoxon signed-rank chart for monitoring location. *Quality and Reliability Engineering International*, *33*(3), 669–685.

Amin, R. W., & Searcy, A. J. (1991). A nonparametric exponentially weighted moving average control scheme. *Communications in Statistics Simulation and Computation*, *20*(4), 1049–1072.

Amin, R. W., Reynolds, M. R, Jr., & Bakir, S. T. (1995). Nonparametric quality control charts based on the sign statistic. *Communications in Statistics Theory and Methods*, *24*(6), 1597–1623.

Bakir, S. T., & Reynolds, M. R, Jr. (1979). A nonparametric procedure for process control based on within-group ranking. *Technometrics*, *21*(2), 175–183.

Capizzi, G. (2015). Recent advances in process monitoring: nonparametric and variable-selection methods for Phase I and Phase II. *Quality Engineering*, *27*(1): 44–67.

Castagliola, P., Tran, K. P., Celano, G., Rakitzis, A. C., & Maravelakis, P. E. (2019). An EWMA-Type Sign Chart with Exact Run Length Properties. *Journal of Quality Technology*, *51*(1), 51–63.

Celano, G., Castagliola, P., Chakraborti, S., & Nenes, G. (2016a). The performance of the Shewhart sign control chart for finite horizon processes. *International Journal of Advanced Manufacturing Technology*, *84*(5), 1497–1512.

Celano, G., Castagliola, P., Chakraborti, S., & Nenes, G. (2016b). On the implementation of the Shewhart sign control chart for low-volume production. *International Journal of Production Research*, *54*(19), 5886–5900.

Chakraborti, S., Van der Laan, P., & Bakir, S. T. (2001). Nonparametric control charts: An overview and some results. *Journal of Quality Technology*, *33*(3), 304–315.

Chakraborty, N., Chakraborti, S., Human, S. W., & Balakrishnan, N. (2016). A generally weighted moving average signed-rank control chart. *Quality and Reliability Engineering International*, *32*(8), 2835–2845.

Gibson, J. D., & Melsa, J. L. *Introduction to nonparametric detection with applications* (Vol. 119, Academic Press).

Graham, M. A., Chakraborti, S., & Human, S. W. (2011). A nonparametric EWMA sign chart for location based on individual measurements. *Quality Engineering*, *23*(3), 227–241.

Li, S., Tang, L., & Ng, S. (2010). Nonparametric CUSUM and EWMA control charts for detecting mean shifts. *Journal of Quality Technology*, *42*(2), 209–226.

Linna, K. W., & Woodall, W. H. (2001). Effect of measurement error on Shewhart control charts. *Journal of Quality Technology*, *33*(2), 213–222.

Lu, S. L. (2015). An extended nonparametric exponentially weighted moving average sign control chart. *Quality and Reliability Engineering International*, *31*(1), 3–13.

Maleki, M. R., Amiri, A., & Castagliola, P. (2017). Measurement errors in statistical process monitoring: A literature review. *Computers & Industrial Engineering*, *103*, 316–329.

McGilchrist, C. A., & Woodyer, K. D. (1975). Note on a distribution-free CUSUM technique. *Technometrics*, *17*(3), 321–325.

Montgomery, D. C. (2013). *Statistical quality control: A modern introduction* (7th ed.). New York: Wiley.

Putter, J. (1955). The treatment of ties in some nonparametric tests. *The Annals of Mathematical Statistics*, *26*(3), 368–386.

Qiu, P. (2018). Some perspectives on nonparametric statistical process control. *Journal of Quality Technology*, *50*(1), 49–65.

Yang, S. F., Lin, J. S., & Cheng, S. W. (2011). A new nonparametric EWMA sign control chart. *Expert Systems with Applications*, *38*, 6239–6243.

Zou, C., & Tsung, F. (2011). A Multivariate Sign EWMA Control Chart. *Technometrics*, *53*(1), 84–97.

Statistical Process Monitoring and the Issue of Assumptions in Practice: Normality and Independence

S. Chakraborti and R. S. Sparks

Abstract Most statistical process monitoring begins with an assumed model (implicitly or otherwise) and further assumptions about the components of the model. These assumptions all play important roles in practice, in the solution that is proposed for the problem at hand. Since the proposal is based on these assumptions, it is important that they are thoroughly investigated and properly validated, so that the results can be depended on. In this paper, we examine two of the important and common assumptions, namely, normality and independence. We provide some examples with real data and illustrate the consequences. It is seen that the nonparametric (distribution-free) approach may be a safer option in many applications in practice.

1 Introduction

In the modern age, large amounts of data are often available from a variety of sources and in a variety of environments that need to be studied and analyzed. This means one needs to make sense of the volume of data and then be able to make efficient decisions based on the data. The study may consist of one time or cross-sectional analysis at a given point in time or a longer term ongoing monitoring of a process. The analysis involves both descriptive and inferential statistics. The descriptive analysis involves visualization and numerical summaries to help understand what is going on. The decision making via prediction, estimation, etc., which is statistical inference, is often based on a confidence interval or a test of hypothesis. While availability of modern software has made this type of work routine and seemingly trivial, one must not forget the assumptions behind the methods that must be satisfied to validate and

S. Chakraborti (✉)
Department of Information Systems, Statistics and Management Science, The University of Alabama, Tuscaloosa, AL 35487, USA
e-mail: schakrab@cba.ua.edu

R. S. Sparks
Data61, CSIRO, Corner of Vimiera and Pembroke Roads, Marsfield, NSW 2122 Sydney, Australia

© Springer Nature Switzerland AG 2020
M. V. Koutras and I. S. Triantafyllou (eds.), *Distribution-Free Methods for Statistical Process Monitoring and Control*,
https://doi.org/10.1007/978-3-030-25081-2_4

justify the end results. Although each statistical inference method requires its own assumptions, some of the most common ones are about randomness, independence and underlying distribution of the data. Violations of one or more of the assumptions might render the decisions invalid and hence useless even though there would seem nothing wrong in terms of crunching the numbers. Much has been written about the importance of checking assumptions during an analysis using statistical methods before decisions are made. In this brief paper, we revisit some of these issues via an example and illustrate some of the challenges associated with data analysis and analytics in general, in practice.

We first consider the assumption about the underlying distribution that may be necessary before a statistical method is applied. This may be necessary simply because the theoretical derivation of the methodology requires such an assumption. However, it is fair to say that assuming (picking) a probability distribution for the data is a daunting task. Whether acknowledged or not, this is one of the most important and challenging aspects of data analysis since the validity of the inference drawn from the application of the method often hinges upon this crucial assumption. The distributions come in all sorts of shapes; the most commonly assumed distribution in the application of statistical methods is the normal distribution which is symmetric and bell-shaped. However, depending on the context, other distributions may also be used, such as the exponential distribution, the uniform distribution, the Weibull distribution, the gamma distribution, to name a few, within the class of continuous distributions. Graphs of some probability density functions of some of the well-known continuous distributions are shown in Fig. 1. The collection includes symmetric and skewed distributions. Note that as shown, even among the symmetric distributions, shapes can vary and this can lead to differences in probability based assessments. The same is true for skewed distributions.

This challenge of making and meeting the distributional assumption is faced by practitioners and data analysts from all areas on a day to day basis. Although it may be possible to use the law of averages and the central limit theorem to by-pass (avoid) the distributional assumption in certain cases (like for large sample sizes) while making statistical inference, it is somewhat of a dicey strategy, particularly in quality control and monitoring applications where the sample sizes are often small. In manufacturing, the typically recommended subgroup size is around five, whereas in real time, individual monitoring of data, the subgroup size is one. Note that in many monitoring settings, data are collected from sensors in a nearly continuous stream and thus it is often more meaningful (and required) to monitor the individual data. Thus applying the central limit theorem to such problems can be risky if not impossible. At times data are aggregated into hourly, four-hourly- or six-hourly intervals to improve the accuracy of measures but such aggregations still will not necessarily allow the central limit to apply.

Statistical process control and monitoring methods originally arose in the context of industrial/manufacturing applications, developed during and after World War Two, in order to produce high quality (and high reliability) items (at a lower cost). This regime involves designing studies (i.e., Design of Experiments), collecting (Sampling), and analyzing data (Analytics). Among the many statistical tools used in

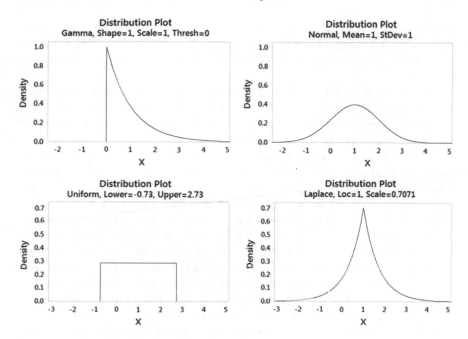

Fig. 1 Some continuous probability distributions

statistical process monitoring the control chart is perhaps the most well-known. This graphic provides a simple and effective visualization of "what's going on" in a process at a given point in time and also over time, in order for the user to make a decision about the quality of the process, including indicating what might not be working and where (and when) so that necessary adjustments may be made. Here the term "quality of the process" is used in a general sense, which is context-dependent. If the process is a manufacturing process producing say copper tubing, the quality of the process may be reflected in the diameter of the tubes (or the average and standard deviation of the diameter of a sample of tubes) and a control chart can describe whether or not the machine (or the process) is producing these tubes that are "in-control" that is meeting specifications for what is expected out of these tubes, so that they can be sold in the market. On the other hand if the context refers to a situation other than in manufacturing, say about the level of pollution in the environment of a certain city, measured in terms of the concentration level of certain matters or chemicals, the quality of the process (whether or not it is in-control) may correspond to what may be dictated by the safety and health considerations.

In any event, if the variables of interest in the outputs of the process are measurable and are monitored with a control chart, over time, and the concept of significance is to be attached to the results, such as, for example, whether or not the process is producing *significantly* more defectives, or, whether or not the pollution level is *significantly* higher than expected, relative to what is "in-control", an assumption about the probability distribution of the variable being measured and monitored may be needed. When this information is available, it is possible to calculate the probability (or the chance) of observing what has been observed, and decide whether or not the observed results are significant, or extreme, compared to some nominal threshold.

In a vast majority of the applications of statistical process monitoring and control, as in many applications of statistics in many other areas, it is fairly common to assume that the underlying probability distribution is of a (given) known form (based on knowledge and experience), but some aspects of the distribution are unknown, say the mean and/or the standard deviation, which are called parameters. This clearly lessens the burden of making the distributional assumption somewhat and provides a bit of flexibility in the choice. For example, one may assume that the diameters of the copper tubes follow a normal distribution with an unknown mean and an unknown standard deviation. Then standard statistical theory and methods may be used with this knowledge of the form of the distribution, to estimate the parameters, and to set up inference procedures. This is the pathway to setting up Shewhart control charts to monitor the mean of the process. Control charts (and statistical methods) developed under the assumption of a known parametric distribution are referred to as parametric control charts (Chakraborti and Graham 2019a, b). Since the normal distribution is among the most commonly assumed distribution sometimes parametric charts are almost synonymous with normal-theory control charts, but we emphasize that many other distributions can be and are used in process monitoring, in a variety of interesting applications.

2 Consequences of the Distributional Assumption

It is clear that if the diameters of the tubes do not follow a normal distribution, but, some other distribution, say a gamma distribution (which is typically skewed to the right) one can set up parametric control charts for the mean taking advantage of that information. This control chart, including the control limits, will not be the same as the one based on the normal distribution. Thus, one may get a different set of results, such as whether or not the process is in control, from an application of each chart. In other words, the statistical inference may be dependent on the assumed distribution for the observed variable(s). Put another way, there is a practical consequence to making the distributional assumption and that consequence, in terms of the probability and the eventual decision, may be slight to severe, depending on how much of the distributional assumption may be violated by the data and how much each decision may end up costing. For example, assuming that the IC distribution of the diameters of copper tubes is exponential with mean 20, the UCL $= 20 + 3 \times 20 = 80$ and LCL $= 20 - 3 \times 20 = -40$, so that the LCL is rounded up to 0. Hence the false alarm rate for the chart is $P(X > 80) = e^{-80/20} = 0.0183$. Thus, where under the normal distribution assumption, for 3-sigma limits, there would be a false alarm, on the average, once in every $1/0.0027 = 370$ samples, under the exponential distribution, there would be a false alarm, on the average, once in every

54.59 (=1/0.0183) samples. This could mean that the manufacturer may soon be out of business because of stopping the process so frequently and looking for a reason that does not exist. Although the example may be somewhat extreme, the point is that there will be a consequence of using the wrong distribution, on the inference or the decision, which can be anywhere, from mild (bearable) to catastrophic.

The issue is that there is usually no way to fully guarantee that the assumed distribution is the correct distribution for the data, or that the data fit the assumed distribution perfectly, since there is always at least a 5% chance of getting it wrong (say based on a goodness of fit test). The bottom line may be that the manufactured copper tubes all pass the quality control check based on the control charts but may be useless in the marketplace or, alternatively, that the tubes do not pass the check and yet may be acceptable.

In the "classical" statistical literature, such consequences are of course well-known and have been examined for some time in terms of what is called robustness. In practice however, the analyst faces a dilemma. The issue is that in practice the consequences of the violations of the assumptions are not always known, advertised, articulated or even appreciated. Or, even if the consequences may be understood, the implementation is not affected or may be delayed, due to lack of training, availability of software, carelessness, ..., just to mention a few reasons. In order to address this dilemma, the area of nonparametric statistics has been developed within statistics. Nonparametric statistical methods provide robust inferential tools (confidence intervals, hypotheses tests) which can be used to make valid statistical inference without assuming a specific parametric form of the underlying distribution. Note that these are not "too good to be true" methods peddled by some suspicious characters at the street corners, but have the backing of a solid theoretical basis. For instance, for the copper tubing example, one can construct a valid 95% nonparametric confidence interval for the median diameter, which does not require the assumption of any particular parametric form of the distribution, except continuity. This is a remarkable result available for many years and should be utilized whenever possible. In short, nonparametric methods apply to a larger class of probability distributions (which may include the one that may have been most commonly used, say the normal). It is true that being applicable to a much broader range of distributions, nonparametric methods may lose some efficiency against parametric methods, for some specific distribution. So if one is sure about the assumption of the distribution, it is perfectly reasonable to proceed along that parametric path. However, it seems fair to say that in most situations, such knowledge is all but nonexistent and one is better off using a nonparametric method.

The same recommendation applies to the area of statistical process monitoring and control. Most of this literature is about parametric charts that are set up assuming a normal (or some other) distribution. However, in the last twenty years, several nonparametric control charts have been proposed in parallel. This area of research has grown rapidly and now a number of software packages are available. In fact, the proliferation of R programming has now reached a state of maturity where it is not entirely unexpected that a user can program a newly proposed chart in a journal article and apply it, even if a packaged solution is not yet available. This is an encouraging development. Several review papers are now available (see for example, Chakraborti, van der Laan and Bakir 2001 and Chakraborti and Graham 2019a, b) and at least two recent books (Qiu 2014 and Chakraborti and Graham 2019a, b) have been written on the subject. A lot of research is currently underway in these areas, both theory and applications.

In this paper our goal is to illustrate the issues with making some of the basic assumptions for a valid statistical analysis with a real dataset arising in a real situation. We focus on this type of a dataset as most datasets used for illustration in the literature, although may be appropriate to illustrate a particular proposed methodology, do not seem to conform to many applications of statistical process monitoring and control in practice. It will be seen that the analysis of real data is hard and our example will show the imperfections in the practical setting, but that is precisely the point. With this is mind, various ways of monitoring such data are considered and a case is made in favor of nonparametric statistical process monitoring.

3   Other Assumptions and Considerations When Designing a Control Chart

The second important issue we consider here is the assumption of independence. In a manufacturing type process monitoring context, it may be reasonable to have data that may be presumed to be independent since it is usually possible to control the monitoring environment quite tightly. However, while monitoring individual data or data monitored over time, it is more often the case that the data are *not* stochastically independent, that is, one data point, in a sequence of data points, influences another, positively or otherwise, that needs to be accounted for in the analysis. A typical scenario involves data collected over time, where the time difference between the successive observations may be small. This could lead to the data being autocorrelated (or serially correlated). Here we consider monitoring applications where the data stream become available (are collected) in near real-time and this could be in the context of a continuous process in manufacturing, social media data streaming in, or pollution data. Given the very nature of these data, there is a high likelihood that the observations are not independent.

The first step in this setting is to define the "common cause" (what is in-control) and "special cause" variation (what is out-of-control) that need to be flagged by the monitoring strategy. This involves understanding the sources of variation by deciding on, for example, what, if any, seasonal adjustments need to be made (should be removed) or what within day influences are to be treated as common cause variation (and should be removed). This would help define common causes of variation that do not need to be signaled (which defines the in-control state). Also, having a clear understanding of what special causes of variation need to be flagged (what is out-of-control) is vital at this design phase. These decisions would have to be made before deciding on the distributional assumptions. These considerations are to be handled before deciding what time series model is going to be fitted and then monitoring the forecast residuals from that model. Deciding on the appropriate rational subgroup also requires some thought.

In parts manufacturing type applications, the basic in-control model is generally taken as

$$y_t = \mu + e_t$$

where μ is the mean process value or the target and the random error is given by $e_t \sim n(0, \sigma^2)$. However, in monitoring applications outside of manufacturing, this is often defined as

$$y_t = \mu_t + e_t$$

where $e_t \sim n(0, \sigma_t^2)$ and the exact form of μ_t and σ_t^2 depend on how common cause variation is defined over time t. A visual example of μ_t and σ_t^2 that are influenced by the seasons is presented in Fig. 1. Notice that in this case, both the mean and the variance of the process are higher in Spring and Summer. We may wish to remove this variation as nothing can be done about it. In such cases we need to be certain about what we wish to control. In addition, the temperature, wind and humidity within a day vary and these influences can be removed if they influence the ozone. However before doing this we need to decide of the purpose of monitoring. If it is to discover what is unusual process behavior, removing all potential influences is recommended, but if the purpose is just to flag high values then no corrections should be made.

4   Applications

The practical example we consider is monitoring the level of ozone in the atmosphere at Chullora, a suburb of the local government areas the Canterbury-Bankstown Council and in the Municipality of Strathfield. It is located 15 km west of the Sydney central business district, in the state of New South Wales, Australia. It is part of the Greater Western Sydney region. The data, plotted in Fig. 2, involve hourly average ozone measures for six years, from June 5, 2013 to June 5, 2019 (inclusive). Monitoring the ozone level is important from a public health perspective since the ozone in the air we breathe can harm our health, especially on hot sunny days when ozone can reach unhealthy levels. Even relatively low levels of ozone can cause health effects. People most at risk from breathing air containing ozone include people with asthma, children, older adults, and people who are active outdoors, especially outdoor workers. In addition, people with certain genetic characteristics, and people with reduced intake of certain nutrients, such as vitamins C and E, are at greater risk from ozone exposure.

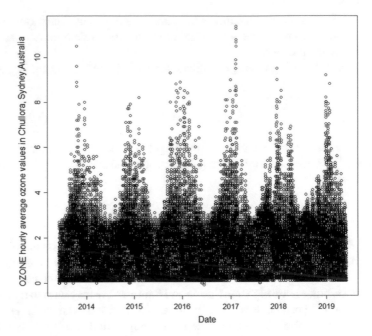

Fig. 2 Hourly ozone measure at Chullora, Sydney, Australia

The first phase in the analysis is to explore the sources of variation in order to understand and define the in- and out-of-control status of the process and the underlying distribution, among other matters. This exploratory phase is called Phase I. We would expect ozone to have a seasonal influence, and Figs. 2 and 3 confirm this influence. It is also evident that the data are not normally distributed given the skewness of the boxplots particularly for months with low values.

There is little we can do about this seasonal influence and so we may wish to remove this variation as common cause. We also assess the hourly influence of ozone measures in Fig. 4. There is a clear influence of hour of the day on the ozone measures. Note that on average lower ozone values are recorded in the afternoon than during other times in the day. It appears as if the hourly average trend could be fitted using a within day harmonic. There is little we can do about this influence and so we may also wish to remove this variation as common cause.

These two sources of variation, the seasonality and the hour of the day (or within day), should be included in the model for the ozone level that can be used to provide one hour-ahead forecasts values. This can be handled as follows. The seasonal influence could be removed by taking first order seasonal differences of the data with season defined as the month (12). The influence of the hour of the day may be removed by fitting an ARIMA(1,1,1) model with "seasonal influence" being 24 hourly values. However, this model failed to remove all the significant autocorrelations and even more complicated ARIMA models failed to fit adequately. As an

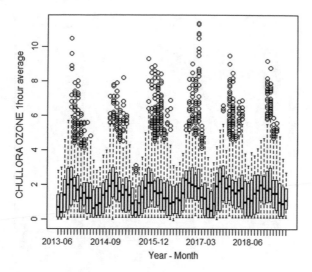

Fig. 3 Monthly boxplots of hourly ozone measures at Chullora

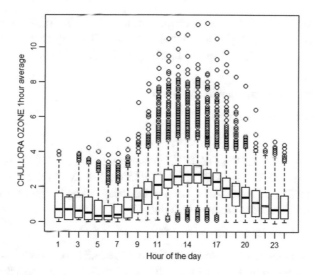

Fig. 4 Boxplots of the hourly ozone measures at Chullora

alternative, the simplest model tried was seasonal harmonics, with day harmonics and up to three lag autoregressive parameters.

Step 1: Transform the data so as to achieve approximate normality.

The transformation that proved closest to normality was (CHUL-LORA.OZONE.1h.average..pphm. $+ 0.101)^{0.77}$

which was obtained using the boxcox in function in R MASS library given as

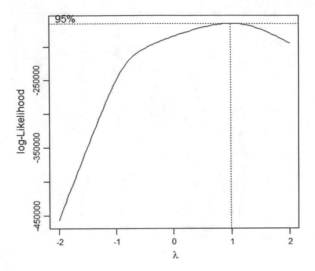

Fig. 5 Box-cox transformation to normality

boxcox(((CHULLORA.OZONE.1h.average..pphm. + 0.101)^0.77 ~ day + cos(2 * pi * day/365.25) + sin(2 * pi * day/365.25) + dw * (Time + cos(2 * pi * Time/24) + sin(2 * pi * Time/24)) + Lag1 + Lag1day + Lag2 + Lag3,data = OZONE.C) (Fig. 5).

Step 2: Fit a model to the transformed data

The fitted model was as follows:

Call:

lm(formula = (CHULLORA.OZONE.1h.average..pphm. + 0.001)^0.77 ~ day + cos(2 * pi * day/365.25) + sin(2 * pi * day/365.25) + dw * (Time + cos(2 * pi * Time/24) + sin(2 * pi * Time/24)) + Lag1 + Lag1day + Lag2 + Lag3 + Temperature + WindSpeed + Humidity, data = OZONE.C)

The output from this is as follows (Fig. 6).

The autocorrelations of the residuals from this model are shown in Fig. 7 which indicate that there is significant autocorrelation up to 24 h but thereafter this autocorrelation is largely small to non-significant. The partial autocorrelations are significant for the first 24 h but the values are low (less than 0.11). Thus there is some evidence that the model does not do a great job at correcting within day variation, and may be improved. Nevertheless, for illustration, we decided to use this model to define the one hour-ahead forecasts, and then use a rational sub-group of hourly average ozone measures in a day. Given that measures were not recorded at 2 a.m. each day and that we are using 3 lagged autoregressive terms in the model this results in mostly 21 measures in a day.

Even though the fitted model does not seem to remove all the influences of the hour of the day since the ACF values are significant for the first 24 h, this model is a

```
Residuals:
     Min       1Q    Median       3Q       Max
-2.27094  -0.14918  -0.01683   0.14246   2.23760


Call:
lm(formula = (CHULLORA.OZONE.1h.average..pphm. + 0.001)^0.77 ~
    day + cos(2 * pi * day/365.25) + sin(2 * pi * day/365.25) +
        dw * (Time + cos(2 * pi * Time/24) + sin(2 * pi * Time/24)) +
        Lag1 + Lag1day + Lag2 + Lag3 + Temperature + WindSpeed +
        Humidity, data = OZONE)


Residuals:
     Min       1Q    Median       3Q       Max
-2.29465  -0.14918  -0.01399   0.14412   2.21179


Coefficients:
                                  Estimate Std. Error t value Pr(>|t|)
(Intercept)                      -5.953e-02  3.852e-02  -1.545 0.122280
day                               1.988e-05  2.005e-06   9.914  < 2e-16 ***
cos(2 * pi * day/365.25)         -8.899e-03  3.112e-03  -2.860 0.004244 **
sin(2 * pi * day/365.25)         -5.037e-02  2.066e-03 -24.374  < 2e-16 ***
dwMonday                          3.484e-02  1.684e-02   2.069 0.038594 *
dwSaturday                        4.373e-02  1.690e-02   2.587 0.009673 **
dwSunday                          1.109e-01  1.685e-02   6.590 4.75e-11 ***
dwThursday                        2.862e-03  1.688e-02   0.170 0.865390
dwTuesday                        -4.222e-03  1.685e-02  -0.251 0.802133
dwWednesday                       1.664e-02  1.686e-02   0.987 0.323844
Time                              2.874e-03  8.365e-04   3.435 0.000592 ***
cos(2 * pi * Time/24)            -7.426e-02  5.462e-03 -13.597  < 2e-16 ***
sin(2 * pi * Time/24)             3.534e-02  7.344e-03   4.812 1.50e-06 ***
Lag1                              7.398e-01  3.921e-03 188.675  < 2e-16 ***
Lag1day                           5.470e-03  2.305e-03   2.373 0.017661 *
Lag2                             -2.444e-01  5.065e-03 -48.250  < 2e-16 ***
Lag3                             -1.091e-02  3.252e-03  -3.355 0.000794 ***
Temperature                       1.783e-02  4.696e-04  37.974  < 2e-16 ***
WindSpeed                         7.052e-02  1.226e-03  57.524  < 2e-16 ***
Humidity                         -3.346e-03  9.357e-05 -35.763  < 2e-16 ***
dwMonday:Time                    -1.768e-03  1.177e-03  -1.502 0.133004
dwSaturday:Time                   1.299e-03  1.180e-03   1.101 0.270949
dwSunday:Time                    -1.349e-03  1.177e-03  -1.146 0.251812
dwThursday:Time                  -2.306e-04  1.180e-03  -0.195 0.845025
dwTuesday:Time                   -1.691e-04  1.177e-03  -0.144 0.885793
dwWednesday:Time                 -1.038e-03  1.178e-03  -0.881 0.378071
dwMonday:cos(2 * pi * Time/24)    3.939e-03  7.329e-03   0.537 0.590930
dwSaturday:cos(2 * pi * Time/24) -1.329e-02  7.343e-03  -1.809 0.070399 .
dwSunday:cos(2 * pi * Time/24)   -1.052e-02  7.321e-03  -1.437 0.150848
dwThursday:cos(2 * pi * Time/24) -2.575e-03  7.369e-03  -0.349 0.726778
dwTuesday:cos(2 * pi * Time/24)   5.561e-03  7.346e-03   0.757 0.449036
dwWednesday:cos(2 * pi * Time/24) -2.508e-03 7.354e-03  -0.341 0.733086
dwMonday:sin(2 * pi * Time/24)   -6.434e-03  9.919e-03  -0.649 0.516547
dwSaturday:sin(2 * pi * Time/24)  2.069e-02  9.927e-03   2.084 0.037161 *
dwSunday:sin(2 * pi * Time/24)    4.507e-02  9.902e-03   4.551 5.34e-06 ***
dwThursday:sin(2 * pi * Time/24) -9.118e-03  9.980e-03  -0.914 0.360916
dwTuesday:sin(2 * pi * Time/24)  -4.839e-04  9.947e-03  -0.049 0.961199
dwWednesday:sin(2 * pi * Time/24) -7.039e-03 9.960e-03  -0.707 0.479706
---
Signif. codes:  0 '***' 0.001 '**' 0.01 '*' 0.05 '.' 0.1 ' ' 1


Residual standard error: 0.2537 on 41983 degrees of freedom
  (10035 observations deleted due to missingness)
Multiple R-squared:  0.9119,    Adjusted R-squared:  0.9118
F-statistic: 1.175e+04 on 37 and 41983 DF,  p-value: < 2.2e-16
```

Fig. 6 Output from model fitting

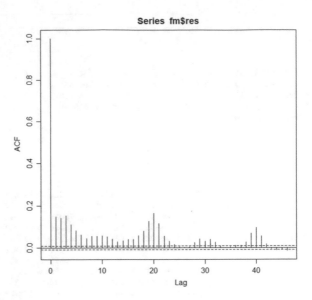

Fig. 7 Autocorrelation of the residuals of the fitted model

reasonably adequate description of in-control data, particularly if we use a rational subgroup of daily measures. However, the other assumption in the model is that the variance of the error term is homogeneous which needs to be checked. In fact, looking at the time series in Fig. 2, there seems to be some questions about the validity of this homoscedasticity assumption. We use the gamlss library in R to assess whether the variance can be assumed homogeneous over time.

The R code is as follows:

••

Family: c("NO", "Normal")

Call: gamlss(formula = (CHULLORA.OZONE.1 h.average..pphm. + 0.001)^0.77 ~ day + cos(2 * pi * day/365.25) + sin(2 * pi * day/365.25) + dw * (Time + cos(2 * pi * Time/24) + sin(2 * pi * Time/24)) + Lag1 + Lag1day + Lag2 + Lag3 + Temperature + WindSpeed + Humidity, sigma.formula = ~ day + cos(2 * pi * day/365.25) + sin(2 * pi * day/365.25) + ((dw == "Wednesday") + (dw == "Sunday") + (dw == "Saturday")) * (Time + cos(2 * pi * Time/24) + sin(2 * pi * Time/24)) + Temperature + WindSpeed + Humidity, data = na.omit(OZONE.C)).

The output is shown in Fig. 8.

Clearly the variances for this process are not homogeneous within days with the variance increasing with the hour of the day and there is a significant harmonic change in variances within the day.

```
Fitting method: RS()
**********************************************************************
Family:  c("NO", "Normal")


Call:  gamlss(formula = (CHULLORA.OZONE.1h.average..pphm. + 0.001)^0.77 ~ day + cos(2 *
pi * day/365.25) + sin(2 * pi * day/365.25) + dw * (Time + cos(2 * pi * Time/24) + sin(2 * pi *
Time/24)) +  Lag1 + Lag1day + Lag2 + Lag3 + Temperature + WindSpeed +            Humidity,
sigma.formula = ~day + cos(2 * pi * day/365.25) + sin(2 * pi * day/365.25) + ((dw == "Wednesday")
+ (dw ==  "Sunday") + (dw == "Saturday")) * (Time + cos(2 * pi * Time/24) +        sin(2 * pi *
Time/24)) + Temperature + WindSpeed + Humidity, data = na.omit(OZONE))


Fitting method: RS()


-----------------------------------------------------------------

Mu link function:  identity
Mu Coefficients:
                                  Estimate Std. Error t value Pr(>|t|)
(Intercept)                      -1.425e-01  3.565e-02  -3.998 6.40e-05 ***
day                               1.916e-05  1.828e-06  10.481  < 2e-16 ***
cos(2 * pi * day/365.25)         -3.318e-02  2.913e-03 -11.389  < 2e-16 ***
sin(2 * pi * day/365.25)         -5.933e-02  1.927e-03 -30.794  < 2e-16 ***
dwMonday                          2.588e-02  1.741e-02   1.487 0.137030
dwSaturday                        3.880e-02  1.796e-02   2.160 0.030779 *
dwSunday                          1.130e-01  1.751e-02   6.453 1.11e-10 ***
dwThursday                       -5.704e-03  1.752e-02  -0.326 0.744781
dwTuesday                        -2.926e-03  1.748e-02  -0.167 0.867069
dwWednesday                       8.781e-03  1.773e-02   0.495 0.620444
Time                              3.820e-03  8.700e-04   4.390 1.13e-05 ***
cos(2 * pi * Time/24)            -7.222e-02  5.329e-03 -13.554  < 2e-16 ***
sin(2 * pi * Time/24)             3.534e-02  6.961e-03   5.077 3.85e-07 ***
Lag1                              7.506e-01  3.958e-03 189.630  < 2e-16 ***
Lag1day                           7.535e-03  2.188e-03   3.444 0.000573 ***
Lag2                             -2.532e-01  5.102e-03 -49.626  < 2e-16 ***
Lag3                             -2.297e-03  3.264e-03  -0.704 0.481558
Temperature                       2.160e-02  4.529e-04  47.688  < 2e-16 ***
WindSpeed                         6.797e-02  1.103e-03  61.642  < 2e-16 ***
Humidity                         -3.370e-03  8.389e-05 -40.174  < 2e-16 ***
dwMonday:Time                    -1.272e-03  1.219e-03  -1.044 0.296707
dwSaturday:Time                   1.564e-03  1.254e-03   1.248 0.212136
dwSunday:Time                    -1.274e-03  1.229e-03  -1.036 0.300266
dwThursday:Time                   4.396e-04  1.226e-03   0.359 0.719913
dwTuesday:Time                   -9.983e-05  1.225e-03  -0.082 0.935033
dwWednesday:Time                 -3.803e-04  1.236e-03  -0.308 0.758215
dwMonday:cos(2 * pi * Time/24)    5.138e-03  7.152e-03   0.718 0.472480
dwSaturday:cos(2 * pi * Time/24) -1.118e-02  7.368e-03  -1.517 0.129307
dwSunday:cos(2 * pi * Time/24)   -3.288e-03  7.106e-03  -0.463 0.643583
dwThursday:cos(2 * pi * Time/24) -2.260e-03  7.219e-03  -0.313 0.754240
dwTuesday:cos(2 * pi * Time/24)   5.812e-03  7.200e-03   0.807 0.419511
dwWednesday:cos(2 * pi * Time/24) -5.290e-03  7.271e-03  -0.728 0.466859
dwMonday:sin(2 * pi * Time/24)   -4.399e-03  9.362e-03  -0.470 0.638469
dwSaturday:sin(2 * pi * Time/24)  2.754e-02  9.557e-03   2.882 0.003954 **
dwSunday:sin(2 * pi * Time/24)    6.082e-02  9.299e-03   6.541 6.17e-11 ***
```

Fig. 8 Model fitting results for the location and variance using gamlss library in R

```
dwThursday:sin(2 * pi * Time/24)   -5.461e-03  9.454e-03  -0.578 0.563501
dwTuesday:sin(2 * pi * Time/24)    -3.383e-03  9.443e-03  -0.358 0.720134
dwWednesday:sin(2 * pi * Time/24)  -4.631e-03  9.441e-03  -0.490 0.623799
---
Signif. codes:  0 '***' 0.001 '**' 0.01 '*' 0.05 '.' 0.1 ' ' 1


----------------------------------------------------------------

Sigma link function:  log
Sigma Coefficients:
                                          Estimate Std. Error t value Pr(>|t|)
(Intercept)                              -2.448e+00  1.033e-01 -23.704  < 2e-16 ***
day                                       7.008e-06  5.573e-06   1.257 0.208626
cos(2 * pi * day/365.25)                 -2.625e-01  8.548e-03 -30.712  < 2e-16 ***
sin(2 * pi * day/365.25)                 -1.363e-01  5.557e-03 -24.519  < 2e-16 ***
dw == "Wednesday"TRUE                     4.484e-02  3.704e-02   1.210 0.226099
dw == "Sunday"TRUE                       -6.864e-02  3.698e-02  -1.856 0.063459 .
dw == "Saturday"TRUE                      4.393e-02  3.719e-02   1.181 0.237424
Time                                     -4.318e-03  1.159e-03  -3.726 0.000195 ***
cos(2 * pi * Time/24)                     2.478e-01  8.231e-03  30.111  < 2e-16 ***
sin(2 * pi * Time/24)                     3.532e-02  1.048e-02   3.371 0.000751 ***
Temperature                               5.746e-02  1.208e-03  47.555  < 2e-16 ***
WindSpeed                                -9.834e-03  3.301e-03  -2.979 0.002894 **
Humidity                                 -1.011e-03  2.553e-04  -3.961 7.47e-05 ***
dw == "Wednesday"TRUE:Time               -3.716e-03  2.589e-03  -1.435 0.151217
dw == "Wednesday"TRUE:cos(2 * pi * Time/24)  3.139e-03 1.614e-02  0.195 0.845773
dw == "Wednesday"TRUE:sin(2 * pi * Time/24) -1.430e-02 2.186e-02 -0.654 0.512879
dw == "Sunday"TRUE:Time                   4.493e-02  2.585e-02   1.738 0.082161 .
dw == "Sunday"TRUE:cos(2 * pi * Time/24)  3.886e-02  1.600e-02   2.428 0.015173 *
dw == "Sunday"TRUE:sin(2 * pi * Time/24)  1.180e-01  2.165e-02   5.451 5.04e-08 ***
dw == "Saturday"TRUE:Time                -6.809e-04  2.597e-03  -0.262 0.793143
dw == "Saturday"TRUE:cos(2 * pi * Time/24) -1.302e-04 1.609e-02 -0.008 0.993546
dw == "Saturday"TRUE:sin(2 * pi * Time/24)  4.744e-02 2.175e-02  2.181 0.029186 *
---
Signif. codes:  0 '***' 0.001 '**' 0.01 '*' 0.05 '.' 0.1 ' ' 1


----------------------------------------------------------------

No. of observations in the fit:  42021
Degrees of Freedom for the fit:  60
        Residual Deg. of Freedom:  41961
                    at cycle:  7


Global Deviance:    -580.6214
          AIC:      -460.6214
          SBC:       58.13409
```

Fig. 8 (continued)

```
Time                               1.301e-02  2.797e-03   4.651 3.31e-06 ***
cos(2 * pi * Time/24)              3.776e-01  2.636e-02  14.322  < 2e-16 ***
sin(2 * pi * Time/24)             2.475e-01  4.220e-02   5.866 4.50e-09 ***
```

The source of variation in the variances includes seasonal variation but no significant increase in variance over time. The changes in variances are also seasonally influenced but the variances are not significantly increasing with each day, this was anticipated as we observed that variance changes significantly with season in Figs. 2 and 3.

day	7.008e-06	5.573e-06	1.257	0.208626
cos(2 * pi * day/365.25)	-2.625e-01	8.548e-03	-30.712	< 2e-16 ***
sin(2 * pi * day/365.25)	-1.363e-01	5.557e-03	-24.519	< 2e-16 ***

Indicating that the ozone values are increasing as the day number increases and the seasonal harmonics for the mean are both statistically significant. During the day the ozone values increase to a maximum at about noon and then decreases thereafter.

Time	3.820e-03	8.700e-04	4.390	1.13e-05 ***
cos(2 * pi * Time/24)	-7.222e-02	5.329e-03	-13.554	< 2e-16 ***
sin(2 * pi * Time/24)	3.534e-02	6.961e-03	5.077	3.85e-07 ***

Note that there is a significant and different Sunday and Saturday influence for day of the week. This model was selected because it delivered the smallest AIC value.

In setting up a Phase II monitoring strategy, we use this model to compute the hour-ahead forecasts for each hour of the day for the last 366 days of data. The data used to provide these forecasts uses a moving window of 5 years of data for each forecast so that the accuracy of the forecast are expected to be the same. The model is used to forecast both the mean and variance of the normally distributed data for a month ahead, and then we calculate the usual month ahead ozone value minus this forecasted value, all divided by the forecast standard deviation. This result is assumed to be approximately normally distributed with mean zero and standard deviation 1 and a three-sigma control chart is used to monitor these standardized residuals. Note, generally speaking, even when the normal distribution fits the training data quite well (in Phase I) it does not guarantee that the forecast errors will be normally distributed (in Phase II—see qqplot in Fig. 9). However we recognize that this may contain out-of-control data and so such judgements are difficult to make. For example, the ozone values could increase significantly to a new steady state, and this new steady state distribution may be different from the normal.

The hour-ahead forecasts and the forecast errors are calculated and used to (Phase II) monitoring changes in the ozone level at Chullora. The assumption when applying the classical x-bar Shewhart chart to the forecast errors using all 23 hourly measures during a day (so the hourly measures within a day is viewed as a subgroup) is likely to be adequate in approximating the normal distribution because of the central limit theorem is likely to apply. Note however, that averaging as many as 23 values is unlikely in the application of most classical x-bar Shewhart Charts in manufacturing.

The x-bar chart with 3-sigma control limits is given for the standardized forecast errors in Fig. 10. It is clear that ozone level is mostly out-of-control on the high side at Chullora with an occasional signal for low ozone measures. This indicates that the ozone measures have been mostly higher than expected during this monitoring period.

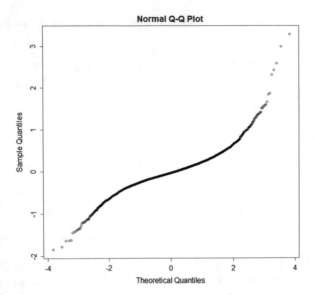

Fig. 9 qq-plot of the one hour-ahead standardized forecasts errors

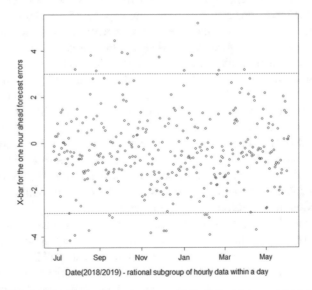

Fig. 10 The x-bar chart or the one hour-ahead standardized forecast errors with 3-sigma control limits

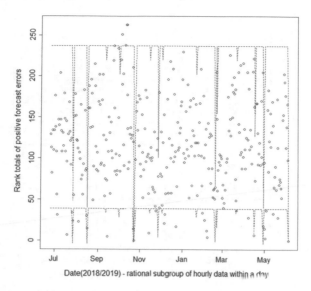

Fig. 11 Wilcoxon signed-rank chart for the negative ranked values for the one hour-ahead standardized forecast errors

Although the x-bar chart in Fig. 10 may be appropriate for this application because it used a relatively large rational subgroup of 23 observations and the Box-Cox transformation is applied to improve the normality assumption. It is likely to provide earlier flags of ozone measures of a health concern. The x-bar chart flags 12 out-of-control high sided ozone days and 14 low sided ozone days. It is unclear whether the transformation to normality will be appropriate in the long term, while the non-parametric Wilcoxon Signed Rank Control chart is always valid. So it may be a safer option if the planned monitoring strategy is going to run for several years. The Wilcoxon Signed Rank Control chart is presented in Fig. 11 and this chart flags four out-of-control high-sided ozone days and 20 low sided ozone days.

Let N be the number of standardized hour-ahead forecast errors in a day. The value of N varies from day to day which explains why the control limits vary according to the number of reported values in the day. Let the rank total for positive standardized forecast errors be T^+, this is the classical Wilcoxon signed-rank statistic. The control limits, the UCL and the LCL are defined by

$$UCL = \frac{N(N+1)}{4} + 0.5 + 3\sqrt{N(N+1)(2n+1)/24}$$

and

$$UCL = \frac{N(N+1)}{4} - 0.5 - 3\sqrt{N(N+1)(2n+1)/24}$$

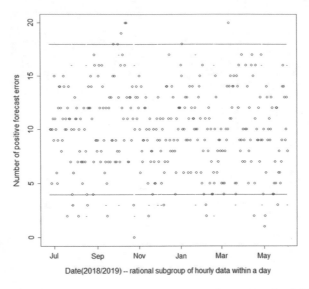

Fig. 12 Sign chart for the number of positive values for the hour-ahead standardized forecast errors

where the mean and the variance of are obtained from, for example, Gibbons and Chakraborti (2010), page 196. Note that since N is not the same every day, the control limits change, and hence the jagged appearance of the control chart.

A required assumption for the signed-rank chart is symmetry which may or may not be satisfied in general. On the other hand, an alternative, namely the sign chart is more general and requires only the assumption of continuity of the data. The control limits for the sign chart can be easily found along the lines for the signed-rank chart using the mean and the variance of the sign statistic, which are available, also in Gibbons and Chakraborti (2010). This chart, shown in Fig. 12, does not seem to lose much power compared to the Wilcoxon Signed Rank control chart.

The sign control chart flags 4–5 out-of-control high sided ozone days and 12 low sided ozone days. Thus, simple nonparametric charts can do the job adequately and one does not need to use a parametric chart and worry about the underlying assumption(s).

5 Concluding Remarks

We have demonstrated the application of some nonparametric control charts in a real monitoring application, using an air pollution data example involving ozone measurements at Chullora, a suburb in Sydney, Australia. The example illustrates how these charts can be applied for monitoring in non-traditional settings, outside of the usual manufacturing situations. It reinforces that there are many facets to monitoring data observed over time in practice and one size fits all recipes using the classical assumptions about independence and normal distribution do not work in many situations. It also shows that in this example very little is lost in applying a nonparametric control chart to the forecast errors (residuals) and it may be a longer term safer option as the distribution of the variables are likely to change over time as the ozone concentrations change. In other words, we can not assume that the plan best at a given point in time is going to remain the best in the future. For this reason, the nonparametric approach may be a safer option unless the parametric chart is periodically reviewed and updated at least annually.

Acknowledgements The authors are grateful to a referee for comments that improved the presentation.

References

Chakraborti, S., & Graham, M. A. (2019a). Nonparametric (distribution-free) control charts: An updated overview and some results. *Quality Engineering*. https://doi.org/10.1080/08982112. 2018.1549330.

Chakraborti, S., & Graham, M. A. (2019b). *Nonparametric statistical process control*. Wiley, New York, New York.

Chakraborti, S., van der Laan, P., & Bakir, S. T. (2001). Nonparametric control charts: An overview and some results. *Journal of Quality Technology, 33*, 304–315.

Gibbons, J. D., & Chakraborti, S. (2010). *Nonparametric statistical inference* (5th ed.). Boca Raton, FL: CRC Press.

Qiu, P. (2014). *Introduction to statistical process control*. Boca Raton, FL: Chapman Hall/CRC.

On Change-Point Analysis-Based Distribution-Free Control Charts with Phase I Applications

Christina Parpoula and Alex Karagrigoriou

Abstract One of the most common challenges in nonmanufacturing control chart Phase I applications is that the underlying process distribution of many of the non-manufacturing quality characteristics is not normal and usually unknown, hence, statistical properties of the most commonly used charts are highly affected. Given these concerns, nonparametric or even distribution-free control charts appear to be ideal candidates for such Phase I applications. However, most of the existing non-parametric control charts are designed for Phase II monitoring, and most Phase I statistical methods can only be applied with subgrouped observations. Nevertheless, nonmanufacturing process-quality characteristics are often observed and recorded as individual observations, and little has been done in developing nonparametric Phase I control charts especially for individual observations that are prevalent in nonmanufacturing applications. Toward this end, in this chapter, existing nonparametric control charts (either for Phase I or Phase II) are briefly discussed and change-point analysis-based distribution-free control charts, designed for Phase I applications especially for individual observations, are constructed for retrospectively detecting single or multiple changes in location and dispersion of univariate variables. A real example is included to unfold the capabilities of the developed methodologies.

Keywords Change-point analysis · Control charts · Distribution-free · Epidemiological surveillance · Statistical process control

Classifications 62-07 · 62-09 · 62G05 · 62G08 · 62G10 · 62P10

C. Parpoula (✉) · A. Karagrigoriou
Lab of Statistics and Data Analysis, Department of Statistics and Actuarial-Financial Mathematics, University of the Aegean, 83200 Karlovasi, Samos, Greece
e-mail: parpoula.ch@aegean.gr

A. Karagrigoriou
e-mail: alex.karagrigoriou@aegean.gr

© Springer Nature Switzerland AG 2020
M. V. Koutras and I. S. Triantafyllou (eds.), *Distribution-Free Methods for Statistical Process Monitoring and Control*,
https://doi.org/10.1007/978-3-030-25081-2_5

157

1 Introduction

Control charts form the core of Statistical Process Control (SPC) and are the main tools of SPC for monitoring the characteristics of a process over time (Montgomery 2013). The application of control charts is usually implemented in two phases, i.e., Phase I and Phase II. In Phase I, process historical data are analyzed retrospectively for characterizing the "in-control" (IC) state (in SPC terminology). In particular, one essentially checks whether the process historically was stable and consistent and a set of "trial" control limits are iteratively revised until no more "out-of-control" (OC) points (in SPC terminology) can be found (Capizzi and Masarotto 2013). Phase II consists of ongoing prospective monitoring with data samples taken successively over time and an assumed underlying probability distribution which is appropriate to the process.

In the past several decades, SPC chart applications can be found predominantly in manufacturing industries for monitoring production lines. Over the past two decades, control chart applications have begun to spread to nonmanufacturing industries and found many applications in areas such as healthcare monitoring, infectious disease surveillance, environment monitoring, banking, and insurance, among others (Ning et al. 2015). One of the most common challenges in nonmanufacturing control chart Phase I applications is that the underlying process distribution of many of the nonmanufacturing quality characteristics is not normal and usually unknown, hence, statistical properties of the most commonly used charts are highly affected. In SPC, it is often reasonable to assume that the IC distribution is (or can reasonably be assumed to be) stationary; observations can be drawn from the process so they are assumed to be independent (or nearly so); monitoring the process mean and standard deviation is usually sufficient; the asymptotic distributions of the statistics being monitored are known and thus can be used to design appropriate control charts; shifts (when they occur) remain until they are detected and corrective action is taken; temporal (as opposed to spatial) detection is the critical problem (Fricker 2011). Although there are manufacturing processes that have characteristics similar to those of the nonmanufacturing industries (e.g., continuous monitoring of processes), for the implementation of quality tools into the nonmanufacturing monitoring problem, in exactly the same fashion as in industrial quality control, many assumptions fail to be satisfied. For example, the distribution of many nonmanufacturing process-quality characteristics is nonstationary; observations are sometimes autocorrelated, and the need for quick detection works against the idea of taking measurements far apart to achieve (near) independence; in some applications there is little information about what types of statistics are useful for monitoring, and one is often looking for anything that seems unusual; individual observations are usually being monitored, thus the idea of asymptotic sampling distributions does not apply and the data often contain significant systematic effects that must be accounted for; processes are subject to step, transient, and even isolated shifts; identifying both spatial and temporal deviations are often critical.

Given these concerns, nonparametric control charts (or even distribution-free control charts in the sense that distributions of their charting statistics do not depend on the true process distribution) appear to be ideal candidates for such Phase I applications. However, most of the existing nonparametric control charts are designed for Phase II monitoring, and most Phase I statistical methods can only be applied with subgrouped observations. Nevertheless, nonmanufacturing process-quality characteristics are often observed and recorded as individual observations, and little has been done in developing nonparametric Phase I control charts especially for individual observations that are prevalent in nonmanufacturing applications. This gap between existing SPC methods and the nonmanufacturing control chart applications provides an area ripe for new research, since as these nonmanufacturing applications continue to spread, new challenges will inevitably arise, which will require developing more effective control charts. Therefore, it is important to develop control charts that are either more robust or do not require normality or any parametric assumption of the quality characteristic being monitored, in order to continue to expand control chart applications into broader sectors. Several authors, for example, Woodall and Montgomery in (1999), Woodall in (2000), and Ning et al. in (2015) have also pointed out the need to develop nonparametric control charts. The interested reader may refer to McCracken and Chakraborti in (2013), Qiu in (2014), and Chakraborti and Graham in (2019) for some comprehensive reviews of both univariate and multivariate nonparametric control charts and for some of the latest developments in this area.

Thus, in this chapter, existing nonparametric control charts (either for Phase I or Phase II) are briefly discussed, and change-point analysis-based distribution-free control charts, designed for Phase I applications especially for individual observations, are constructed for retrospectively detecting single or multiple changes in location and dispersion of univariate variables. A nonmanufacturing real example is included, especially aiming to adequately serve epidemiological surveillance and healthcare monitoring purposes, in which the few recently developed competing charts are applied and compared, and their ability to detect the true and correct amount of change points is evaluated in terms of several performance evaluation metrics.

The rest of the chapter is organized as follows. In Sect. 2, existing nonparametric control charts, and major issues and developments in Phase I and Phase II analysis are discussed. In Sect. 3, the statistical reference framework is introduced, and technical details for Phase I distribution-free control charts for individual observations are provided. In Sect. 4, an application to nonmanufacturing real data is illustrated. Within this framework, several distribution-free control charting methods are constructed (with emphasis on change-point analysis-based methods), and the output of Phase I analysis is interpreted (from both a statistical and an epidemiological perspective). In Sect. 5, the performance of the developed competing charts is compared and evaluated in terms of retrospectively (Phase I) detecting change points (outbreaks) with scientific and methodological adequacy. Finally, in Sect. 6, some concluding remarks are made.

2 Background and Literature Review

The successful performance of SPC applications in Phase II mostly depends on
the correct characterization of the IC state attained in Phase I (Jensen et al. 2006).
However, despite its crucial importance, the vast majority of research on process
monitoring has considered the development and performance of Phase II control
charting methods. Mukherjee in (2016) gave an overview of Phase II monitoring
of the probability distributions of univariate continuous processes. Further, most of
the currently available nonparametric control charts are for Phase II applications
(with the vast majority of them focusing on monitoring the location parameter). The
interested reader may refer to, among others, Chakraborti et al. in (2008), Chatterjee
and Qiu in (2009), Capizzi and Masarotto in (2012), Qiu and Zhang in (2015), Qiu
in (2018) and (2019), Malela-Majika et al. in (2019), and Triantafyllou in (2019) and
(2019) for some perspectives on nonparametric control charts with main focus on
Phase II. Moreover, some nonparametric control charts are based on a change-point
model, such as those studied by Zou and Tsung in (2010), Qiu and Li in (2011), and
Ross and Adams in (2012), and are capable of detecting changes in the distribution
function that could be a result of changes in the location parameter or scale parameter
or both. Further, almost all of the existing charts are designed for Phase II monitoring.
As such, they typically assume that the IC value of the parameter to be monitored is
either known or can be reasonably estimated based on Phase I data that were collected
when the process was IC (Ning et al. 2015). It should be noted though that in those
cases that such an assumption is not made, they assume instead either the existence
of an IC Phase I sample or that the process was IC (Ning et al. 2015). Note that any of
these assumptions makes most of the aforementioned nonparametric Phase II control
charts (see Zhou et al. in (2009), Hawkins and Deng in (2010), Zou and Tsung in
(2010), Qiu and Li in (2011), and Liu et al. in (2013)) not suitable for retrospective
analysis in Phase I control.

The past decade, Phase I analysis (with emphasis on distribution-free techniques)
has received increasing attention. Chakraborti et al. in (2009) gave an overview
of the available literature on the retrospective use of univariate control charts in
Phase I, providing also important technical details about performance measures and
comparisons of various control charts, and Jones-Farmer et al. in (2014) discussed
some of the important aspects of univariate Phase I analysis and reviewed some of the
recent developments in this field. The aforementioned reviews show that most Phase I
control charts are based on the assumption of normally distributed observations
or that the underlying process follows some parametric distribution. The statistical
properties of commonly employed Phase I control charts, such as the Shewhart-
type (Shewhart 1939) and the CUSUM-type (Page 1955) control charts or the charts
based on binary and multiple segmentation (see Sullivan and Woodall in (1996) and
Sullivan in (2002)), are exact only if the normality assumption is satisfied.

Moreover, the ability of parametric Phase I control charts to correctly distin-
guish between IC and OC observations is directly connected to the correct specification of the IC probability model. However, during Phase I, little information

on IC distribution is available to practitioners. When distributional assumptions underlying a parametric control chart are not satisfied, or cannot be verified, the performance and sensitivity of parametric Phase I methods deteriorate, as pointed out by Capizzi and Masarotto in (2017). It is worth to be noted that due to the fact that the underlying distributional form cannot be checked (because the process may not be stable) further complications arise in Phase I, as discussed by Woodall in (2000). Thus, several researchers (see, for example, Chakraborti et al. in (2009) and Jones-Farmer et al. in (2014)) recommended verifying the form of the underlying IC distribution only after process stability has been established using a suitable distribution-free control chart. Further, in Phase I, it is very common for practitioners to face the objective difficulty in distinguishing points coming from an OC distribution from those coming from either skewed or heavy-tailed IC distributions, as emphasized by Capizzi and Masarotto in (2013).

Conclusively, most SPC applications assume that the quality of a process can be adequately represented by the distribution of a quality characteristic, and the IC and OC distributions are the same with only differing parameters. However, in many nonmanufacturing applications, the underlying process distribution is not normal and usually unknown, hence, statistical properties of the most commonly used charts are violated. It is also important to note that most Phase I statistical methods can only be applied with subgrouped observations. However, process-quality characteristics are often observed and recorded as individual observations ($n = 1$). There is little guidance in the literature on the benefits or drawbacks analyzing Phase I data as individuals (Jones-Farmer et al. 2014). When individual data are used, the central limit theorem cannot be invoked, and as a result the statistical properties of such charts are highly affected (even for slight deviations from the specified parametric model) (Capizzi and Masarotto 2013).

Given all the aforementioned concerns, nonparametric control charts appear to be ideal candidates for such Phase I applications. Some of them could even be distribution-free in the sense that distributions of their charting statistics do not depend on the true process distribution. Qiu in (2018) pointed out that the terminologies "nonparametric control charts" and "distribution-free control charts" are both referred to charts that can be used when the process distribution does not have a parametric form. In order to make a clear distinction between these two terminologies, it is worth to be noted that "nonparametric control charts" may not be distribution-free, in the sense that their design may still depend on the process distribution (although a parametric form for the process distribution is not required). Jones-Farmer et al. in (2009), Jones-Farmer and Champ in (2010), and Graham et al. in (2010), among others, presented nonparametric Shewhart-type control charts for Phase I applications. In particular, these Phase I univariate nonparametric control charts have been developed for monitoring either the location (see Jones-Farmer et al. 2009; Graham et al. 2010) or the scale of a continuous variable (see Jones-Farmer and Champ 2010). Further, these nonparametric Shewhart-type control charts cannot be implemented in practical situations where only individual observations can be gathered in Phase I.

Little has been done in developing nonparametric Phase I control charts especially for individual observations that are prevalent in nonmanufacturing applications (Ning et al. 2015). Recently, Ning et al. in (2015) developed a new nonparametric Phase I control chart (for individual observations) whose construction is essentially based on the empirical likelihood ratio test. However, the proposed approach is designed for monitoring only the location parameter. Nonparametric Phase I control charts for monitoring both location and scale have been introduced by Capizzi and Masarotto (see Capizzi and Masarotto 2013; Capizzi and Masarotto 2018) and implemented in practical situations where either individual or subgrouped observations are gathered in Phase I. More specifically Capizzi and Masarotto in (2013) they introduced a new distribution-free strategy based on a recursive segmentation and permutation approach (RS/P) for detecting shifts in process location and/or scale during Phase I. The RS/P procedure is advantageous since it easily implementable with individual (or subgrouped) observations and assuming no prior knowledge of the underlying process. Further, the idea of using a permutation approach to hypothesis testing seems to be unexplored for Phase I analysis in SPC. According to Capizzi's and Masarotto's derived simulation results, the RS/P approach leads to an effective Phase I distribution-free procedure. However, Schmid in (2015) discusses that the RS/P approach of Capizzi and Masarotto illustrates the importance of change-point analysis, however, it also reflects its limits. Schmid in (2015) points out that it would be very interesting to know how well it really identifies the positions of the change points. Therefore, in this chapter, the ability of RS/P method to detect the true (and correct amount of) change points is tested through benchmarking (by conducting a retrospective analysis of real epidemiological time series), and the RS/P approach is compared in terms of its performance with some competitors.

3   Change-Point Analysis-Based Distribution-Free Control Charts

Let $x_1, x_2, \ldots, x_m$, be m independent observations collected from the distribution of a characteristic X, either continuous or discrete. When the process is IC, these observations are assumed to be independent with an unknown but common cumulative distribution function (c.d.f.), $F_0(x)$, whereas the OC state can be described by the following multiple change-point model:

$$x_i \sim \begin{cases} F_0(x) & \text{if } 0 < i \leq \tau_1, \\ F_1(x) & \text{if } \tau_1 < i \leq \tau_2, \\ \vdots & \vdots \\ F_k(x) & \text{if } \tau_k < i \leq m, \end{cases} \tag{1}$$

where $0 < \tau_1 < \tau_2 < \cdots < \tau_k < m$ denote k change points, $F_r(\cdot), r = 0, \ldots, k$, are unknown c.d.f. which, at one or several times, may shift in position and the shift times are also assumed to be unknown. Model (1) includes a variety of OC situations and can describe processes subject to step, and even transient shifts. This Phase I analysis procedure provides a statistical test for verifying the hypothesis system

$$\begin{cases} H_0 : & \text{the process was IC } (k = 0) \\ H_1 : & \text{the process was OC } (k > 0), \end{cases} \tag{2}$$

and some graphical diagnostic aids useful for identifying the time and the type of the changes when the hypothesis of an IC process is rejected. It is worth to be noted here that the aforementioned hypothesis testing system (performed in Phase I) requires the specification of a nominal False Alarm Probability (FAP). Here, the typical difference between parametric and nonparametric tests takes place. A parametric test can guarantee a prescribed FAP only if $F_0(\cdot)$ belongs to a particular family of probability distribution, whereas a distribution-free or a nonparametric procedure enables controlling the FAP without knowing the specific distribution from which individual observations (or samples) are drawn. Following the RS/P approach of Capizzi and Masarotto in (2013), choosing an acceptable FAP value, say α (that is, a probability of false alarm equals α), then p-values are computed using a permutation approach and can be used to test the stability over time of the level and scale parameters. In applications, it is important to check the stability of the level and the scale of the quality characteristic under study. This corresponds to the traditional practice of using two Shewhart-type control charts, with one chart designed to detect mean changes and the other designed to detect scale changes. Capizzi and Masarotto in (2013) also designed two control charts to detect separately location and scale shifts and the FAP is "balanced" between the separate control charts. In particular, if either one or both p-values are less than α, an alarm is signaled. Indeed, Capizzi and Masarotto in (2013) adjusted p-values so that comparing both p-values with the same threshold α results in an overall $FAP \leq \alpha$. Following the RS/P approach of Capizzi and Masarotto in (2013), three distinct steps need to be executed for combined level-and scale-changes detection, i.e., 1. Detection of single or multiple level shifts; 2. Detection of scale changes; 3. Adjustment of the p-values obtained during the previous two steps.

3.1 Detection of Single or Multiple Level Shifts

Let us consider the problem of testing the null hypothesis that the process was IC against the alternative hypothesis that the process mean experienced an unknown number of step shifts. In such a case, a set of test (control) statistics is needed for detecting $1, 2, \ldots, K$ step shifts. Here, K denotes the maximum number of hypothetical change points. The test statistics T_k, $k = 1, \ldots, K$ are designed for testing H_0 against the alternatives

$$H_{1,k} : E(x_i) = \begin{cases} \mu_0 & \text{if } 0 < i \leq \tau_1, \\ \mu_1 & \text{if } \tau_1 < i \leq \tau_2, \\ \vdots & \vdots \\ \mu_k & \text{if } \tau_k < i \leq m. \end{cases} \tag{3}$$

The mean values $\mu_0, \ldots, \mu_k$, and the change points $0 < \tau_1 < \cdots < \tau_k < m$ are assumed to be unknown. Further, defining $\tau_0 = 0$ and $\tau_{k+1} = m$, it is also assumed that $\tau_r - \tau_{r-1} \geq l_{MIN}$, $r = 1, \ldots, k+1$, where l_{MIN} is a (user prespecified) constant giving the minimum number of successive observations allowed between two change points. Implementing the approach of Capizzi and Masarotto in (2013) for a sequence of individual observations, the control statistic and the possible change points are computed using a simple forward recursive segmentation approach. The algorithm starts with $k = 0$ and then proceeds in K successive stages. At the beginning of stage k, the interval $[1, m]$ is partitioned into k subintervals, each having a length greater or equal to l_{MIN}. At stage k, one of these subintervals is split, adding a new potential change point. The new change point is selected maximizing

$$\sum_{i=1}^{k+1} (\hat{\tau}_i - \hat{\tau}_{i-1})(\bar{x}(\hat{\tau}_{i-1}, \hat{\tau}_i) - \bar{x}_{om})^2, \tag{4}$$

conditionally on the results of the previous stages. Here $\bar{x}_{om}$ represents the overall mean (om) of observations, $\bar{x}(a, b) = \frac{1}{b-a}\sum_{i=a+1}^{b} x_i$, and $0 = \hat{\tau}_0 < \hat{\tau}_1 < \cdots < \hat{\tau}_k < \hat{\tau}_{k+1} = m$ are the boundaries of the new partition. The control statistic T_k, $k = 1, \ldots, K$ is equal to the attained maximum value of Eq. (4). The number of elementary test statistics for the case of individual observations equals K. The IC probability distribution function of $T_1, \ldots, T_K$ depends on the unknown distribution $F_0(\cdot)$, and thus is unknown as well. Let $\mathbf{X}_{(\cdot)} = (x_{(1)}, \ldots, x_{(N)})$ be the order statistic associated with $N(= m)$ Phase I observations. Then, it is well known (see Lehmann and Romano in (2005)) that under the null (IC) hypothesis

$$P\{\mathbf{X} = (\alpha_1, \ldots, \alpha_N) \mid \mathbf{X}_{(\cdot)}\} = \begin{cases} \frac{1}{N!} & \text{if } (\alpha_1, \ldots, \alpha_N) \text{ is a permutation of } \mathbf{X}_{(\cdot)}, \\ 0 & \text{otherwise.} \end{cases} \tag{5}$$

Observe that Eq. (5) does not depend on $F_0(\cdot)$. Therefore, given a test statistic, its p-value can be calculated, conditionally on $\mathbf{X}_{(\cdot)}$, as the proportion of permutations under which the statistic value exceeds or is equal to the statistic computed from the original sample of observations. It is worth to be noted here that taking into account that the number of permutations, N, can be extremely large and that, in our case, we have K elementary test statistics, we followed the permutation procedure as described by Capizzi and Masarotto in (2013) analogously treated for our case since we deal with a sequence of individual observations ($n = 1$). From now and on, the mean estimates are denoted by $\hat{\mu}_i$, $i = 1, \ldots, m$. Let now assume that choosing an acceptable FAP, say α, then, two possible cases can be distinguished here.

If p-value $\geq \alpha$, the hypothesis of a constant mean is accepted and the overall mean is used to estimate the level, i.e., we set $\hat{\mu}_i = \bar{x}_{om}$ for $i = 1, \ldots, m$, whereas if p-value $< \alpha$ the null hypothesis that the process was IC is rejected and the test signals an OC condition that can be attributed to the presence of k^* step shifts for $0 < k^* \leq K$.

3.2 Detection of Scale Changes

A modified version of the aforementioned procedure is applied for detecting scale changes in the centered observations $x_i - \hat{\mu}_i$. In this case, the objective function of Eq. (4) used for the recursive segmentation is replaced by

$$\sum_{i=1}^{k+1} (\hat{\tau}_i - \hat{\tau}_{i-1}) \log \left(\frac{s^2}{s^2(\hat{\tau}_{i-1}, \hat{\tau}_i)} \right), \tag{6}$$

where $s^2 = \frac{1}{m} \sum_{i=1}^{m} s_i^2, s^2(\alpha, b) = \frac{1}{b-\alpha} \sum_{i=\alpha+1}^{b} s_i^2$, and $s_i = x_i - \hat{\mu}_i$. The other steps of the procedure remain unchanged.

3.3 Adjustment of the p-Values Obtained During the Previous Two Steps

Since we execute a combined level-and scale-changes detection we need to adjust the p-values obtained during the previous two steps (following the Holm-Bonferroni methods in Holm (1979) and the suggestions of Capizzi and Masarotto in (2013)). Let p_{level}^* and p_{scale}^* be the original p-values and $\bar{p} = \min \left(1, 2 \min(p_{level}^*, p_{scale}^*) \right)$, then the adjusted p-values are

$$(p_{level}, p_{scale}) = \begin{cases} \left(\bar{p}, max(\bar{p}, p_{scale}^*) \right) & \text{if } p_{level}^* \leq p_{scale}^*, \\ \left(max(\bar{p}, p_{level}^*), \bar{p} \right) & \text{if } p_{level}^* > p_{scale}^*. \end{cases} \tag{7}$$

The adjusted p-values in Eq. (7) are advantageous since they are directly comparable with the desired FAP, as pointed out by Capizzi and Masarotto in (2013).

4 Empirical Study

In the modern world, the timely and accurate detection of epidemics has been recognized as an extremely important problem of epidemiological surveillance, as also evidenced by the implementation of statistical routine methods to detect outbreaks in epidemiological surveillance systems in several European countries (European Centre for Disease Prevention and Control-ECDC) and in U.S.A. (Centers for Disease Control and Prevention-CDC). In particular, the study of the evolution of the Influenza-Like Illness (ILI) syndrome is a major public health concern, since despite the fact that it belongs to the epidemiological surveillance priorities in the European region, it is also high in terms of international interest due to its potential for widespread transmission (with ILI also representing a potential pandemic risk) as discussed by Parpoula et al. in (2017). The past decades, public health practitioners have turned their attention to the field of SPC. Indeed, several proposed approaches for the detection of outbreaks of infectious diseases not only are inspired by, or related to, methods of SPC, but also quality tools are sometimes applied and adapted directly to epidemiological surveillance. However, the general epidemiological surveillance problem violates many of the assumptions that are reasonable to assume in SPC. For example, there is little to no control over disease incidence and thus the distribution of disease incidence is usually nonstationary, observations are autocorrelated, individual observations are usually being monitored, and outbreaks are sometimes transient, with disease incidence returning to its original state once an outbreak has run its course (Fricker 2011).

An effective approach to overcome these difficulties and limitations is to consider nonparametric or even distribution-free control charts which offer the benefits of not depending on the shape of the distribution and the amount of available data. Toward this end, in this empirical study, the previously described control charts (especially for individual observations) are constructed for retrospectively (Phase I) detecting single or multiple changes in location and dispersion of univariate variables (especially aiming at the detection of ILI outbreaks, and in this way adequately serving epidemiological surveillance and healthcare monitoring purposes). In particular, we perform the RS/P approach in the case of a sequence of individual observations, i.e., weekly ILI rate data, between September 29, 2014 (week40/2014) and October 2, 2016 (week39/2016). Therefore, we conduct a retrospective (Phase I) analysis for the period from 2014 to 2016, based on these two seasons' historical data. It is important to note that in correspondence with the epidemiological surveillance terminology, the retrospective analysis of historical data is referred to as Phase I, whereas the prospective monitoring of future data is referred to as Phase II.

Data values are assumed to be independent, are individual (one observation at each instant of time), and Phase I control limits are computed, so that a prescribed FAP is guaranteed without making any parametric assumptions on the stable (IC) distribution. Further, the ability of RS/P method to detect the true (and correct amount of) change points is compared, in terms of several performance evaluation metrics, with some competitors suggested in the literature, appropriately adjusted for our study, that is the use of distribution-free control limits is emphasized. Firstly, we construct control charts for which the control statistic is based on a Generalized Likelihood Ratio Test (GLRT) computed under a Gaussian assumption (see Sullivan and Woodall in (1996)); however, the control limits are computed by permutation following the suggestions of Cappizi and Masarotto in (2018), thus the resulting control charts are indeed distribution-free. Then, we construct analogous control charts based on a rank transformation of the original observations, i.e., rank-based GLRT control charts. This rank transformation is applied to possibly improve the performance of control charts in the case of nonnormal data. In the latter case, the control limits are also computed by permutation and the resulting control charts are also distribution-free. It is important to note here that the decomposition of the likelihood ratio test statistic suggested by Sullivan and Woodall in (1996) for diagnostic purposes is also computed and plotted in both cases (with and without a rank transformation).

4.1 RS/P-Based Phase I Analysis

Capizzi and Masarotto in (2013) suggest that $L = 1000$ random permutations of the data are sufficient to get stable and accurate enough estimates of the p-value to be used for hypothesis testing when the usual significance levels (e.g., $\alpha = 0.01$, 0.05, and 0.1) are used. In this study, $L = 100,000$ permutations are adapted allowing us to maintain a relatively low complexity of the algorithm due to the relatively small dimension of our data (a total of 105 observations). Following the Capizzi's and Masarotto's suggestions in Capizzi and Masarotto (2013), the maximum number of hypothetical change points $K = \max\left(3, \min\left(50, \left[\frac{m}{15}\right]\right)\right)$, and the minimum number of successive observations allowed between two change points $l_{MIN} = 5$, since this constant value was found to reduce the IC variability of the test statistics without hindering the power to detect short, transient OC phases (Capizzi and Masarotto 2013). Figure 1 illustrates the application of the R/SP-based Phase I analysis to each influenza period under study (first period: week40/2014–week39/2015 and second period: week40/2015–week39/2016). Assuming that a probability of false alarm

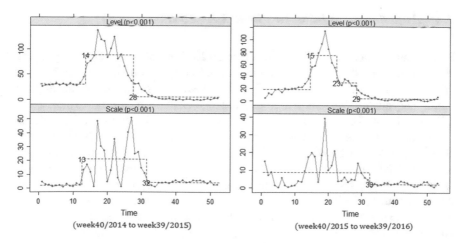

Fig. 1 RS/P-based Phase I analysis of ILI rate data (week40/2014–week39/2016)

equal to $\alpha = 0.05$ is acceptable, our procedure signals both a possible change of the mean (p-value < 0.001 for a change in level) and of scale (p-value < 0.001 for a change in scale). Here, graphics can be used for diagnostic purposes. The solid line in the upper panels shows the observed ILI rate values, whereas the solid line in the lower panels represents the s_i of the process, for each influenza period under study.

The behavior of the dashed line in the upper panels representing an estimate, $\hat{\mu}_i$, of the time-varying process mean, suggests an increase in ILI rate after the first 13 weeks and a decrease after the 27th week, for the *first period* under study (a case of "two-steps" level change); it also suggests an increase in ILI rate after the first 14 weeks, a partial decrease after the 22nd week as well a significant decrease after the 28th week, for the *second period* under study (a case of "three-steps" level change). Note here that in the latter case, from an epidemiological perspective, the RS/P approach also succeeded to identify short, transient shifts ($\hat{\tau}_{23}$: week09/2016 to $\hat{\tau}_{28}$: week14/2016) since the number of successive observations between these change points is almost equal to the minimum number of observations allowed between two change points. In the lower panels, the behavior of the dashed line representing an estimate, $\hat{s}_i$, of the time-varying process scale, suggests a dispersion shift starting on week 13, and no additional shift is detected after week 31, for the *first period* under study; it also suggests that no scale changes are signaled for the first 32 weeks, and a dispersion shift is detected from week 33, for the *second period* under study. Conclusively, for the *first period*, the hypothesis of a stable process is not rejected from week40/2014 to week52/2014 and from week15/2015 to week39/2015, whereas location or/and dispersion shifts are detected from week01/2015 to week14/2015, whereas for the *second period*, the hypothesis of a stable process is not rejected from week40/2015 to week53/2015 and from week09/2016 to week39/2016, whereas location or/and dispersion shifts are detected from week01/2016 to week08/2016.

4.2 Adjusted GLRT-Based Phase I Analysis

Sullivan and Woodall in (1996) and Sullivan in (2002) introduced preliminary control charts based on individual observations, for detecting a shift either in the mean or variance, or both, developed from a Likelihood Ratio Test (LRT), and developed a preliminary analysis based on a time-ordered segmentation of the original data. These LRT-based control charts offer the advantage of improved detection of OC conditions compared to combined X and moving range control charts, as well as CUSUM charts. However, these LRT-based control charts are exact only if the normality assumption is satisfied. In such cases, a normal transformation of the observations is certainly feasible; however, it will be done at the expense of the interpretability of the analysis that is particularly important to control chart users in nonmanufacturing applications.

For these reasons, in this empirical study, adjusted GLRT-based distribution-free control charts are developed for detecting a step shift either in the mean or variance (or both) of a sequence of individual observations. In particular, we adopt a change-point approach by constructing preliminary control charts based on individual observations, developed from a LRT, for verifying whether the mean and dispersion are the same before and after τ, with permutation-based limits. Further, we compute the LRT statistics, with or without a rank transformation, for $\tau = 2, \ldots, m$, and control limits proportional to the in-control mean of the LRT statistics are used, as suggested by Sullivan and Woodall in (1996). Further, the control statistics divided by the time-varying control limits are plotted with a "pseudo-limit" equal to one, and the decomposition of the LRT statistics (i.e., decomposition of the control statistics in the two parts due to changes in the mean and dispersion, respectively) for diagnostic purposes is also computed and plotted. It is worth to be noted that following the approach of Capizzi and Masarotto in (2018), the control limits are computed by permutation so that the desired FAP is guaranteed. In particular, the adjusted GLRT-based distribution-free control charts have been designed evenly balancing the FAP values of the location- and scale-control schemes, i.e., the single schemes have the same probability of giving a false alarm and the probability that at least one of the charts signal is equal to the specified FAP. Thus, the resulting adjusted GLRT-based control charts (with permutation-based control limits) are distribution-free.

4.2.1 GLRT Approach

Figure 2 illustrates the application of the adjusted GLRT-based Phase I analysis (without transformation) to each influenza period under study. The control statistics are based on a GLRT computed under a Gaussian assumption; however, the control limits are computed by permutation ($L = 100,000$), and an alarm is signaled if the statistics are greater than a positive control limit. Assuming that a probability of false alarm equal to $\alpha = 0.05$ is acceptable, the upper panels show the control statistics for verifying the presence of a shift either in the location or dispersion. Since for many weeks (in both periods), the values are greater than the permutation-based control

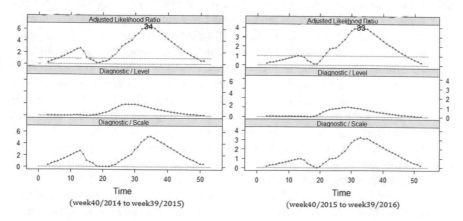

Fig. 2 GLRT-based Phase I analysis of ILI rate data (week40/2014–week39/2016)

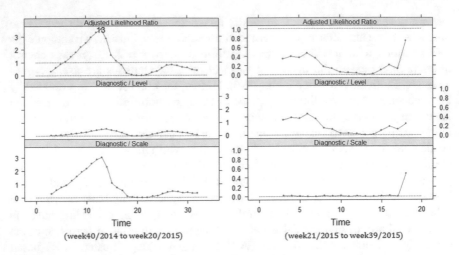

Fig. 3 First round of recursive GLRT approach (First period)

limit, the hypothesis of a stable process is rejected, and, in particular, the graph points to a possible shift on week **34** (week21/2015) and on week **33** (week19/2016), for the *first* and *second* period under study, respectively. The middle and lower panels show the decomposition of the control statistics in the two parts due to changes in the location and dispersion, respectively. These diagnostic graphs clearly point to a shift in the dispersion for the employed ILI rate data in this study (Fig. 2).

Having divided the observations into two subperiods, that is, before and after week **34** (week21/2015) for the *first period*, as well as before and after week **33** (week19/2016) for the *second period*, it is also useful to see if there is evidence of other shifts *within these periods*. First round (see Fig. 3) of recursive application of GLRT approach for the *first period*, suggests that another dispersion shift was probably present starting on week **13** (week52/2014), and no additional shift is detected after week 33 (week20/2015), whereas second round (see Fig. 4) suggests that the process was stable up to week 12 (week51/2014), but another location shift was probably present starting on week **14** (week13/2015). Conclusively, for the *first period*, the hypothesis of a stable process cannot be rejected from week 40/2014 to week51/2014 and from week13/2015 to week39/2015, whereas location or/and dispersion shifts are detected from week52/2014 to week12/2015. First round (see Fig. 5) of recursive application of GLRT approach for the *second period*, suggests that another dispersion shift was probably present starting on week **13** (week52/2015), and no additional shift is detected after week 32 (week18/2016), whereas second round (see Fig. 6) suggests that the process was stable up to week 12 (week51/2015), but another location shift was probably present starting on week **11** (week09/2016). Conclusively, for the *second period*, the hypothesis of a stable process is not rejected from week40/2015 to week51/2015 and from week 09/2016 to week39/2016, whereas location or/and dispersion shifts are detected from week52/2015 to week08/2016.

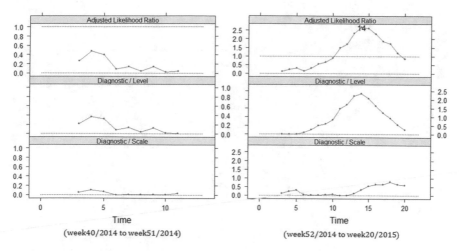

Fig. 4 Second round of recursive GLRT approach (First period)

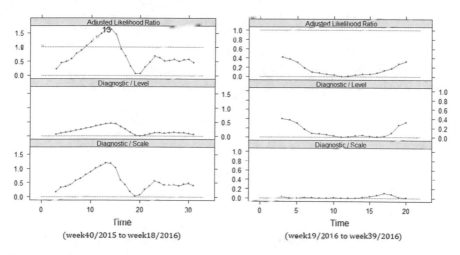

Fig. 5 First round of recursive GLRT approach (Second period)

4.2.2   Rank-Based GLRT Approach

A rank transformation of the original data is now used to possibly improve the performance of GLRT-based control charts (as in our case of nonnormal data). Figure 7 illustrates the application of the adjusted rank-based GLRT-based Phase I analysis (with transformation) to each influenza period under study. The upper panels show the control statistics for verifying the presence of a shift either in the location or dispersion. Since for many weeks (in both periods) the values are greater than the permutation rank-based control limit, the hypothesis of a stable process is rejected, and, in particular, the graph points to a possible shift on week **30** (week 17/2015) and

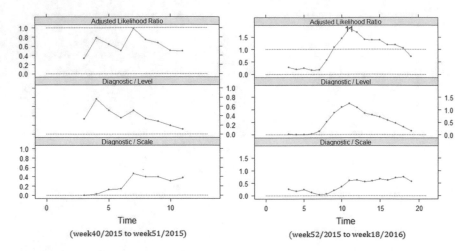

Fig. 6 Second round of recursive GLRT approach (Second period)

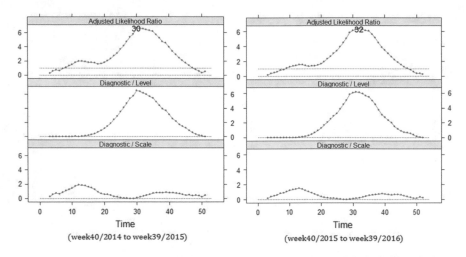

Fig. 7 Rank-based GLRT-based Phase I analysis of ILI rate data (week40/2014–week39/2016)

on week **32** (week18/2016), for the *first* and *second* period under study, respectively. The middle and lower panels show the decomposition of the control statistics in the two parts due to changes in the location and dispersion, respectively. The rank-based diagnostic graphs clearly point to a shift in the location for the employed ILI rate data in this study (Fig. 7).

Having divided the observations into two subperiods, that is, before and after week **30** (week17/2015) for the *first period*, as well as before and after week **32** (week18/2016) for the *second period*, it is also useful to see if there is evidence of other shifts *within these periods*. First round (see Fig. 8) of recursive application of rank-based GLRT approach for the *first period*, suggests that another

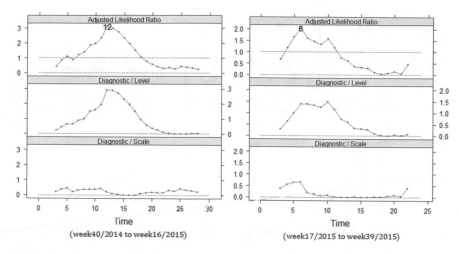

Fig. 8 First round of recursive rank-based GLRT approach (First period)

location shift was probably present starting on week **12** (week51/2014) and on week **6** (week22/2015), whereas second round (see Fig. 9) suggests that the process was stable up to week 11 (week2014/50), but another location shift was probably present starting on week **16** (week14/2015), and no other shift was detected from week22/2015 and onwards. Conclusively, for the *first period*, the hypothesis of a stable process is not rejected from week40/2014 to week50/2014 and from week14/2015 to week39/2015, whereas location or/and dispersion shifts are detected from week51/2014 to week13/2015. First round (see Fig. 10) of recursive application of rank-based GLRT change point method for the *second period*, suggests that another location shift was probably present starting on week **11** (week50/2015), and no additional shift is detected after week 31 (week17/2016), whereas second round (see Fig. 11) suggests the process was stable up to week 10 (week49/2015), but another location shift was probably present starting on **17** (week13/2016). Conclusively, for the *second period*, the hypothesis of a stable process is not rejected from week40/2015 to week49/2015 and from week13/2016 to week39/2016, whereas location or/and dispersion shifts are detected from week50/2015 to week12/2016.

5 Comparative Performance Evaluation

In this section we proceed with a comparative study, in order to evaluate the performance of the developed change-point analysis-based control charting methods in terms of retrospectively (Phase I) detecting outbreaks with scientific and methodological adequacy (from both a statistical and an epidemiological perspective). As discussed earlier, Schmid in (2015) pointed out that the RS/P approach of Capizzi and Masarotto in (2013) illustrates the importance of change-point analysis,

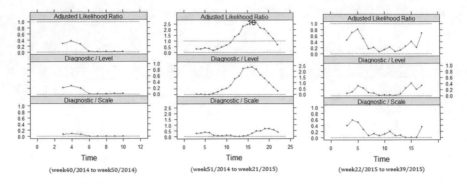

Fig. 9 Second round of recursive rank-based GLRT approach (First period)

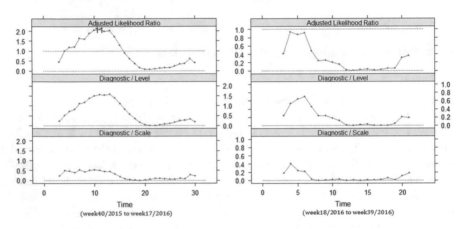

Fig. 10 First round of recursive rank-based GLRT approach (Second period)

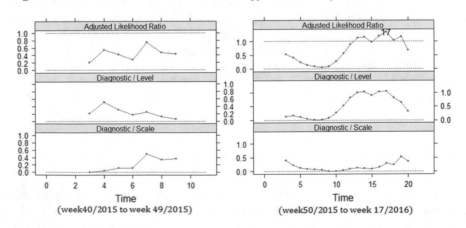

Fig. 11 Second round of recursive rank-based GLRT approach (Second period)

however, it also reflects its limits since it would be very interesting to know how well it really identifies the positions of the change points. Therefore, in this comparative analysis, the ability of RS/P method to detect the true (and correct amount of) change points is tested through benchmarking. The RS/P approach is compared, in terms of several performance evaluation metrics, with its competitors, i.e., the GLRT as well as the rank-based GLRT approaches. More specifically, the derived change points from RS/P, GLRT and rank-based GLRT approaches are compared with those derived after executing the "gold standard" approach to influenza surveillance (CDC and ECDC flu detection algorithm, that is, Serfling's model).

The current approach to influenza surveillance is based on Serfling's method (Serfling 1963), i.e., a commonly used fully parametric regression model for outbreak detection (based on a trigonometric function with linear trend, assuming Gaussian white noise errors in order to model historical baselines) that can be described by

$$X(t) = \alpha_0 + \alpha_1 t + \gamma_1 \cos \left(\frac{2\pi t}{m} \right) + \delta_1 \sin \left(\frac{2\pi t}{m} \right) + \varepsilon(t), \qquad (8)$$

where $X(t)$ are the observed time-series values (weekly ILI rate), $\varepsilon(t)$ are centered zero-mean random variables with variance σ^2, m denotes the number of observations within one year, and model coefficients are estimated by least squares regression. Parpoula et al. in (2017) developed extended Serfling-type periodic regressions models, and then compared their performance to typical forecasting models (including Serfling's method). The aforementioned procedure allowed Parpoula et al. in (2017) to extract the signaled start weeks (sw) and End Weeks (ew) of the epidemics, i.e., sw01-ew13/2015 and sw01-ew08/2016 for the first and second period under study, respectively. It is important to note that these signaled start and end weeks of the epidemics were extracted considering either Serfling's model or an extended Serfling's model (identified as the optimal one). The interested reader may refer to Parpoula et al. in (2017) for more details.

Therefore, we then examine the ability of the RS/P, GLRT and rank-based GLRT approaches to detect the true change points, using Receiver Operating Characteristic (ROC) curve analysis and its related metric, namely the Area under the ROC curve (AUC). The ROC curve is a visual illustration of the success and error observed in a classification model. The AUC is a measure of predictive accuracy of a classification model and its score can be interpreted as either the average value of sensitivity for all possible values of specificity or as the average value of specificity for all possible values of sensitivity, and is always bounded between zero and one (there is no realistic classifier with an AUC lower than 0.5). For more detailed discussions on the theory and practice of ROC curves, the interested reader may refer to, among others, Pepe in (2003) and Zhou et al. in (2011). Here, we present only some discussion of the details of calculation of the ROC curves and its related statistics/metrics particular to our problem. In our study, we performed pairwise comparison of ROC curves for all approaches examined, that is, RS/P, GLRT, and rank-based GLRT. In particular, we implemented the method of DeLong et al. in (1988), as well as of Hanley and McNeil in (1982) and (1983), for the calculation of the Standard Error (SE) of the estimated

AUCs, and we also estimated the exact binomial Confidence Interval (CI) for each derived AUC. Table 1 presents the estimated AUCs, the SEs, and the 95% CIs. Tables 2 and 3 present the pairwise comparison of ROC curves, and Fig. 12 displays the ROC curves that summarize the accuracy of the RS/P, GLRT and rank-based GLRT approaches for detection of outbreaks. Note here that the aforementioned calculations and ROC graphs are obtained for each employed method as well as for both periods under study.

Table 1 Estimated AUC, SE, 95% CI for RS/P, GLRT and rank-based GLRT approaches

Period[1]	Method	AUC[2]	AUC[3]	SE[2]	SE[3]	95% CI[2]	95% CI[3]
First	RS/P	0.987	0.987	0.0128	0.0151	0.908–1.000	0.908–1.000
First	GLRT	0.949	0.949	0.0405	0.0457	0.849–0.991	0.849–0.991
First	Rank-based GLRT	0.974	0.974	0.0179	0.0213	0.887–0.999	0.887–0.999
Second	RS/P	0.989	0.989	0.0111	0.0133	0,.912–1.000	0.912–1.000
Second	GLRT	0.978	0.978	0.0155	0.0189	0.894–0.999	0.894–0.999
Second	Rank-based GLRT	0.911	0.911	0.0288	0.0393	0.800–0.972	0.800–0.972

[1] First period: week40/2014–week39/2015, Second period: week40/2015–week39/2016
[2] DeLong et al. (1988)
[3] Hanley and McNeil (1982)

Table 2 Pairwise comparison of ROC curves (First period: week40/2014–week39/2015)

RS/P ∼ GLRT	DeLong et al. (1988)	Hanley and McNeil (1983)
Difference between AUCs	0.0385	0.0385
SE	0.0426	0.0480
95% CI	−0.0451 to 0.122	−0.0557 to 0.133
Z-statistic	0.902	0.801
Significance level	P = 0.3669	P = 0.4233
RS/P ∼ Rank-based GLRT	DeLong et al. (1988)	Hanley and McNeil (1983)
Difference between AUCs	0.0128	0.0128
SE	0.0224	0.0261
95% CI	−0.0311 to 0.0567	−0.0383 to 0.0639
Z-statistic	0.572	0.491
Significance level	P = 0.5671	P = 0.6231
GLRT ∼ Rank-based GLRT	DeLong et al. (1988)	Hanley and McNeil (1983)
Difference between AUCs	0.0256	0.0256
SE	0.0405	0.0460
95% CI	−0.0538 to 0.105	−0.0646 to 0.116
Z-statistic	0.632	0.557
Significance level	P = 0.5271	P = 0.5775

Table 3 Pairwise comparison of ROC curves (Second period: week40/2015-week39/2016)

RS/P ~ GLRT	DeLong et al. (1988)	Hanley and McNeil (1983)
Difference between AUCs	0.0111	0.0111
SE	0.0194	0.0231
95% CI	−0.0269 to 0.0491	−0.0342 to 0.0564
Z-statistic	0.573	0.480
Significance level	P = 0.5666	P = 0.6310
RS/P ~ Rank-based GLRT	DeLong et al. (1988)	Hanley and McNeil (1983)
Difference between AUCs	0.0778	0.0778
SE	0.0273	0.0403
95% CI	0.0242 to 0.131	−0.00126 to 0.157
Z-statistic	2.847	1.929
Significance level	**P = 0.0044**	P = 0.0538
GLRT ~ Rank-based GLRT	DeLong et al. (1988)	Hanley and McNeil (1983)
Difference between AUCs	0.0667	0.0667
SE	0.0256	0.0410
95% CI	0.0164 to 0.117	−0.0137 to 0.147
Z-statistic	2.602	1.626
Significance level	**P = 0.0093**	P = 0.1040

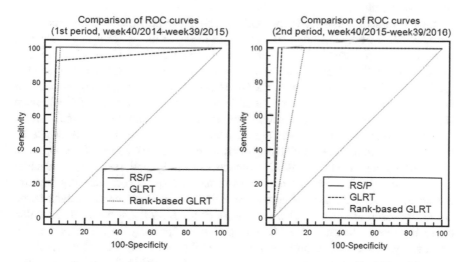

Fig. 12 Comparative ROC curve analysis

Table 1 and Fig. 12 indicate that RS/P approach was found to be superior, for both periods under study, compared to GLRT and rank-based GLRT approaches, in terms of higher AUC and smaller SE values. However, Tables 2 and 3 indicate that statistically significant differences between AUCs were identified as regards the RS/P and rank-based GLRT approaches (p-value = 0.0044) as well as the GLRT and rank-based GLRT approaches (p-value = 0.0093), only for the second period under study. Therefore, in order to clarify if the RS/P approach does indeed compare favorably with its competitors (GLRT and rank-based GLRT approaches), we further executed an additional comparative analysis, taking now into account several commonly accepted accuracy measures of a diagnostic test (Zhou et al. 2011), that is, sensitivity, specificity, and accuracy. Sensitivity (SENS) is the probability that a test result will be positive when the epidemic is present (true positive rate, larger-the-better), specificity (SPEC) is the probability that a test result will be negative when the epidemic is not present (true negative rate, larger-the-better), and accuracy (ACC) is the overall probability that a case will be correctly classified (larger-the-better). We estimated these metrics (expressed as percentages) along with their exact Clopper–Pearson 95% CIs (see Clopper and Pearson in (1934)) for each method examined as well as for both periods under study, as shown in Table 4. Further, we computed the False Alarm Rate (FAR), that is, the probability of false detection, defined as the number of negative events wrongly categorized as positive (false positives-false alarms) per the total number of actual negative events regardless of classification (smaller-the-better).

Table 4 indicates that RS/P approach was found to be superior compared to its competitors, for both periods under study, in terms of all performance evaluation metrics considered, that is, higher (or equal in some cases) ACC, SENS, and SPEC values. Further, the RS/P approach was also found to be superior in terms of smaller (or equal) FAR values compared to the GLRT and rank-based GLRT approaches, for both periods under study. It is important to note here that only the RS/P approach shows a quite robust and satisfactory performance for both periods under study, thus it can be recommended to be used in Phase I applications in which there is no prior information on the IC underlying distribution. The rank-based GLRT approach seems to be dominated by the GLRT approach, especially for the second period under study, despite the fact that the rank transformation was used to possibly improve the performance in the case of nonnormal data. The FAR values of the rank-based GLRT approach are much higher than those of the GLRT approach. In particular, the FAR value of the rank-based GLRT approach, especially for the second period under study, seems unacceptable for practical applications, whereas GLRT approach seems to approximately guarantee an acceptable FAR value. The aforementioned issues reveal that the "best" chart based on GLRT strongly depends on the different shift patterns, and the GLRT-based approaches could be recommended to be used only when the shape of the distribution is known.

Therefore, the derived results provide sufficient enough evidence that the RS/P approach seems to detect successfully the "true" and "correct" amount of change points (recall here that RS/P change points are compared with those derived after executing the "gold standard" CDC and ECDC flu detection algorithm), was found

Table 4 Metrics of RS/P, GLRT and rank-based GLRT approaches

First period: week40/2014–week39/2015

Metric	RS/P	GLRT	Rank-based GLRT
SENS (95% CI)	100.0% (75.29–100.0%)	92.31% (63.97–99.81%)	100.0% (75.29–100.0%)
SPEC (95% CI)	97.44% (86.52–99.94%)	97.44% (86.52–99.94%)	94.87% (82.68–99.37%)
ACC (95% CI)	98.08% (89.74–99.95%)	96.15% (86.79–99.53%)	96.15% (86.79–99.53%)
FAR	0.025	0.025	0.051

Second period: week40/2015–week39/2016

Metric	RS/P	GLRT	Rank-based GLRT
SENS (95% CI)	100.0% (63.06–100.0%)	100.0% (63.06–100.0%)	100.0% (63.06–100.0%)
SPEC (95% CI)	97.78% (88.23–99.94%)	95.56% (84.85–99.46%)	82.22% (67.95–91.99%)
ACC (95% CI)	98.11% (89.93–99.95%)	96.23% (87.02–99.54%)	84.91% (72.41–93.25%)
FAR	0.022	0.044	0.177

to be the "best" scheme in the presence of multiple change points, as in the case of "two-" and "three-steps" patterns, and can be considered as an ideal retrospective (Phase I) change-point analysis-based control charting method for outbreak detection. However, the most important feature of RS/P is that it is a distribution-free approach; that is, it is able to guarantee a prescribed FAP without any knowledge about the IC underlying distribution. This feature is particularly relevant and useful for any Phase I application since it is well known that it is not possible to check a distributional assumption before verifying the stability of a nonmanufacturing process, and the use of a misspecified IC distribution can result in an unacceptably high false alarm probability (Woodall 2000).

6 Concluding Remarks

The SPC techniques have a long history of application to nonmanufacturing industries such as healthcare monitoring, among others. One of the most common challenges in nonmanufacturing control chart Phase I applications is that the underlying process distribution of many of the nonmanufacturing quality characteristics is not normal and usually unknown, hence, statistical properties of the most commonly used charts are highly affected. Given these concerns, nonparametric or even distribution-free control charts appear to be ideal candidates for such Phase I applications. However, most of the existing nonparametric control charts are designed for Phase II monitoring, and most Phase I statistical methods can only be applied with sub-grouped observations. However, nonmanufacturing process-quality characteristics are often observed and recorded as individual observations, and little has been done in developing nonparametric Phase I control charts especially for individual observations that are prevalent in nonmanufacturing applications. Toward this end, in this chapter, change-point analysis-based distribution-free control charts designed for Phase I applications (especially for individual observations) were constructed for retrospectively detecting single or multiple changes in location and dispersion of univariate variables, and some of the statistical issues involved in the construction and evaluation of such control charts were discussed. A nonmanufacturing real example was included to unfold the capabilities of the developed methodologies. The interpretation of these charts, from a statistical (and an epidemiological) perspective, facilitated the better understanding of process variability and highlighted important issues for an SPC practitioner (and a healthcare practitioner).

The empirical comparative study provided sufficient evidence that RS/P control charting method was found to be superior compared to its competitors in terms of several performance evaluation metrics considered, and seems to detect successfully the true change points compared to a "gold standard" approach. The GLRT-based approaches were found to strongly depend on different shift patterns, and could be recommended to be used only when the shape of the distribution is known (since the probability of false detection in some cases was found to be unacceptable for practical applications). It is important to note here that only the RS/P approach indicated a quite robust and satisfactory performance, thus it could be recommended to be used in Phase I applications in which there is no prior information on the IC underlying distribution. However, despite these appealing properties, as its competitors, RS/P, requires the independence of the observations (at least in the IC state). Hence, more research is still needed to account for autocorrelation in historical process data.

Conclusively, the present chapter provided general recommendations to serve critical needs of Phase I nonmanufacturing applications (especially for individual observations). It is worth to be noted that the process knowledge and insights gained from Phase I data and analysis can be too important for one to move quickly into Phase II (even if this is possible in theory). A Phase I analysis should encompass more than simply applying a control chart to data to determine which observations are in control, such as including visualization of the process data as well as applying statistical methods in data in order to gain richer insights into the process and determine the appropriate model for process improvement and monitoring, and in this sense, the current study, through benchmarking among competing change-point analysis-based distribution-free control charts and conducting a retrospective analysis of real time-series data, is filling up the gap in the relevant state-of-the-art literature.

Acknowledgements The work was carried out at the Lab of Statistics and Data Analysis of the University of the Aegean. The authors would like to thank the Department of Epidemiological Surveillance and Intervention of the Hellenic Center for Disease Control and Prevention (HCDCP) for providing the Influenza-Like Illness (ILI) rate data, collected weekly through the sentinel surveillance system. The authors would also like to thank the anonymous reviewer(s) for their valuable comments and suggestions for improving the clarity and presentation of the chapter.

References

Capizzi, G., & Masarotto, G. (2012). Adaptive generalized likelihood ratio control charts for detecting unknown patterned mean shifts. *Journal of Quality Technology, 44*, 281–303.

Capizzi, G., & Masarotto, G. (2013). Phase I distribution-free analysis of univariate data. *Journal of Quality Technology, 45*, 273–284.

Capizzi, G., & Masarotto, G. (2017). Phase I distribution-free analysis of multivariate data. *Technometrics, 59*, 484–495.

Capizzi, G., & Masarotto, G. (2018). Phase I distribution-free analysis with the R package dfphase1. *Frontiers in Statistical Quality Control, 12*, 3–19.

Chakraborti, S., Eryilmaz, S., & Human, S. W. (2008). A Phase II non parametric control chart based on precedence statistics with runs-type signaling rules. *Computational Statistics and Data Analysis, 53*, 1054–1065.

Chakraborti, S., & Graham, M. A. (2019). Nonparametric (distribution-free) control charts: An updated overview and some results. *Quality Engineering*. https://doi.org/10.10180/08982112.2018.1549330.

Chakraborti, S., Human, S. W., & Graham, M. A. (2009). Phase I statistical process control charts: An overview and some results. *Quality Engineering, 21*, 52–62.

Chatterjee, S., & Qiu, P. (2009). Distribution-free cumulative sum control chart using bootstrap-based control limits. *Annals of Applied Statistics, 3*, 349–369.

Clopper, C., & Pearson, E. S. (1934). The use of confidence or fiducial limits illustrated in the case of the binomial. *Biometrika, 26*, 404–413.

DeLong, E. R., DeLong, D. M., & Clarke-Pearson, D. L. (1988). Comparing the areas under two or more correlated receiver operating characteristic curves: A nonparametric approach. *Biometrics, 44*, 837–845.

Fricker, R. D. (2011). Methodological issues in biosurveillance. *Statistics in Medicine, 30*, 403–415.

Graham, M. A., Human, S. W., & Chakraborti, S. (2010). A phase I nonparametric Shewhart-type control chart based on the median. *Journal of Applied Statistics, 37*, 1795–1813.

Hanley, J. A., & McNeil, B. J. (1982). The meaning and use of the area under a receiver operating characteristic (ROC) curve. *Radiology, 143*, 29–36.

Hanley, J. A., & McNeil, B. J. (1983). A method of comparing the areas under receiver operating characteristic curves derived from the same cases. *Radiology, 148*, 839–843.

Hawkins, D. M., & Deng, Q. (2010). A nonparametric change-point control chart. *Journal of Quality Technology, 42*, 165–173.

Holm, S. (1979). A simple sequentially rejective multiple test procedure. *Scandinavian Journal of Statistics, 6*, 65–70.

Jensen, W. A., Jones-Farmer, L. A., Champ, C. W., & Woodall, W. H. (2006). Effects of parameter estimation on control chart properties: A literature review. *Journal of Quality Technology, 38*, 349–364.

Jones-Farmer, L. A., & Champ, C. W. (2010). A distribution-free Phase I control chart for subgroup scale. *Journal of Quality Technology, 42*, 373–387.

Jones-Farmer, L. A., Jordan, V., & Champ, C. W. (2009). Distribution-free Phase I control charts for subgroup location. *Journal of Quality Technology, 41*, 304–317.

Jones-Farmer, L. A., Woodall, W. H., Steiner, S. H., & Champ, C. W. (2014). An overview of Phase I analysis for process improvement and monitoring. *Journal of Quality Technology, 46*, 265–280.

Lehmann, E. L., & Romano, J. P. (2005). *Testing Statistical Hypotheses* (3rd ed.). New York: Springer.

Liu, L., Zou, C., Zhang, J., & Wang, Z. (2013). A sequential rank-based nonparametric adaptive EWMA control chart. *Communications in Statistics: Simulation and Computation, 42*, 841–859.

Malela-Majika, J. C., Rapoo, E. M., Mukherjee, A., & Graham, M. A. (2019). Distribution-free precedence schemes with a generalized runs-rule for monitoring unknown location. *Communications in Statistics-Theory and Methods*. https://doi.org/10.1080/03610926.2019.1612914.

McCracken, A. K., & Chakraborti, S. (2013). Control charts for joint monitoring of mean and variance: An overview. *Quality Technology and Quantitative Management, 10,* 17–36.

Montgomery, D. C. (2013). Introduction to statistical quality control (7th ed.). Hoboken, New Jersey: Wiley.

Mukherjee, P. S. (2016). On Phase II monitoring of the probability distributions of univariate continuous processes. *Statistical Papers, 57,* 539–562.

Ning, W., Yeh, A. B., Wu, X., & Wang, B. (2015). A nonparametric Phase I control chart for individual observations based on empirical likelihood ratio. *Quality and Reliability Engineering International, 31,* 37–55.

Page, E. S. (1955). A test for change in a parameter occurring at an unknown point. *Biometrika, 42,* 523–527.

Parpoula, C., Karagrigoriou, A., & Lambrou, A. (2017). Epidemic intelligence statistical modelling for Biosurveillance, In J. Blömer, et al. (Eds.), MACIS 2017, Lecture notes in computer science (LNCS) (Vol. 10693, pp. 349–363). Cham: Springer International Publishing AG.

Pepe, M. S. (2003). *The statistical evaluation of medical tests for classification and prediction.* Oxford: Oxford University Press.

Qiu, P. (2014). *Introduction to statistical process control.* Boca Raton, FL: Chapman & Hall/CRC.

Qiu, P. (2018). Some perspectives on nonparametric statistical process control. *Journal of Quality Technology, 50,* 49–65.

Qiu, P. (2019). Some recent studies in statistical process control. In Y. Lio, H. Ng, T. R. Tsai, & D. G. Chen (Eds.), *Statistical quality technologies* (pp. 3–19). Cham: ICSA Book Series in Statistics, Springer.

Qiu, P., & Li, Z. (2011). On nonparametric statistical process control of univariate processes. *Technometrics, 53,* 390–405.

Qiu, P., & Zhang, J. (2015). On Phase II SPC in cases when normality is invalid. *Quality and Reliability Engineering International, 31,* 27–35.

Ross, G. J., & Adams, N. M. (2012). Two nonparametric control charts for detecting arbitrary distribution changes. *Journal of Quality Technology, 44,* 102–116.

Schmid, W. (2015). Discussion on recent advances in process monitoring: Nonparametric and variable-selection methods for Phase I and Phase II. *Quality Engineering, 27,* 68–72.

Serfling, R. (1963). Methods for current statistical analysis of excess pneumonia-influenza deaths. *Public Health Reports, 78,* 494–506.

Shewhart, W. A. (1939). *Statistical method from the viewpoint of quality control.* New York, NY: Dover Publications.

Sullivan, J. H. (2002). Detection of multiple change points from clustering individual observations. *Journal of Quality Technology, 34,* 371–383.

Sullivan, J. H., & Woodall, W. H. (1996). A control chart for preliminary analysis of individual observations. *Journal of Quality Technology, 28,* 265–278.

Triantafyllou, I. S. (2019). A new distribution-free control scheme based on order statistics. *Journal of Nonparametric Statistics, 31,* 1–30.

Triantafyllou, I. S. (2019). Wilcoxon-type rank-sum control charts based on progressively censored reference data. *Communications in Statistics-Theory and Methods.* https://doi.org/10.1080/03610926.2019.1634816.

Woodall, W. H. (2000). Controversies and contradictions in statistical process control. *Journal of Quality Technology, 32,* 341–350.

Woodall, W. H., & Montgomery, D. C. (1999). Research issues and ideas in statistical process control. *Journal of Quality Technology, 31,* 376–386.

Zhou, X. H., Obuchowski, N. A., & McClish, D. K. (2011). *Statistical methods in diagnostic medicine* (2nd ed.). New Jersey: Wiley.

Zhou, C., Zou, C., Zhang, Y., & Wang, Z. (2009). Nonparametric control chart based on change-point model. *Statistical Papers, 50,* 13–28.

Zou, C., & Tsung, F. (2010). Likelihood ratio-based distribution-free EWMA control chart. *Journal of Quality Technology, 42,* 174–196.

A Class of Distribution-Free Exponentially Weighted Moving Average Schemes for Joint Monitoring of Location and Scale Parameters

Zhi Song, Amitava Mukherjee, Marco Marozzi and Jiujun Zhang

Abstract In this chapter, we investigate and compare six distribution-free exponentially weighted moving average (EWMA) schemes for simultaneously monitoring the location and scale parameters of a univariate continuous process. More precisely, we consider a well-known distribution-free EWMA scheme based on the Lepage statistic, and we propose five new EWMA schemes for the same purpose. One of the five new schemes is based on the maximum of EWMA of two individual components, one for the location parameter and the other for the scale parameter, of the Lepage statistic. Such a component-wise combined EWMA is referred to as the cEWMA. Further, we consider an EWMA scheme based on the Cucconi test statistic. We show that it is possible to express the Cucconi statistic as a quadratic combination of two orthogonal statistics, one of which is useful for monitoring the location parameter and the other for monitoring the scale parameter. Such decomposition of the Cucconi statistic is not unique, and one can split it into three different ways. Therefore, we design three more cEWMA schemes corresponding to the decompositions of the Cucconi statistic. We discuss the implementation steps along with an illustration. We perform a detailed comparative study based on Monte Carlo simulation. We observe that the three cEWMA-Cucconi schemes perform very well for various location–scale models.

Z. Song
College of Science, Shenyang Agricultural University,
Shenyang 110866, People's Republic of China

A. Mukherjee (✉)
Production, Operations and Decision Sciences Area,
XLRI-Xavier School of Management, XLRI Jamshedpur, India
e-mail: amitmukh2@yahoo.co.in

M. Marozzi
Department of Environmental Sciences, Informatics and Statistics,
Ca' Foscari University of Venice, Venezia, Italy

Z. Song · J. Zhang
Department of Mathematics, Liaoning University,
Shenyang 110036, People's Republic of China

© Springer Nature Switzerland AG 2020
M. V. Koutras and I. S. Triantafyllou (eds.), *Distribution-Free
Methods for Statistical Process Monitoring and Control*,
https://doi.org/10.1007/978-3-030-25081-2_6

Keywords Cucconi statistic · Distribution-free · Exponentially weighted moving average (EWMA) · Lepage statistic · Process monitoring

1 Introduction

Early evidence of researches in Nonparametric Statistical Process Monitoring (NSPM) mainly capitalized the asymptotic theory of sequential analysis, see, for example, Bhattacharya and Frierson Jr (1981), Park et al. (1987). About 20 years ago, Woodall and Montgomery (1999) anticipated an increasing role of nonparametric approaches in process monitoring in the twenty-first century. In the past two decades, not only their vision came true, but also the application of nonparametric process control reached beyond the production line of manufacturing industries. From monitoring healthcare system to service quality in a call center, various NSPM schemes are playing a transformative role even in the Industry 4.0 era. For some interesting applications of the NSPM schemes, see, for example, Stromberg (2005), Mukherjee and Marozzi (2017b), and Mukherjee and Sen (2018), among others.

Bakir (2004, 2006), respectively, proposed some NSPM schemes based on the signed ranks and some signed-rank type statistics. Among the most notable early works on the distribution-free approaches of process monitoring, we recommend Chakraborti et al. (2004) that introduced a class of NSPM schemes based on the precedence statistic; Li et al. (2010) designed cumulative sum (CUSUM) and exponentially weighted moving average (EWMA) schemes for Phase-II monitoring of unknown location parameter of a process using the Wilcoxon statistic. Qiu and Li (2011a, b) studied some NSPM schemes primarily focusing on the location parameter of univariate processes. For the same purpose, Graham et al. (2012) and Mukherjee et al. (2013), respectively, designed EWMA and CUSUM schemes based on the precedence statistic in the distribution-free setup. For some notable contributions in the multivariate nonparametric setup, we recommend Qiu and Hawkins (2001, 2003). Chatterjee et al. (2009) proposed some distribution-free CUSUM schemes using bootstrap-based control limits. From a statistical perspective, we often see that the distributional assumption of various parametric approaches is unrealistic or untenable in practice. Therefore, we strongly recommend practitioners to rely on nonparametric and distribution-free procedures. While the book by Montgomery (2009) provides a sound understanding of the fundamentals of statistical process control, Qiu (2014) offers a systematic treatment of some NSPM schemes and is worth reading to learn and understand the advantages of the NSPM schemes over traditional parametric approaches. For a recent review on various perspectives of NSPM schemes, we recommend Qiu (2018) and the book by Chakraborti and Graham (2019b).

Statistical process monitoring is broadly based on two phases. In the first phase, practitioners collect a sample from an in-control (IC) process, systematically analyze and examine if there is any signal due to an assignable cause. Phase-I or retrospective analysis is essential for establishing a reference sample. Such a reference sample is used as a benchmark in the course of monitoring incoming series of observations in

Phase-II. Establishment of Phase-I sample is itself an exciting area of research and is addressed by several researchers in recent times. However, in the current chapter, we focus on the Phase-II NSPM schemes only. Interested readers may see Jones-Farmer et al. (2014), Capizzi and Masarotto (2018) and Li et al. (2019) for more details of Phase-I analysis. We assume that a reference sample of fixed size is available a priori from an IC process. For Phase-II analysis, most of the NSPM schemes discussed before 2012–2013 are designed either for monitoring the location parameter of the process or for the process range or variability separately. For example, Bakir (2004, 2006), Li et al. (2010), Qiu and Li (2011a, b), Graham et al. (2012), and Mukherjee et al. (2013) are effective in monitoring the location parameter. On the other hand, Stromberg (2005) is useful for monitoring the process range.

Nevertheless, several researchers recommend using a combined charting scheme that can simultaneously monitor both location and scale parameters of a process. Under parametric setup, researches on joint monitoring started during the last decade of the twentieth century, whereas there was no NSPM scheme for joint monitoring for a long time. Mukherjee and Chakraborti (2012) first addressed nonparametric joint monitoring of location and scale parameters using a single plotting statistic, and, more precisely, developed a Shewhart-type NSPM scheme based on the Lepage statistic, introduced by Lepage (1971), in connection to the classical two-sample location–scale problem of testing of hypothesis.

The Lepage statistic is a rank-based statistic defined as a quadratic combination of the standardized Wilcoxon statistic for location and the standardized Ansari–Bradley statistic for scale. Eventually, a large number of rank statistics have been proposed for the two-sample location–scale problem. Most of these statistics are of the Lepage-type, that is, a combination of a location test statistic and a scale test statistic. However, there was a different test statistic designed by Cucconi (1968) in connection to a similar problem in an Italian national journal, remained unnoticed by the scientific community outside Italy before Marozzi (2009) provided new results about it. Marozzi (2013) reviewed and compared the performance of several two-sample location–scale tests of the Lepage-type as well as the Cucconi statistics. Chowdhury et al. (2014) used this statistic to construct a Shewhart-type NSPM scheme for joint monitoring of location and scale parameters.

Bonnini et al. (2014) and Marozzi (2014) emphasized that the Cucconi test is of historical interest as it was proposed some years before the Lepage test. Moreover, the genesis of the Cucconi statistic is not through a combination of a test statistic for location and a test statistic for scale as the various other Lepage-type location–scale tests. In this chapter, we show that even the Cucconi statistic can be decomposed as a quadratic combination of a location statistic and a scale statistic. Interestingly, such decomposition is not unique, and here we present three different decompositions. Nevertheless, one should note that the original Cucconi statistic was not proposed as a combination of a location statistic and a scale statistic. Marozzi (2013) showed that the Cucconi statistic compares favorably with Lepage-type statistic in connection to two-sample testing in terms of power and type one error probability. Likewise, Chowdhury et al. (2014) established that the Shewhart-type NSPM scheme based on the Cucconi statistic competes well with the Shewhart–Lepage scheme as in

Mukherjee and Chakraborti (2012) in terms of run length characteristics. The Cucconi statistic is slightly more straightforward and faster as far as computational time and complexity are concerned, because it uses only the ranks of one sample in the combined sample, whereas the other statistics also require scores of various types. In some cases, one needs to estimate the mean and variance of test statistics via permutation as their analytic formulae are not available.

After Mukherjee and Chakraborti (2012) and Chowdhury et al. (2014), a large volume of literature has been introduced for joint monitoring of location and scale parameters. Chowdhury et al. (2015) and Mukherjee and Marozzi (2017b) designed two CUSUM-type NSPM schemes using the Lepage and Cucconi statistics, respectively. Further, Mukherjee (2017a) discussed various EWMA schemes based on the Lepage statistic. Chong et al. (2017) designed a new NSPM scheme, referred to as the premier monitoring scheme, using the Lepage statistic and both the maximum and distance metrics. Mukherjee and Marozzi (2017a) introduced the circular-grid version of the Shewhart–Lepage scheme. Noting that the Lepage statistic is by default designed for two-sided shifts in location or scale parameters, Chong et al. (2018) proposed several one-sided Shewhart-type Lepage schemes with appropriate modification of the Lepage statistic. Song et al. (2019) introduced a new distribution-free adaptive Shewhart–Lepage-type scheme for simultaneous monitoring using information about symmetry and tail weights of the process distribution. Among other notable works, Celano et al. (2016) proposed joint monitoring schemes for finite horizon processes; Mahmood et al. (2017) studied the performance of some joint monitoring schemes, and Zafar et al. (2018) introduced some progressive approaches of joint monitoring.

Roberts (1959) first introduced the notion of EWMA schemes. Various EWMA schemes are widely used in industry and are very popular among practitioners as they are very efficient in detecting small to moderate shifts. In fact, in the context of joint monitoring, Mukherjee (2017a) established that if the smoothing parameter is appropriately tuned, the EWMA Lepage scheme will be as good as the Shewhart scheme even for significant shifts in location or scale parameters. However, discussion of the EWMA type schemes for joint monitoring of location and scale is somewhat limited. Apart from Mukherjee (2017a), a recent work by Li et al. (2018) considered an EWMA scheme based on ranks. Moreover, Mukherjee (2017b) presented the EWMA scheme based on the Cucconi statistic. Nevertheless, researchers have never explored the max-type combination of individual EWMAs based on the location and scale components of the Lepage and Cucconi statistics. We refer to these new types of EWMA schemes as cEWMA schemes. The notion of distribution-free cEWMA is one of the pivotal contributions of this chapter.

The rest of the chapter is organized as follows. In Sect. 2, we describe the six distribution-free EWMA schemes for joint monitoring based on the Lepage and Cucconi statistics. In Sect. 3, we outline the implementation design of various EWMA and cEWMA schemes. In Sect. 4, we investigate and compare the performance of the six schemes. Their application is illustrated with real data in Sect. 5. We offer concluding remarks and some directions for further research in Sect. 6.

2 Distribution-Free EWMA Schemes

Consider a stable univariate and continuous process that is unaffected by any assignable or special causes of variation. In connection to statistical process monitoring, such a process is called IC process. An IC process is subjected to random variation only. In the present chapter, we assume the availability of a reference sample of size m from the IC process. To be precise, we consider that a reference sample $U_1, U_2, \ldots, U_m$ is collected from an IC process with a continuous cumulative distribution function (cdf) $F(x)$. We also assume that $U_i's$ are independently and identically distributed (i.i.d) random variables each having cdf F. As mentioned earlier, one must establish a reference sample through appropriate Phase-I analysis. We omit details on this to save space. We assume that the functional form of F is unknown. At each stage of Phase-II monitoring, we collect a test sample of size n. Let $V_1, V_2, \ldots, V_n$, be the test sample from an unknown continuous distribution with cdf $G(x)$. The test sample is mutually independent of the reference sample. We further assume that $V_i's$ are i.i.d for every i, ($1 \le i \le n$). Ideally, two distribution functions F and G should be identical in all respect when the process is in IC. Nevertheless, when there is a shift in process location or scale or both at Phase-II, we often see that $G(x) = F(\frac{x-\theta}{\delta})$, $\theta \in \Re$, $\delta > 0$, where the constants θ and δ represent the unknown shift in the location and scale parameters, respectively. In classical statistical literature, such a framework is referred to as the general location–scale model. See, for example, Hájek et al. (1999). When $\theta = 0$ and $\delta = 1$, we observe an IC setup. Note that $\theta \neq 0$ but $\delta = 1$, representing a pure location shift, while $\theta = 0$ and $\delta \neq 1$ indicate a pure scale shift. Finally, if both $\theta \neq 0$ and $\delta \neq 1$, we observe a shift in both the location and scale parameters.

2.1 Lepage-Type Monitoring Schemes

The well-known Lepage statistic is the sum of squares of the standardized Wilcoxon rank-sum (WRS) statistic for location and the standardized Ansari–Bradley (AB) statistic for scale. Introduce an indicator variable $I_k = 0$ or 1 accordingly as the kth order statistic of the combined $N(= m + n)$ sample is a U observation or V. The WRS statistic for testing the equality of two location parameters is the sum of ranks of $V_i's$ in the combined sample of size N and is given by

$$T_1 = \sum_{k=1}^{N} k I_k.$$

The AB test is a nonparametric test for the two-sample scale problem based on the statistic T_2, defined as

$$T_2 = \sum_{k=1}^{N} \left| k - \frac{1}{2}(N+1) \right| I_k.$$

The combination of the standardized WRS and AB statistics is the Lepage statistic and is given by

$$L = \left(\frac{T_1 - E(T_1)}{\sqrt{\mathrm{Var}(T_1)}} \right)^2 + \left(\frac{T_2 - E(T_2)}{\sqrt{\mathrm{Var}(T_2)}} \right)^2. \tag{1}$$

It is well known (see Mukherjee and Chakraborti (2012)) that

$$E(T_1) = \frac{1}{2} n(N+1),$$

$$\mathrm{Var}(T_1) = \frac{1}{12} mn(N+1).$$

Further,

$$E(T_2) = \begin{cases} \frac{n(N^2-1)}{4N} & \text{when } N \text{ is odd} \\ \frac{nN}{4} & \text{when } N \text{ is even}, \end{cases}$$

$$\mathrm{Var}(T_2) = \begin{cases} \frac{mn(N+1)(N^2+3)}{48N^2} & \text{when } N \text{ is odd} \\ \frac{mn(N^2-4)}{48(N-1)} & \text{when } N \text{ is even}. \end{cases}$$

Mukherjee and Chakraborti (2012) recommended using this statistic for joint monitoring of location and scale parameters of an unknown but continuous univariate process. They designed a Phase-II Shewhart-type scheme. Interested readers may see Chowdhury et al. (2015) for the Phase-II CUSUM Lepage scheme. A Phase-I Lepage-type scheme is also available in literature. See Li et al. (2019) for more details.

2.1.1 The EWMA Lepage Scheme

The EWMA schemes are usually more effective in detecting persistent shifts than Shewhart schemes. Mukherjee (2017a) proposed some distribution-free EWMA schemes based on the Lepage statistic for jointly monitoring the location and the scale parameters. These schemes are referred to as the EL procedures. Let L_j denote the Lepage statistic L computed by using m Phase-I observations from the jth test sample. Note that $E(L_j|\mathrm{IC}) = 2$. Then the EL scheme is given by

$$R^*_{\mathrm{EL},j} = \max\{2, \lambda L_j + (1 - \lambda) R_{\mathrm{EL},j-1}\}, \quad j = 1, 2, \ldots,$$

with the starting value $R_{\mathrm{EL},0} = 2$. Here, $0 < \lambda \le 1$ is the smoothing parameter. Note that we get the Shewhart-type scheme when $\lambda = 1$. This type of EWMA structure helps in reducing inertia effect of the EWMA. We refer p. 423. of Montgomery (2009) for better understanding of the inertia effect of an EWMA scheme. Here, we consider standard (traditional) EWMA structure. We consider

$$R_{EL,j} = \lambda L_j + (1 - \lambda)R_{EL,j-1}, \quad j = 1, 2, \ldots,$$

with the starting value $R_{EL,0} = 2$. Chakraborti and Graham (2019a) mentioned about this type of traditional ways of designing the EL scheme. The EWMA Schemes considered in this chapter, are all based on the traditional design. We recommend future study of the EWMA design with correction of inertia effect.

2.1.2 The cEWMA Lepage Scheme

Unlike, the traditional EL schemes discussed in Mukherjee (2017a), practitioners may find a max-type variant of the EL scheme more interesting. The idea is to combine two individual EWMA statistics, one based on the location component of the Lepage statistic and the other based on the scale component. We refer to this new scheme as the cEWMA Lepage (cEL) scheme. Let $T_{1,j}$ and $T_{2,j}$ denote the WRS statistic T_1 and the AB statistic T_2, respectively, computed by using m Phase-I observations and the jth test sample. Writing $S_{1,j} = \frac{T_{1,j} - E(T_{1,j})}{\sqrt{Var(T_{1,j})}}$ and $S_{2,j} = \frac{T_{2,j} - E(T_{2,j})}{\sqrt{Var(T_{2,j})}}$, we may define the following EWMA statistics:

$$\begin{aligned} R_{1,j} &= \lambda S_{1,j}^2 + (1 - \lambda)R_{1,j-1}, \\ R_{2,j} &= \lambda S_{2,j}^2 + (1 - \lambda)R_{2,j-1}, \qquad j = 1, 2, \ldots, \end{aligned} \tag{2}$$

with $R_{1,0} = R_{2,0} = 1$, as it is easy to see that $E\left(S_{1,i}^2\right) = E\left(S_{2,j}^2\right) = 1$. The plotting statistic of the cEL scheme is defined by

$$R_{cEL,j} = \max\{R_{1,j}, R_{2,j}\} \text{ for } j = 1, 2, \ldots$$

2.2 Cucconi-Type Monitoring Schemes

The Cucconi (1968) statistic is another important statistic for the classical distribution-free two-sample location–scale problems. Marozzi (2009, 2013) popularized this statistic and established that the test based on the Cucconi statistic is preferable to the familiar Lepage test and several other tests for various distributions and different sizes of shifts. Now, we shall briefly discuss the Cucconi statistic. Consider the sum of the squares of the ranks of $V_i's$ in the combined sample of size N as

$$S_1 = \sum_{k=1}^{N} k^2 I_k.$$

Further, let the sum of squares of anti-ranks (alias contrary ranks) of $V_i's$ in the combined sample be S_2. Cucconi (1968) showed that

$$S_2 = \sum_{k=1}^{N} (N + 1 - k)^2 I_k = n(N + 1)^2 - 2(N + 1)T_1 + S_1,$$

and that

$$E(S_l|IC) = \frac{n(N + 1)(2N + 1)}{6},$$

and

$$\mathrm{Var}(S_l|IC) = \frac{mn(N + 1)(2N + 1)(8N + 11)}{180}, \; l = 1, 2.$$

Consider the standardized S_l statistics for $l = 1, 2$ as $W = \frac{S_1 - E(S_1|IC)}{\sqrt{\mathrm{Var}(S_1|IC)}}$ and $Z = \frac{S_2 - E(S_2|IC)}{\sqrt{\mathrm{Var}(S_2|IC)}}$ where

$$\frac{S_l - E(S_l|IC)}{\sqrt{\mathrm{Var}(S_l|IC)}} = \frac{6S_l - n(N + 1)(2N + 1)}{\sqrt{\frac{1}{5}(mn(N + 1)(2N + 1)(8N + 11))}}, \; l = 1, 2,$$

so that $E(W|IC) = E(Z|IC) = 0$ and $\mathrm{Var}(W|IC) = \mathrm{Var}(Z|IC) = 1$. Moreover, W and Z are negatively correlated with correlation coefficient $\rho \in \left(-1, -\frac{7}{8}\right)$. Explicitly,

$$\mathrm{Corr}(W, Z) = \rho = \frac{2(N^2 - 4)}{(2N + 1)(8N + 11)} - 1.$$

The minimum -1 occurs in the trivial situation where $N = 2$, while the supremum is reached when N diverges to infinity with $\lim_{N \to \infty} \rho = -7/8$.

The process is out-of-control (OOC) if either of the following three situations arises in course of Phase-II monitoring: (i) when $\theta \neq 0$ and $\delta = 1$, (ii) when $\theta = 0$ and $\delta \neq 1$, and (iii) when $\theta \neq 0$ and $\delta \neq 1$. In these situations, one or both of $E(W)$ and $E(Z)$ are nonzero; the various scenarios are reported in details by Marozzi (2009). In order to combine the information provided by both W and Z regarding the presence of a difference in location as well as in scale, Cucconi (1968) proposed the following rank-based statistic:

$$C = \frac{W^2 + Z^2 - 2\rho WZ}{2(1 - \rho^2)}. \tag{3}$$

The higher the deviation of θ from 0 and/or δ from 1, the larger is the value of C, and therefore C plays an important role in detecting any possible shifts in θ and/or δ.

Chowdhury et al. (2014) developed a Phase-II Shewhart-type joint monitoring schemes using the Cucconi statistic as an alternative to the Shewhart–Lepage scheme. Further, Mukherjee and Marozzi (2017b) designed the Phase-II CUSUM Cucconi scheme.

2.2.1 The EWMA Cucconi Scheme

Mukherjee (2017b) proposed a single distribution-free EWMA scheme for jointly monitoring the location and the scale parameters, which is based on the Cucconi statistic and referred to as the EC procedure. Let C_j denote the Cucconi statistic C computed by using m Phase-I observations from the jth test sample. Note that $E(C_j|\text{IC}) = 1$. Then the EC scheme is given by

$$E_j = \lambda C_j + (1 - \lambda)E_{j-1}, \ j = 1, 2, \ldots,$$

with the starting value $E_0 = 1$.

2.2.2 The cEWMA-Cucconi Schemes

In the literature, nonparametric tests for jointly detecting location and scale changes are based on the combination of two tests, one for location and another for scale. Generally, the combination is achieved through the sum of the squared standardized test statistics, and this is just the case of the Lepage test. The traditional viewpoint is that the Cucconi statistic is not based on a combination of a test statistic for location and a test statistic for scale as the other location–scale tests. However, in this chapter, we show that even the Cucconi statistic can be decomposed as a quadratic combination of a location statistic and a scale statistic. Interestingly, such decomposition is not unique, and here we present three different decompositions. It is easy to see that

$$\text{Case 1: } C = \frac{W^2}{2} + \frac{(Z - \rho W)^2}{2(1 - \rho^2)};$$

$$\text{Case 2: } C = \frac{(W - \rho Z)^2}{2(1 - \rho^2)} + \frac{Z^2}{2};$$

$$\text{Case 3: } C = \frac{(W - Z)^2}{4(1 - \rho)} + \frac{(W + Z)^2}{4(1 + \rho)}.$$

Therefore, we can construct three cEWMA-Cucconi (abbreviated by cEC) schemes as follows:

cEC-Case 1(cEC1): Suppose $T_{11}^2 = W^2$ and $T_{12}^2 = \frac{(Z-\rho W)^2}{1-\rho^2}$. Note that $E(T_{11}^2|\mathrm{IC}) = E(T_{12}^2|\mathrm{IC}) = 1$. We consider

$$E_{11,j} = \lambda T_{11,j}^2 + (1-\lambda)E_{11,j-1},$$

$$E_{12,j} = \lambda T_{12,j}^2 + (1-\lambda)E_{12,j-1},$$

with the starting value $E_{11,0} = E_{12,0} = 1$. The plotting statistic is given by

$$E_{1,j} = \max\{E_{11,j}, E_{12,j}\} \text{ for } j = 1, 2, \ldots$$

cEC-Case 2(cEC2): Suppose $T_{21}^2 = \frac{(W-\rho Z)^2}{1-\rho^2}$ and $T_{22}^2 = Z^2$. We consider the EWMA statistics:

$$E_{21,j} = \lambda T_{21,j}^2 + (1-\lambda)E_{21,j-1},$$

$$E_{22,j} = \lambda T_{22,j}^2 + (1-\lambda)E_{22,j-1},$$

with the starting value $E_{21,0} = E_{22,0} = 1$, as $E(T_{21}^2|\mathrm{IC}) = E(T_{22}^2|\mathrm{IC}) = 1$. The plotting statistic is given by

$$E_{2,j} = \max\{E_{21,j}, E_{22,j}\} \text{ for } j = 1, 2, \ldots$$

cEC-Case 3(cEC3): Suppose $T_{31}^2 = \frac{(W-Z)^2}{2(1-\rho)}$ and $T_{32}^2 = \frac{(W+Z)^2}{2(1+\rho)}$. We consider

$$E_{31,j} = \lambda T_{31,j}^2 + (1-\lambda)E_{31,j-1},$$

$$E_{32,j} = \lambda T_{32,j}^2 + (1-\lambda)E_{32,j-1},$$

with the starting value $E_{31,0} = E_{32,0} = 1$, as $E(T_{31}^2|\mathrm{IC}) = E(T_{32}^2|\mathrm{IC}) = 1$. The plotting statistic is defined by

$$E_{3,j} = \max\{E_{31,j}, E_{32,j}\} \text{ for } j = 1, 2, \ldots$$

3   Implementation of the Proposed Monitoring Procedures

For various practical purposes, it is essential to follow specific implementation steps for various schemes discussed in Sect. 2. Note that, irrespective of type and direction of the shift, either in location or in scale or in both, the statistic of each scheme is expected to be larger when the process in OOC. Thus, each of the six monitoring schemes requires only an upper control limit (UCL).

3.1 Implementation of the EL and cEL Schemes

The EL and cEL schemes may be implemented in practice via the following steps:

Step 1: Collect the reference sample from IC process, $X_m = (X_1, X_2, \ldots, X_m)$.

Step 2: Let $Y_j = (Y_{j1}, Y_{j2}, \ldots, Y_{jn})$ be the jth test sample of size n, $j = 1, 2, \ldots$

Step 3: Identify the U's with the X's and the V's with the Y's, respectively. Calculate $T_{1,j}$ and $T_{2,j}$ between the reference sample and the jth test sample, and obtain their means and variances according to whether $N = m + n$ is even or odd.

Step 4: (i) For the EL scheme, calculate the Lepage statistic L_j using the reference sample and the jth test sample as in (1). Then compute the plotting statistic $R_{EL,j} = \lambda L_j + (1 - \lambda) R_{EL,j-1}, j = 1, 2, \ldots$, sequentially for the jth test sample of the EL scheme starting with $R_{EL,0} = 2$.

(ii) For the cEL scheme, calculate the EWMA statistics $R_{1,j}$ and $R_{2,j}$ for the jth subgroup ($j = 1, 2, \ldots$) as in (2). Then compute the plotting statistic $R_{cEL,j} = \max\{R_{1,j}, R_{2,j}\}$, $j = 1, 2, \ldots$, sequentially for the jth test sample of the cEL scheme.

Step 5: Let H_{EL} and H_{cEL} be the UCL of the EL and cEL schemes, respectively. For $j = 1, 2, \ldots$, compare the plotting statistic $R_{[S],j}$ with the corresponding UCL, $H_{[S]}$, where $[S] = EL$ or cEL, as the case may be.

Step 6: Obtain an OOC signal at the jth stage of inspection if $R_{[S],j}$ exceeds $H_{[S]}$, $[S] = EL$ or cEL, and a search for assignable cause begins. Otherwise, the process is thought to be IC, and monitoring continues to the next test sample.

3.2 Implementation of the EC, cEC1, cEC2, and cEC3 Schemes

The EC, cEC1, cEC2, and cEC3 monitoring schemes may be constructed as follows:

Step 1: Same as in Sect. 3.1.

Step 2: Same as in Sect. 3.1.

Step 3: Identify the U's with the X's and the V's with the Y's, respectively. Calculate the Cucconi statistic C_j between the reference sample and the jth test sample as in (3).

Step 4:   (i) For the EC scheme, calculate the plotting statistic $E_j = \lambda C_j + (1 - \lambda)E_{j-1}$, sequentially for the jth test sample of the EC scheme starting with $E_0 = 1$.

(ii) For the cEC[I] schemes, calculate the EWMA statistics $E_{[I]1,j}$ and $E_{[I]2,j}$ for the jth subgroup ($j = 1, 2, \ldots$), where $[I] = 1, 2, 3$, using the formulae presented in Sect. 2.2.2. Then compute the plotting statistic $E_{[I],j}$, $[I] = 1, 2, 3$, sequentially for the jth test sample of the cEL[I] scheme, respectively.

Step 5:   Let H_{EC} and $H_{cEC[I]}$ be the UCL of the EC and cEC[I] ($[I] = 1, 2, 3$) schemes, respectively. For $j = 1, 2, \ldots$, compare the plotting statistic E_j with the UCL, H_{EC}, and similarly, the plotting statistic $E_{[I],j}$ with the corresponding UCL $H_{cEC[I]}$, where $[I] = 1, 2, 3$.

Step 6:   Obtain an OOC signal at the jth stage of inspection if E_j exceeds H_{EC}, or $E_{[I],j}$ exceeds $H_{cEC[I]}$, $[I] = 1, 2, 3$, and a search for assignable cause begins. Otherwise, the process is thought to be IC, and monitoring continues to the next test sample.

4   Numerical Results and Comparisons

We analyze IC and OOC performance of various schemes discussed in Sect. 2 via Monte Carlo simulation. The average run length (ARL) and the standard deviation of the run length (SDRL) are popular performance indicators, but since the run length distribution is right-skewed, it is also worthwhile to study the 5th, 25th, 50th, 75th, and the 95th percentiles of the run length.

4.1   Determination of Control Limits

For the implementation of the aforementioned schemes, we need to determine the UCL that provides certain target IC ARL (ARL$_0$). We perform a Monte Carlo simulation study in FORTRAN to determine the UCL values on the basis of 50,000 replications. Because of the distribution-free nature of the proposed class of monitoring schemes, without loss of generality, we generate m observations from a standard normal distribution for the Phase-I sample and n observations from the same distribution for each test sample. In Table 1, we provide the UCL values for some selected (m, n, λ) and nominal ARL$_0$ values. We select $m = 50, 100, 300$, and 500, to cover small to moderate reference sample sizes. Further, we select $n = 5, 10$ and 15 as test sample size. Finally, we consider the values of λ as 0.05 and 0.1 and the nominal ARL$_0$ as 250, 370, and 500.

Table 1 The UCL values of six schemes for some standard (target) values of ARL_0

Chart parameter			EL scheme ARL_0			cEL scheme ARL_0			EC scheme ARL_0			cEC1 scheme ARL_0			cEC2 scheme ARL_0			cEC3 scheme ARL_0		
λ	m	n	250	370	500	250	370	500	250	370	500	250	370	500	250	370	500	250	370	500
0.05	50	5	2.393	2.465	2.530	1.473	1.523	1.565	1.185	1.217	1.245	1.459	1.509	1.547	1.459	1.509	1.546	1.453	1.500	1.539
	50	10	2.329	2.396	2.458	1.445	1.494	1.538	1.164	1.197	1.224	1.443	1.494	1.533	1.442	1.494	1.532	1.443	1.490	1.534
	50	15	2.248	2.309	2.362	1.383	1.438	1.481	1.125	1.158	1.182	1.395	1.444	1.486	1.396	1.444	1.485	1.394	1.443	1.483
	100	5	2.505	2.587	2.641	1.518	1.570	1.609	1.245	1.278	1.308	1.509	1.561	1.603	1.509	1.560	1.603	1.501	1.548	1.585
	100	10	2.496	2.565	2.622	1.521	1.575	1.618	1.239	1.275	1.301	1.514	1.568	1.609	1.517	1.569	1.611	1.513	1.566	1.607
	100	15	2.439	2.517	2.575	1.499	1.551	1.597	1.224	1.256	1.282	1.497	1.551	1.593	1.501	1.552	1.592	1.500	1.550	1.593
	300	5	2.600	2.678	2.736	1.547	1.600	1.639	1.302	1.340	1.371	1.558	1.616	1.661	1.560	1.616	1.662	1.541	1.594	1.633
	300	10	2.612	2.689	2.746	1.569	1.630	1.670	1.299	1.341	1.370	1.570	1.628	1.670	1.570	1.628	1.672	1.564	1.620	1.661
	300	15	2.605	2.682	2.741	1.575	1.632	1.674	1.297	1.336	1.368	1.572	1.629	1.672	1.572	1.629	1.673	1.569	1.625	1.668
	500	5	2.619	2.698	2.758	1.552	1.607	1.645	1.316	1.356	1.387	1.628	1.630	1.676	1.570	1.631	1.675	1.552	1.606	1.646
	500	10	2.631	2.715	2.773	1.580	1.638	1.679	1.317	1.358	1.387	1.580	1.640	1.685	1.579	1.639	1.684	1.575	1.629	1.674
	500	15	2.632	2.718	2.776	1.585	1.645	1.688	1.312	1.354	1.385	1.581	1.642	1.686	1.583	1.644	1.688	1.579	1.639	1.683
0.1	50	5	2.843	2.947	3.028	1.860	1.938	1.998	1.399	1.447	1.489	1.840	1.912	1.974	1.840	1.911	1.974	1.833	1.901	1.956
	50	10	2.786	2.881	2.958	1.351	1.927	1.991	1.382	1.428	1.467	1.859	1.910	1.973	1.839	1.910	1.973	1.838	1.915	1.974
	50	15	2.778	2.852	2.857	1.785	1.862	1.925	1.339	1.385	1.423	1.791	.862	1.923	1.792	1.862	1.924	1.790	1.864	1.926
	100	5	2.991	3.093	3.173	1.919	1.996	2.050	1.479	1.537	1.580	1.916	2.001	2.066	1.916	2.002	2.066	1.892	1.965	2.024
	100	10	2.975	2.977	3.157	1.943	2.029	2.091	1.471	1.524	1.563	1.932	2.014	2.077	1.932	2.014	2.077	1.931	2.009	2.070
	100	15	2.925	3.030	3.111	1.929	2.008	2.075	1.455	1.505	1.545	1.918	1.997	2.061	1.918	1.996	2.062	1.919	1.997	2.059
	300	5	3.087	3.210	3.295	1.953	2.030	2.088	1.558	1.621	1.669	1.997	2.088	2.159	1.997	2.088	2.159	1.950	2.033	2.095
	300	10	3.110	3.225	3.317	2.009	2.093	2.158	1.552	1.612	1.655	2.009	2.096	2.166	2.009	2.095	2.166	1.993	2.080	2.145
	300	15	3.108	3.220	3.310	2.019	2.107	2.173	1.547	1.608	1.649	2.011	2.102	2.172	2.011	2.102	2.173	2.004	2.093	2.159
	500	5	3.118	3.236	3.321	1.958	2.035	2.095	1.579	1.642	1.691	2.017	2.113	2.187	2.017	2.113	2.187	1.967	2.049	2.114
	500	10	3.144	3.253	3.349	2.014	2.103	2.169	1.573	1.633	1.680	2.026	2.120	2.192	2.026	2.121	2.192	2.010	2.099	2.163
	500	15	3.133	3.257	3.468	2.032	2.121	2.189	1.568	1.629	1.674	2.034	2.125	2.196	2.034	2.125	2.197	2.028	2.114	2.182

4.2   Performance of the Proposed Procedures at Microlevel

We conduct a detailed Monte Carlo simulation to evaluate the IC and OOC performance of the proposed class of schemes. As the competitive schemes, we also include two EWMA schemes based on two well-known distribution-free statistics, namely, the Cramér–Von Mises (CvM) statistic and the Kolmogorov–Smirnov (KS) statistic, which are widely used to detect a general shift in the process distribution. We consider three well-known symmetric distributions and two asymmetric distributions from the class of the general location–scale family. Among symmetric distributions, we consider

i. Thin-tailed Uniform distribution over support $(\theta - \delta, \theta + \delta)$, denoted by Uniform$(\theta, \delta)$.
ii. Medium-tailed normal distribution with mean θ and standard deviation δ, denoted by $N(\theta, \delta)$.
iii. Heavy-tailed Laplace distribution with mean θ and standard deviation $\delta\sqrt{2}$, denoted by Laplace(θ, δ).

Among asymmetric distributions, we consider the following:

i. The Rayleigh distribution, denoted by Rayleigh(θ, δ) and having probability density function (pdf)

$$f(x) = \frac{1}{\delta}\left(\frac{x - \theta}{\delta}\right) e^{-\frac{1}{2}\left(\frac{x-\theta}{\delta}\right)^2}, \ x \in [\theta, \infty).$$

ii. The shifted exponential distribution abbreviated as SE(θ, δ) and having pdf

$$f(x) = \frac{1}{\delta} e^{-\frac{1}{\delta}(x-\theta)}, \ x \in [\theta, \infty).$$

To examine the effect of shifts in location and scale parameters, we consider three shift scenarios:

i. A pure location shift case, where only θ changes, $\delta = 1$, with $\theta = 0, \pm 0.5, \pm 1, \pm 1.5$, and ± 2.
ii. A pure scale shift case, where only δ changes, $\theta = 0$, with $\delta = 0.5, 1, 1.25, 1.5, 1.75$, and 2.
iii. Three mixed location–scale shift cases. First, we consider $(\theta, \delta = e^{\theta})$, $\theta = 0, \pm 0.5, \pm 1, \pm 1.5$, and ± 2. Further, we consider $(\theta, \delta = e^{\theta/2})$, $\theta = 0, \pm 0.5, \pm 1, \pm 1.5$, and ± 2. Finally, we consider $(\theta, \delta = e^{2\theta})$, $\theta = 0, \pm 0.25, \pm 0.5$, and ± 1.

Note that the situation $\theta = 0$ and $\delta = 1$ corresponds to the IC situation and the remain cases correspond to the OOC situations. For brevity, we only tabulate the results for $m = 100$, $n = 5$, $\lambda = 0.05$, and ARL$_0 = 500$. We display the run length properties for different distributions with various parameter settings in Tables 2, 3, 4, 5, and 6. The first row of each cell in Tables 2, 3, 4, 5, and 6 shows the ARL followed by the corresponding SDRL in parentheses, whereas the second row shows the values of the 5th, 25th, 50th, 75th, and 95th percentiles (in this order). We highlight the schemes with the best OOC performance in the sense that the least OOC-ARL (ARL$_1$) is observed for a given shift with a dark gray shade.

From Tables 2, 3, 4, 5, and 6, we may reach at the following conclusions:

i. When the underlying process distribution is Uniform, the EC and cEC2 schemes perform well for detecting downward shifts in location under scale invariance, whereas the EC and cEC1 schemes are better for upward shifts in location with no shift in scale. When only the scale shift takes place and the location parameter remains invariant, cEC3 and EC perform better than their competitors, respectively, for downward and upward scale shifts. For mixed location and scale shifts, EC has overall better performance. Interestingly, for $\theta = -0.5$, we see an interesting feature. Here, for a small to moderate downward scale shift, the cEC3 is superior and for a moderate to large downward scale shift, the EWMA-CvM is better.

ii. When the process follows the normal distribution, the general pattern for a pure location shift is somewhat similar to that of the Uniform distribution. The advantage of cEC2 for small downward location shift and cEC1 for small upward location shift is more prominent. The cEL scheme is the best when only a pure downward scale shift takes place. The EC scheme performs well for an upward scale shift. For mixed location and scale shifts, EC performs well for an upward location and scale shift. For a downward location and scale shift, there is no clear winner. Nevertheless, for $\theta = -1$ along with some small to large downward scale shifts, EWMA-CvM offers better result, whereas for $\theta = -0.5$ accompanied by downward shift in scale, EWMA-KS is preferable. For $\theta = -0.25$, and $\delta = e^{2\theta}$, EWMA-CvM displays ARL bias, whereas performance of cEL is exceptionally good. Overall, for a downward location and scale shift, cEL is very effective.

iii. For the Laplace distribution, again, there is no clear winner. The EC and cEC2 schemes are good in detecting large downward pure location shifts ($\theta \leq -1$). However, for a small pure downward location shift ($\theta = -0.5$), EWMA-CvM is the best. For small upward pure location shifts ($0 < \theta \leq 1$), EWMA-CvM is again better. When the magnitude of θ increases, the EC and cEC1 perform better. Further, for the pure downward scale shift, cEL offers the best result where EWMA-CvM and EWMA-KS have significant ARL bias. When there is a pure upward scale shift, EC performs better than its competitors. For mixed upward location and scale shifts, EC displays better OOC performance. For moderate to large downward location shifts ($\theta \leq -1$) along with some downward scale shift, EWMA-CvM is the best, while for a small downward location shift ($\theta = -0.5$) accompanied by some downward scale shift, EWMA-KS is the best.

iv. For the Rayleigh distribution and the shifted exponential distribution, overall EC is the scheme of choice. The cEC1 scheme is also very competitive. Apart from that, for the Rayleigh distribution, EWMA-CvM is the best in detecting a small upward pure location shift ($\theta = 0.5$). Similarly, EWMA-CvM also detects a pure downward scale shift ($\delta = 0.5$) faster than its competitors. Further, for

Table 2 Comparison of the eight schemes under Uniform(θ, δ) distribution for $m = 100$, $n = 5$, $\lambda = 0.05$, and $\mathrm{ARL}_0 = 500$

θ	δ		EC	cEC1	cEC2	cEC3	EL	cEL	EWMA-CvM	EWMA-KS
0	0		501.90(882.53) 12, 56, 172, 507, 2247	503.44(819.34) 16, 72, 201, 549, 2093	498.12(817.92) 16, 70, 199, 535, 2079	501.91(605.31) 17, 73, 206, 551, 2049	502.43(837.59) 15, 69, 198, 531, 2118	502.00(762.89) 21, 88, 232, 571, 1926	500.01(768.16) 18, 82, 227, 566, 1996	511.70(711.16) 23, 98, 257, 617, 1888
						Pure location shifts				
−2	1		1.00(0.00) 1, 1, 1, 1, 1	1.00(0.00) 1, 1, 1, 1, 1	1.00(0.00) 1, 1, 1, 1, 1	1.00(0.00) 1, 1, 1, 1, 1	1.00(0.00) 1, 1, 1, 1, 1	1.00(0.00) 1, 1, 1, 1, 1	1.00(0.20) 1, 1, 1, 1, 1	2.00(0.00) 2,2,2,2,2
−1.5	1		1.03(0.17) 1, 1, 1, 1, 1	1.12(0.33) 1, 1, 1, 1, 2	1.05(0.22) 1, 1, 1, 1, 1	1.37(0.48) 1, 1, 1, 2, 2	1.06(0.23) 1, 1, 1, 1, 2	1.60(0.49) 1, 1, 1, 2, 2	1.17(0.37) 1, 1, 1, 1, 2	2.04(0.19) 2,2,2,2,2
−1	1		1.72(0.80) 1, 1, 2, 2, 3	2.22(1.15) 1, 1, 2, 3, 4	1.77(0.73) 1, 1, 2, 2, 3	2.22(0.72) 1, 2, 2, 3, 3	1.96(0.90) 1, 1, 2, 2, 4	2.32(0.70) 1, 2, 2, 3, 4	2.08(0.74) 1, 2, 2, 2, 3	3.06(0.77) 2,3,3,3,4
−0.5	1		5.59(3.89) 1, 3, 5, 7, 13	8.04(5.28) 2, 4, 7, 10, 18	5.33(3.32) 2, 3, 5, 7, 12	6.75(4.07) 2, 4, 6, 8, 14	6.61(4.46) 2, 4, 6, 8, 15	7.11(4.24) 2, 4, 6, 9, 15	6.67(4.31) 2, 4, 6, 8, 15	8.95(5.11) 4,6,8,11,18
0.5	1		5.57(3.86) 1, 3, 5, 7, 13	5.36(3.34) 2, 3, 5, 7, 12	8.03(5.31) 2, 4, 7, 11, 18	6.72(4.00) 2, 4, 6, 8, 14	6.54(4.44) 2, 3, 6, 8, 15	7.08(4.23) 2, 4, 6, 9, 15	6.69(4.36) 2, 4, 6, 8, 15	8.93(5.03) 4,6,8,11,18
1	1		1.72(0.80) 1, 1, 2, 2, 3	1.78(0.74) 1, 1, 2, 2, 3	2.23(1.15) 1, 1, 2, 3, 4	2.22(0.72) 1, 2, 2, 3, 3	1.96(0.90) 1, 1, 2, 2, 4	2.32(0.70) 1, 2, 2, 3, 4	2.07(0.74) 1, 2, 2, 2, 3	3.06(0.78) 2,3,3,4
1.5	1		1.03(0.17) 1, 1, 1, 1, 1	1.05(0.22) 1, 1, 1, 1, 1	1.12(0.33) 1, 1, 1, 1, 1	1.37(0.48) 1, 1, 1, 2, 2	1.06(0.23) 1, 1, 1, 1, 2	1.60(0.49) 1, 1, 1, 2	1.17(0.37) 1, 1, 1, 1, 2	2.04(0.19) 2,2,2,2
2	1		1.00(0.00) 1, 1, 1, 1, 1	1.00(0.00) 1, 1, 1, 1, 1	1.00(0.00) 1, 1, 1, 1, 1	1.00(0.00) 1, 1, 1, 1, 1	1.00(0.00) 1, 1, 1, 1, 1	1.00(0.00) 1, 1, 1, 1, 1	1.00(0.00) 1, 1, 1, 1, 1	2.00(0.00) 2,2,2,2
						Pure scale shifts				
0	0.5		11.43(40.69) 6, 7, 9, 12, 21	7.75(5.08) 4, 5, 7, 9, 14	7.83(22.79) 4, 5, 7, 9, 14	6.66(3.85) 4, 5, 6, 7, 12	14.04(100.56) 4, 6, 8, 12, 27	6.76(7.29) 3, 4, 6, 8, 14	1303.74(1756.02) 25, 85, 357, 1840, 5000	261.88(570.47) 15,33,75,218,1136
0	1.25		10.77(8.55) 2, 5, 8, 14, 27	13.03(10.13) 2, 6, 10, 17, 32	13.08(10.13) 2, 6, 11, 17, 32	12.53(9.63) 2, 6, 10, 16, 31	17.76(15.07) 3, 8, 14, 23, 47	23.19(19.34) 5, 10, 18, 30, 60	42.47(40.05) 6, 16, 30, 55, 121	56.22(56.2) 9,21,39,71,162
0	1.5		4.98(3.24) 1, 3, 4, 6, 11	5.73(3.65) 1, 3, 5, 7, 13	5.78(3.69) 1, 3, 5, 7, 12	5.51(3.48) 1, 3, 5, 7, 12	7.70(5.14) 2, 4, 6, 10, 18	9.18(5.86) 3, 5, 8, 12, 20	18.39(13.48) 4, 9, 15, 24, 44	21.22(14.77) 6,11,17,27,49
0	1.75		3.44(2.03) 1, 2, 3, 4, 7	3.87(2.26) 1, 2, 3, 5, 8	3.87(2.27) 1, 2, 3, 5, 8	3.69(2.13) 1, 2, 3, 5, 8	5.15(3.01) 2, 3, 4, 7, 11	5.90(3.37) 2, 3, 5, 8, 12	12.01(7.67) 3, 7, 10, 16, 27	13.25(7.49) 5,8,12,17,28
0	2		2.75(1.50) 1, 2, 3, 4, 7	3.05(1.68) 1, 2, 3, 5, 8	3.06(1.70) 1, 2, 3, 5, 8	2.92(1.57) 1, 2, 3, 5, 8	4.03(2.21) 1, 2, 3, 5, 8	4.50(2.42) 2, 3, 4, 6, 9	9.30(5.48) 3, 5, 8, 12, 19	10.09(5.04) 4,6,9,13,20
						Mixed location and scale shifts (at $\delta = e^{\theta}$)				
−2	0.14		1.00(0.00) 1, 1, 1, 1, 1	1.00(0.00) 1, 1, 1, 1, 1	1.00(0.00) 1, 1, 1, 1, 1	1.00(0.00) 1, 1, 1, 1, 1	1.00(0.00) 1, 1, 1, 1, 1	1.00(0.00) 1, 1, 1, 1, 1	1.00(0.02) 1, 1, 1, 1, 1	2.00(0.00) 2,2,2,2
−1.5	0.22		1.00(0.00) 1, 1, 1, 1, 1	1.00(0.00) 1, 1, 1, 1, 1	1.00(0.00) 1, 1, 1, 1, 1	1.00(0.00) 1, 1, 1, 1, 1	1.00(0.00) 1, 1, 1, 1, 1	1.00(0.00) 1, 1, 1, 1, 1	1.00(0.00) 1, 1, 1, 1, 1	2.00(0.00) 2,2,2,2
−1	0.37		1.07(0.25) 1, 1, 1, 1, 2	1.25(0.46) 1, 1, 1, 1, 2	1.10(0.30) 1, 1, 1, 1, 1	1.60(0.49) 1, 1, 2, 2	1.10(0.30) 1, 1, 1, 1, 2	1.84(0.37) 1, 2, 2, 2	1.27(0.44) 1, 1, 1, 2	2.01(0.11) 2,2,2,2
−0.5	0.6		5.93(3.41) 2, 4, 5, 7, 12	6.99(3.20) 4, 5, 6, 8, 12	5.85(3.83) 2, 3, 5, 7, 13	5.86(3.27) 3, 4, 5, 7, 12	5.96(2.99) 3, 4, 5, 7, 12	5.94(3.09) 3, 4, 5, 7, 12	4.89(2.65) 2, 3, 4, 6, 10	5.44(2.13) 3,4,5,6,9
0.5	1.65		3.44(2.10) 2, 4, 5, 7, 12	4.45(2.62) 1, 3, 4, 6, 9	3.64(2.19) 1, 2, 3, 5, 8	4.22(2.46) 1, 2, 4, 5, 9	4.63(2.85) 1, 3, 4, 6, 10	6.20(3.53) 2, 4, 5, 8, 13	7.45(4.67) 2, 4, 6, 10, 16	8.97(4.76) 3,6,8,11,18

(continued)

Table 2 (continued)

θ	δ	EC	cEC1	cEC2	cEC3	EL	cEL	EWMA-CvM	EWMA-KS
1	2.72	1.84(0.90) 1, 1, 1, 2, 2, 4	2.24(1.11) 1, 1, 1, 2, 3, 4	1.88(0.93) 1, 1, 1, 2, 3, 4	2.08(0.98) 1, 1, 2, 3, 4	2.44(1.21) 1, 2, 2, 3, 5	2.91(1.41) 1, 2, 3, 4, 5	4.14(2.21) 1, 3, 4, 5, 8	4.82(1.97) 2, 3, 4, 6, 8
1.5	4.48	1.44(0.59) 1, 1, 1, 2, 2	1.64(0.71) 1, 1, 2, 2, 3	1.44(0.51) 1, 1, 1, 2, 3	1.54(0.63) 1, 1, 1, 2, 3	1.93(0.79) 1, 1, 2, 2, 3	2.01(0.92) 1, 1, 2, 3, 4	3.51(1.64) 1, 2, 3, 4, 7	3.93(1.43) 2, 3, 4, 5, 7
2	7.39	1.25(0.46) 1, 1, 1, 1, 2	1.36(0.54) 1, 1, 1, 2, 2	1.26(0.56) 1, 1, 1, 1, 2	1.31(0.49) 1, 1, 1, 2, 2	1.75(0.62) 1, 1, 2, 2, 3	1.59(0.70) 1, 1, 1, 2, 3	3.31(1.37) 1, 2, 3, 4, 6	3.58(1.17) 2, 3, 3, 4, 6
				Mixed location and scale shifts ($\theta, \delta = e^{\frac{\theta}{4}}$)						
-2	0.37	1.00(0.00) 1, 1, 1, 1, 1	1.00(0.00) 1, 1, 1, 1, 1	1.00(0.03) 1, 1, 1, 1, 1	1.00(0.00) 1, 1, 1, 1, 1	1.00(0.00) 1, 1, 1, 1,	1.00(0.00) 1, 1, 1, 1, 1	1.00(0.01) 1, 1, 1, 1, 1	2.00(0.00) 2, 2, 2, 2
-1.5	0.47	1.00(0.00) 1, 1, 1, 1, 1	1.00(0.00) 1, 1, 1, 1, 1	1.00(0.00) 1, 1, 1, 1, 1	1.00(0.00) 1, 1, 1, 1, 1	1.00(0.00) 1, 1, 1, 1, 1	1.00(0.00) 1, 1, 1, 1, 1	1.00(0.00) 1, 1, 1, 1, 1	2.00(0.00) 2, 2, 2, 2
-1	0.61	1.30(0.49) 1, 1, 1, 2, 2	1.64(0.72) 1, 1, 2, 2, 3	1.36(0.50) 1, 1, 1, 2, 2	1.82(0.45) 1, 2, 2, 2, 2	1.40(0.53) 1, 1, 1, 2, 2	1.94(0.36) 1, 2, 2, 2, 2	1.62(0.50) 1, 1, 2, 2, 2	2.36(0.49) 2, 2, 2, 3, 3
-0.5	0.78	5.71(3.71) 1, 3, 5, 7, 13	7.99(4.52) 2, 5, 7, 10, 16	5.40(3.40) 2, 3, 5, 7, 12	6.05(3.44) 2, 4, 5, 7, 12	6.23(3.79) 2, 4, 5, 8, 13	6.29(3.53) 3, 4, 5, 8, 13	5.63(3.33) 2, 3, 5, 7, 12	7.04(3.49) 3, 5, 6, 8, 13
0.5	1.28	5.00(3.47) 1, 3, 4, 7, 12	5.41(3.40) 1, 3, 5, 7, 12	6.01(3.98) 1, 3, 5, 8, 14	6.96(4.26) 2, 4, 6, 9, 15	6.21(4.25) 1, 3, 5, 8, 14	8.08(4.94) 2, 5, 7, 10, 17	7.56(5.09) 2, 4, 6, 10, 17	9.82(5.63) 4, 6, 9, 12, 21
1	1.65	2.29(1.31) 1, 1, 2, 3, 5	2.53(1.28) 1, 2, 2, 3, 5	2.68(1.57) 1, 2, 3, 6	3.17(1.54) 1, 2, 3, 4, 6	2.74(1.52) 1, 2, 2, 4, 6	3.51(1.63) 1, 2, 3, 4, 7	3.17(1.60) 1, 2, 3, 4, 6	4.29(1.70) 2, 3, 4, 5, 7
1.5	2.12	1.69(0.85) 1, 1, 2, 3, 5	1.91(0.87) 1, 1, 2, 2, 3	1.89(1.01) 1, 1, 2, 2, 4	2.34(1.01) 1, 2, 2, 3, 4	2.00(0.99) 1, 1, 2, 2, 4	2.59(1.09) 1, 2, 2, 3, 5	2.32(1.06) 1, 2, 2, 3, 4	3.19(1.08) 2, 3, 4, 5
2	2.72	1.48(0.69) 1, 1, 1, 2, 3	1.73(0.76) 1, 1, 2, 2, 3	1.59(0.79) 1, 1, 2, 2, 3	2.01(0.86) 1, 1, 2, 2, 4	1.73(0.82) 1, 1, 2, 2, 3	2.30(0.96) 1, 2, 2, 3, 4	2.06(0.92) 1, 1, 2, 3, 4	2.81(0.87) 2, 3, 3, 4
				Mixed location and scale shifts ($\theta, \delta = e^{2\theta}$)						
-1	0.14	1.00(0.02) 1, 1, 1, 1, 1	1.01(0.09) 1, 1, 1, 1, 1	1.00(0.03) 1, 1, 1, 1, 1	1.12(0.32) 1, 1, 1, 1, 2	1.00(0.02) 1, 1, 1, 1, 1	1.42(0.49) 1, 1, 1, 2, 2	1.00(0.04) 1, 1, 1, 1, 1	2.00(0.00) 2, 2, 2, 2
-0.5	0.37	6.64(2.20) 3, 5, 7, 8, 10	5.76(1.87) 4, 5, 5, 6, 9	7.74(4.46) 3, 4, 7, 10, 17	6.23(3.30) 3, 4, 5, 7, 13	7.04(2.56) 3, 5, 7, 9, 11	6.34(3.10) 3, 4, 5, 8, 13	3.86(1.50) 2, 3, 4, 4, 7	3.87(0.93) 3, 3, 4, 4, 5
-0.25	0.61	32.81(155.97) 8, 12, 17, 26, 68	32.72(67.75) 8, 14, 21, 34, 84	21.98(66.77) 6, 10, 15, 23, 50	25.13(58.36) 7, 11, 18, 28, 60	30.75(169.14) 6, 10, 14, 22, 63	22.46(60.88) 7, 9, 15, 24, 55	144.63(567.46) 6, 11, 21, 50, 513	48.71(215.66) 6, 10, 15, 28, 134
0.25	1.65	3.77(2.29) 1, 2, 3, 5, 8	4.67(2.81) 1, 3, 4, 6, 10	4.02(2.43) 1, 2, 4, 5, 9	4.24(2.51) 1, 2, 4, 5, 9	5.51(3.40) 2, 3, 5, 7, 12	6.70(3.91) 2, 4, 6, 9, 14	1.39(7.59) 3, 6, 10, 15, 26	13.04(7.56) 4, 8, 11, 16, 27
0.5	2.72	1.93(0.93) 1, 1, 2, 2, 4	2.24(1.10) 1, 1, 2, 3, 4	2.00(0.99) 1, 1, 2, 2, 4	2.07(0.98) 1, 1, 2, 3, 4	2.71(1.27) 1, 2, 2, 3, 5	2.97(1.44) 1, 2, 3, 4, 6	5.58(2.93) 2, 3, 5, 7, 11	6.12(2.58) 3, 4, 6, 8, 11
1	7.39	1.27(0.47) 1, 1, 1, 2, 2	1.35(0.53) 1, 1, 1, 2, 2	1.29(0.49) 1, 1, 1, 2, 2	1.30(0.48) 1, 1, 1, 2, 2	1.82(0.60) 1, 1, 2, 2, 3	1.59(0.70) , 1, 1, 2, 3	3.65(1.46) 1, 3, 3, 5, 6	3.85(1.25) 2, 3, 4, 5, 6

Table 3 Comparison of the eight schemes under $N(\theta, \delta)$ distribution for $m = 100$, $n = 5$, $\lambda = 0.05$, and $ARL_0 = 500$

θ	δ	EC	cEC1	cEC2	cEC3	EL	cEL	EWMA-CvM	EWMA-KS
0	0	503.22(890.88) 12, 57, 172, 503, 2254	499.95(818.02) 17, 70, 197, 541, 2062	504.99(823.77) 16, 72, 203, 549, 2095	499.26(798.52) 18, 75, 208, 550, 2027	503.69(849.11) 15, 69, 196, 524, 2130	502.54(762.52) 21, 87, 224, 573, 1936	501.54(769.33) 17, 81, 224, 573, 1940	506.92(709.16) 22, 98, 263, 632, 1893
Pure location shifts									
−2	1	1.27(0.48) 1, 1, 1, 2, 2	1.54(0.66) 1, 1, 1, 2, 3	1.36(0.51) 1, 1, 1, 2, 2	1.80(0.53) 1, 1, 1, 2, 3	1.38(0.54) 1, 1, 1, 2, 2	1.96(0.44) 1, 2, 2, 2, 3	1.56(0.60) 1, 1, 2, 2, 2	2.37(0.53) 2, 2, 2, 3, 3
−1.5	1	1.90(0.91) 1, 1, 2, 2, 4	2.58(1.30) 1, 2, 2, 3, 5	1.96(0.82) 1, 1, 2, 2, 3	2.45(0.82) 1, 2, 2, 3, 4	2.08(0.96) 1, 1, 2, 3, 4	2.56(0.82) 2, 2, 2, 3, 4	2.21(0.86) 2, 2, 2, 3, 4	3.15(0.91) 2, 3, 3, 4, 5
−1	1	3.99(2.52) 1, 2, 3, 5, 9	5.79(3.27) 2, 3, 5, 7, 12	3.80(2.10) 1, 2, 3, 5, 8	4.54(2.23) 2, 3, 4, 6, 9	4.38(2.67) 1, 3, 4, 6, 9	4.76(2.33) 2, 3, 4, 6, 9	4.15(2.20) 2, 3, 4, 5, 8	5.46(2.41) 3, 4, 5, 6, 10
−0.5	1	24.10(44.35) 3, 7, 14, 26, 74	37.17(73.75) 5, 12, 21, 39, 113	19.14(37.08) 3, 7, 12, 21, 54	24.41(38.95) 4, 9, 15, 27, 69	25.83(42.93) 3, 8, 15, 29, 78	26.19(41.46) 5, 10, 16, 29, 75	21.30(45.89) 4, 8, 13, 24, 60	26.14(39.34) 5, 10, 17, 29, 74
0.5	1	23.87(46.63) 3, 7, 14, 26, 73	19.08(28.06) 5, 12, 21, 39, 54	37.52(41.67) 5, 12, 21, 39, 112	24.74(41.67) 4, 9, 15, 27, 71	25.82(46.35) 3, 8, 15, 29, 79	26.46(43.89) 5, 10, 16, 29, 77	21.16(31.38) 4, 8, 13, 24, 60	26.21(38.74) 5, 10, 17, 29, 74
1	1	4.01(2.54) 1, 2, 3, 5, 9	3.78(2.09) 1, 2, 3, 5, 8	5.78(3.27) 2, 3, 5, 7, 12	4.56(2.26) 2, 3, 4, 6, 9	4.39(2.68) 1, 3, 4, 6, 9	4.78(2.34) 2, 3, 4, 6, 9	4.15(2.18) 2, 3, 4, 5, 8	5.46(2.41) 3, 4, 5, 7, 10
1.5	1	1.89(0.91) 1, 1, 2, 2, 3	1.96(0.82) 1, 1, 2, 2, 3	2.58(1.30) 1, 2, 2, 3, 4	2.45(0.83) 1, 2, 2, 3, 4	2.08(0.96) 1, 1, 2, 3, 4	2.57(0.81) 2, 2, 2, 3, 4	2.21(0.85) 1, 2, 2, 3, 4	3.15(0.9) 2, 3, 3, 4, 5
2	1	1.27(0.48) 1, 1, 1, 2, 2	1.35(0.51) 1, 1, 1, 2, 2	1.54(0.66) 1, 1, 1, 2, 3	1.79(0.53) 1, 1, 1, 2, 3	1.39(0.54) 1, 1, 1, 2, 2	1.96(0.44) 1, 2, 2, 2, 3	1.55(0.56) 1, 1, 2, 2, 2	2.37(0.53) 2, 2, 2, 3, 3
Pure scale shifts									
0	0.5	47.76(231.65) 8, 12, 18, 31, 105	17.67(67.17) 6, 8, 11, 17, 39	17.39(61.41) 6, 8, 11, 17, 39	13.41(40.49) 5, 7, 10, 14, 29	36.40(196.41) 5, 8, 14, 24, 81	12.87(47.98) 3, 6, 9, 13, 30	3034.88(2107.62) 53, 607, 4194, 5000, 5000	533.6(915.36) 20, 61, 172, 540, 2452
0	1.25	31.85(46.09) 4, 10, 19, 36, 101	41.03(58.20) 5, 13, 24, 47, 129	41.03(57.89) 5, 13, 24, 47, 131	41.56(63.60) 5, 13, 24, 47, 131	47.15(66.85) 5, 14, 27, 55, 151	68.71(99.96) 8, 20, 39, 78, 227	86.55(107.62) 8, 26, 52, 107, 279	107.43(141.94) 11, 30, 62, 129, 354
0	1.5	10.07(8.51) 2, 5, 8, 13, 26	12.33(10.30) 2, 6, 10, 16, 31	12.28(10.21) 2, 6, 10, 16, 31	11.93(10.12) 2, 6, 9, 15, 30	14.81(12.91) 3, 7, 11, 19, 39	19.24(16.95) 4, 9, 15, 24, 50	34.04(32.93) 5, 13, 24, 43, 96	39.16(38.74) 8, 16, 28, 48, 107
0	1.75	5.85(4.02) 1, 3, 5, 7, 13	6.82(4.58) 2, 4, 6, 9, 16	6.90(4.63) 2, 4, 6, 9, 16	6.56(4.34) 2, 4, 6, 8, 15	8.32(5.74) 2, 4, 7, 11, 19	10.09(6.78) 3, 5, 8, 13, 23	20.03(15.79) 4, 9, 16, 26, 50	21.73(16.1) 6, 11, 17, 27, 52
0	2	4.20(2.58) 1, 2, 4, 5, 9	4.82(2.93) 1, 3, 4, 6, 10	4.80(2.93) 1, 3, 4, 6, 10	4.61(2.78) 1, 3, 4, 6, 10	5.93(3.63) 2, 3, 5, 8, 13	6.91(4.03) 2, 4, 6, 9, 15	14.16(9.80) 3, 7, 12, 18, 33	15.15(9.75) 5, 9, 13, 19, 33
Mixed location and scale shifts ($\theta, \delta = e^{\theta}$)									
−2	0.14	1.00(0.02) 1, 1, 1, 1, 1	1.02(0.13) 1, 1, 1, 1, 1	1.00(0.03) 1, 1, 1, 1, 1	1.27(0.45) 1, 1, 1, 1, 2	1.00(0.02) 1, 1, 1, 1, 1	1.65(0.48) 1, 1, 2, 2, 2	1.00(0.03) 1, 1, 1, 1, 1	2.00(0.00) 2, 2, 2, 2, 2
−1.5	0.22	1.24(0.44) 1, 1, 1, 1, 2	1.79(0.66) 1, 1, 2, 2, 3	1.31(0.47) 1, 1, 1, 2, 2	1.95(0.23) 1, 1, 2, 2, 2	1.27(0.45) 1, 1, 1, 2, 2	2.00(0.11) 2, 2, 2, 2, 2	1.36(0.48) 1, 1, 1, 1, 2	2(0.06) 2, 2, 2, 2, 2
−1	0.37	3.36(1.45) 2, 2, 3, 4, 6	3.84(0.78) 3, 3, 4, 4, 5	3.00(1.38) 2, 2, 3, 3, 5	3.03(0.96) 2, 2, 3, 3, 4	3.35(1.57) 2, 2, 3, 4, 6	3.15(1.00) 2, 3, 3, 4, 5	2.35(0.60) 2, 2, 2, 3, 3	2.94(0.6) 2, 3, 3, 3, 4
−0.5	0.61	23.81(44.68) 5, 11, 16, 26, 60	22.96(41.24) 6, 10, 14, 23, 62	25.13(34.69) 5, 10, 18, 30, 67	23.16(36.59) 5, 9, 15, 26, 64	22.38(46.78) 5, 10, 16, 25, 54	21.70(27.85) 5, 9, 15, 25, 58	20.21(97.55) 4, 6, 10, 17, 47	14.66(88.53) 4, 7, 10, 15, 35
0.5	1.65	5.41(3.70) 1, 3, 5, 7, 12	7.07(4.68) 2, 4, 6, 9, 16	6.00(4.04) 2, 3, 5, 8, 14	6.68(4.39) 2, 4, 6, 9, 15	7.28(5.04) 2, 4, 6, 9, 17	9.66(6.35) 3, 5, 8, 12, 22	12.86(9.75) 3, 6, 10, 17, 31	15.15(10.43) 5, 8, 12, 19, 35

(continued)

Table 3 (continued)

Table 3 (Continue)

θ	δ	EC	cEC1	cEC2	cEC3	EL	cEL	EWMA-CvM	EWMA-KS
1	2.72	2.23(1.15) 1, 1, 2, 3, 4	2.76(1.39) 1, 2, 3, 3, 5	2.32(1.8) —, 1, 2, 3, 5	2.50(1.25) 1, 2, 2, 3, 5	2.99(1.49) 1, 2, 3, 4, 6	3.57(1.67) 2, 2, 3, 4, 7	5.57(3.14) 2, 3, 5, 7, 11	6.42(2.83) 3, 4, 6, 8, 12
1.5	4.48	1.58(0.70) 1, 1, 1, 2, 3	1.83(0.81) 1, 1, 2, 2, 3	—.62(0.71) 1, 1, 2, 2, 3	1.68(0.73) 1, 1, 2, 2, 3	2.15(0.88) 1, 2, 2, 3, 4	2.32(0.96) 1, 2, 2, 3, 4	4.24(2.06) 3, 4, 5, 8	4.70(1.74) 2, 3, 4, 6, 8
2	7.39	1.32(0.51) 1, 1, 1, 2, 2	1.45(0.59) 1, 1, 1, 2, 2	1.35(0.55) 1, 1, 1, 2, 2	1.37(0.53) 1, 1, 1, 2, 2	1.85(0.64) 1, 1, 2, 2, 3	1.80(0.72) 1, 1, 2, 2, 3	3.71(1.58) 1, 3, 4, 5, 6	4.01(1.33) 2, 3, 4, 5, 6
Mixed location and scale shifts ($\theta, \delta = e^{\frac{w}{2}}$)									
-2	0.37	1.00(0.06) 1, 1, 1, 1, 1	1.06(0.25) 1, 1, 1, 1, 2	1.01(0.04) 1, 1, 1, 1, 1	1.41(0.49) 1, 1, 1, 2, 2	1.01(0.09) 1, 1, 1, 1, 1	1.75(0.43) 1, 1, 2, 2, 2	1.03(0.17) 1, 1, 1, 1, 1	2.00(0.01) 2, 2, 2, 2, 2
-1.5	0.47	1.43(0.55) 1, 1, 1, 2, 2	2.00(0.79) 1, 1, 2, 2, 3	1.89(0.53) 1, 1, 1, 2, 2	1.98(0.30) 1, 2, 2, 2, 2	1.48(0.55) 1, 1, 1, 2, 2	2.04(0.24) 2, 2, 2, 2, 3	1.64(0.49) 1, 1, 2, 2, 2	2.23(0.42) 2, 2, 2, 2, 3
-1	0.61	3.65(1.92) 1, 2, 3, 5, 7	4.56(1.41) 2, 4, 4, 5, 7	3.28(1.65) 2, 3, 3, 4, 6	3.49(1.33) 2, 3, 3, 4, 6	3.76(1.96) 2, 3, 3, 5, 7	3.61(1.38) 2, 3, 3, 4, 6	2.94(1.12) 2, 2, 3, 3, 5	3.75(1.10) 2, 3, 4, 4, 6
-0.5	0.78	39.19(109.86) 4, 10, 19, 37, 121	45.52(142.88) 7, 13, 20, 37, 133	33.91(88.65) 4, 9, 16, 32, 109	32.48(84.75) 5, 9, 16, 30, 99	35.48(91.02) 5, 10, 18, 35, 104	29.95(66.97) 5, 10, 16, 30, 88	23.41(88.86) 4, 7, 12, 22, 63	22.48(51.58) 5, 9, 13, 23, 62
0.5	1.28	10.31(9.58) 2, 4, 8, 13, 27	11.89(10.35) 2, 6, 9, 15, 30	13.25(12.82) 2, 6, 10, 17, 35	13.86(12.18) 3, 6, 11, 17, 35	13.05(12.05) 3, 6, 10, 16, 34	17.68(15.57) 4, 8, 13, 22, 45	16.88(15.64) 3, 7, 12, 21, 45	21.66(20.67) 5, 10, 16, 26, 56
1	1.65	3.33(2.03) 1, 2, 3, 4, 7	4.09(2.36) 1, 2, 4, 5, 8	3.85(2.33) 1, 2, 5, 8	4.43(2.49) 1, 3, 4, 6, 9	4.12(2.46) 1, 2, 4, 5, 9	5.61(3.01) 2, 3, 5, 7, 11	5.58(3.35) 2, 3, 5, 7, 12	7.06(3.55) 3, 5, 6, 9, 14
1.5	2.12	2.13(1.12) 1, 2, 3, 4, 7	2.68(1.37) 1, 2, 2, 3, 5	2.32(1.21) 1, 1, 2, 3, 5	2.72(1.34) 1, 2, 2, 3, 5	2.61(1.35) 1, 2, 2, 3, 5	3.55(1.64) 2, 2, 3, 4, 7	3.68(1.97) 1, 2, 3, 5, 7	4.74(1.94) 2, 3, 4, 6, 8
2	2.72	1.71(0.82) 1, 1, 2, 2, 3	2.13(1.00) 1, 1, 2, 3, 4	1.81(0.85) 1, 1, 2, 2, 3	2.09(0.95) 1, 1, 2, 3, 4	2.12(1.01) 1, 1, 2, 3, 4	2.77(1.18) 1, 2, 3, 3, 5	3.08(1.58) 1, 2, 3, 4, 6	3.93(1.47) 2, 3, 4, 5, 7
Mixed location and scale shifts ($\theta, \delta = e^{2w}$)									
-1	0.14	3.24(1.28) 2, 2, 3, 4, 6	3.42(0.60) 3, 3, 3, 4, 4	2.87(1.35) 2, 2, 3, 3, 4	2.80(0.82) 2, 2, 3, 3, 4	3.15(1.56) 2, 2, 3, 4, 6	2.91(0.85) 2, 2, 3, 3, 4	2.03(0.26) 2, 2, 2, 2, 2	2.29(0.45) 2, 2, 2, 3, 3
-0.5	0.37	9.66(2.75) 6, 8, 9, 11, 15	10.39(4.89) 5, 7, 9, 13, 20	11.54(5.59) 5, 7, 10, 15, 22	11.84(5.73) 5, 8, 11, 15, 22	10.60(3.86) 5, 8, 10, 13, 17	12.37(6.56) -, 7, 11, 15, 24	8.11(8.67) 3, 5, 6, 9, 18	6.47(3.69) 3, 5, 6, 7, 12
-0.25	0.61	178.23(548.06) 12, 24, 45, 106, 636	135.00(392.80) 12, 24, 44, 100, 452	63.34(223 55) 8, 15, 25, 43, 181	72.18(217.09) 9, 17, 30, 60, 223	121.01(437.98) 8, 17, 32, 70, 376	31.32(208.79) 6, 13, 24, 49, 181	816.10(1532.94) 10, 29, 90, 559, 5000	250.20(653.70) 9, 20, 46, 150, 1224
0.25	1.65	6.53(4.68) 2, 3, 5, 8, 15	8.32(5.89) 2, 4, 7, 11, 19	7.35(5.22) 2, 4, 6, 9, 17	7.69(5.43) 2, 4, 6, 10, 18	9.19(5.74) 2, 5, 7, 12, 22	11.71(8.38) 3, 6, 10, 15, 27	19.64(16.12) 4, 9, 15, 25, 51	22.4(17.47) 6, 11, 18, 28, 55
0.5	2.72	2.46(1.28) 2, 2, 3, 5	2.90(1.51) 1, 2, 3, 4, 6	2.61(1.36) 1, 2, 3, 5	2.65(1.35) 1, 2, 3, 5	3.39(1.71) 1, 2, 3, 4, 7	3.85(1.84) 2, 3, 5, 7	7.43(4.23) 2, 4, 7, 10, 15	8.08(3.78) 3, 5, 7, 10, 15
1	7.39	1.34(0.53) 1, 1, 1, 2, 2	1.45(0.59) 1, 1, 1, 2, 2	1.39(0.56) 1, 1, 1, 2, 2	1.38(0.54) 1, 1, 1, 2, 2	1.91(0.64) 1, 2, 2, 2, 3	-.82(0.73) 1, 1, 2, 2, 3	3.97(1.65) 2, 3, 4, 5, 7	4.27(1.4) 2, 3, 4, 5, 7

Table 4 Comparison of the eight schemes under Laplace(θ, δ) distribution for $m = 100$, $n = 5$, $\lambda = 0.05$, and $ARL_0 = 500$

θ	δ		EC	cEC1	cEC2	cEC3	EL	cEL	EWMA-CvM	EWMA-KS
0	1		506.39(892.29) 12, 56, 173, 509, 2269	498.35(807.69) 16, 72, 203, 542, 2055	499.10(812.78) 16, 71, 200, 538, 2082	496.50(791.98) 17, 74, 206, 549, 2009	512.04(861.03) 15, 69, 198, 535, 2177	505.66(779.22) 21, 87, 228, 571, 1959	493.44(760.45) 17, 82, 222, 559, 1939	506.86(710.07) 22, 95, 259, 626, 1874
						Pure location shifts				
-2	1		1.70(0.76) 1, 1, 2, 2, 3	2.27(1.07) 1, 2, 2, 3, 4	1.80(0.75) 1, 1, 2, 2, 3	2.32(0.72) 1, 2, 2, 3, 4	1.81(0.79) 1, 1, 2, 2, 3	2.45(0.72) 2, 2, 2, 3, 4	1.97(0.79) 1, 1, 2, 2, 3	2.73(0.73) 2, 2, 3, 3, 4
-1.5	1		2.66(1.44) 1, 2, 2, 3, 5	3.79(1.92) 1, 2, 3, 5, 7	2.65(1.25) 1, 2, 3, 3, 5	3.24(1.33) 2, 2, 3, 4, 6	2.77(1.42) 1, 2, 2, 3, 5	3.39(1.37) 2, 2, 3, 4, 6	2.79(1.20) 1, 2, 3, 3, 5	3.59(1.20) 2, 3, 3, 4, 6
-1	1		6.08(5.02) 2, 3, 5, 7, 15	8.32(5.28) 2, 3, 5, 7, 10, 18	5.41(3.81) 2, 3, 5, 7, 12	6.27(3.91) 2, 4, 5, 8, 13	6.29(5.15) 2, 3, 5, 8, 15	6.60(4.16) 2, 4, 6, 8, 14	5.21(3.14) 2, 3, 4, 6, 11	6.13(3.05) 3, 4, 5, 7, 12
-0.5	1		52.94(152.91) 4, 10, 20, 45, 183	64.56(154.81) 7, 16, 29, 60, 213	41.89(114.60) 4, 9, 17, 36, 139	44.44(110.28) 5, 11, 20, 40, 144	59.02(164.07) 4, 11, 23, 50, 206	47.76(115.42) 6, 12, 22, 43, 155	28.86(68.04) 4, 9, 16, 29, 86	30.13(56.89) 5, 10, 17, 32, 90
0.5	1		51.81(139.25) 4, 10, 20, 46, 182	41.71(115.38) 4, 9, 17, 35, 138	64.21(151.26) 6, 16, 29, 60, 216	44.67(117.80) 5, 11, 20, 40, 141	55.28(161.90) 4, 11, 23, 51, 207	48.17(120.12) 6, 12, 22, 44, 158	29.31(65.47) 4, 9, 16, 29, 87	29.88(65.6) 5, 10, 17, 31, 89
1	1		6.10(5.06) 2, 3, 5, 8, 14	5.43(4.03) 2, 3, 5, 7, 12	8.28(5.18) 2, 5, 7, 10, 18	6.26(3.82) 2, 4, 5, 8, 13	6.31(5.33) 2, 3, 5, 8, 15	6.62(3.96) 2, 4, 6, 8, 14	5.14(3.00) 2, 3, 4, 6, 11	6.13(3.06) 3, 4, 5, 7, 12
1.5	1		2.66(1.45) 1, 2, 2, 3, 5	2.66(1.26) 1, 2, 2, 3, 5	3.79(1.92) 1, 2, 3, 5, 7	3.24(1.34) 2, 2, 3, 4, 6	2.78(1.43) 1, 2, 2, 3, 5	3.38(1.37) 2, 2, 3, 4, 6	2.80(1.22) 1, 2, 3, 3, 5	3.59(1.2) 2, 3, 3, 4, 6
2	1		1.70(0.77) 1, 1, 2, 2, 3	1.80(0.74) 1, 1, 2, 2, 3	2.27(1.07) 1, 2, 3, 3, 4	2.32(0.72) 1, 2, 2, 3, 4	1.81(0.79) 1, 1, 2, 2, 3	2.45(0.71) 2, 2, 2, 3, 4	1.97(0.77) 1, 1, 2, 2, 3	2.74(0.74) 2, 2, 3, 3, 4
						Pure scale shifts				
0	0.5		235.14(688.91) 12, 24, 48, 124, 986	61.93(263.18) 8, 13, 21, 39, 163	58.42(237.79) 8, 13, 21, 39, 155	39.61(167.61) 7, 11, 18, 31, 98	109.85(434.80) 6, 13, 25, 57, 334	30.02(130.35) 4, 8, 14, 25, 76	3464.51(1980.80) 91, 1299, 5000, 5000, 5000	903.59(1239.9) 27, 112, 370, 1126, 4177
0	1.25		60.60(126.22) 5, 13, 29, 63, 207	76.85(135.55) 6, 18, 37, 82, 270	78.30(147.55) 6, 18, 38, 83, 267	80.09(141.47) 6, 18, 38, 86, 281	86.22(160.60) 6, 19, 41, 92, 299	122.46(208.13) 10, 28, 60, 134, 424	123.58(173.33) 9, 31, 69, 148, 418	154.59(221.58) 13, 38, 84, 181, 524
0	1.5		17.98(20.85) 3, 7, 12, 22, 52	29.16(27.48) 4, 9, 16, 28, 67	23.10(27.10) 3, 9, 16, 28, 67	22.92(27.63) 4, 9, 16, 27, 66	25.70(30.50) 4, 9, 17, 31, 76	37.18(49.58) 6, 13, 24, 44, 111	52.50(58.51) 6, 18, 35, 66, 157	62.16(74.22) 9, 21, 39, 75, 188
0	1.75		9.60(8.32) 2, 4, 7, 12, 24	11.79(9.98) 2, 5, 9, 15, 30	11.81(10.10) 2, 5, 9, 15, 30	11.42(9.42) 2, 5, 9, 14, 29	13.48(11.51) 3, 6, 10, 17, 34	17.41(15.06) 4, 8, 13, 22, 44	30.53(28.14) 5, 12, 22, 39, 85	33.61(31.61) 7, 14, 24, 42, 90
0	2		6.56(4.76) 2, 3, 5, 8, 15	7.83(5.60) 2, 4, 6, 10, 18	7.83(5.70) 2, 4, 6, 10, 18	7.53(5.41) 2, 4, 6, 10, 18	9.15(6.73) 2, 5, 7, 12, 22	11.13(7.88) 3, 6, 9, 14, 26	21.47(17.50) 4, 10, 17, 28, 54	22.85(17.77) 6, 11, 18, 29, 56
						Mixed location and scale shifts (θ, $\delta = e^{\theta}$)				
-2	0.14		1.22(0.42) 1, 1, 1, 1, 2	1.77(0.66) 1, 1, 2, 2, 3	1.27(0.45) 1, 1, 1, 2, 2	1.95(0.23) 1, 2, 2, 2, 2	1.22(0.42) 1, 1, 1, 1, 2	2.00(0.11) 2, 2, 2, 2, 2	1.21(0.41) 1, 1, 1, 1, 2	2.00(0.01) 2, 2, 2, 2, 2
-1.5	0.22		2.02(0.76) 1, 2, 2, 2, 3	2.76(0.65) 2, 2, 3, 3, 4	1.96(0.56) 1, 2, 2, 2, 3	2.17(0.40) 2, 2, 2, 2, 3	1.98(0.67) 1, 2, 2, 2, 3	2.22(0.44) 2, 2, 2, 2, 3	1.84(0.38) 1, 2, 2, 2, 2	2.11(0.32) 2, 2, 2, 2, 3
-1	0.37		4.83(2.21) 3, 4, 4, 6, 9	4.55(1.11) 3, 4, 4, 5, 7	4.39(2.81) 2, 3, 4, 5, 10	3.96(1.69) 2, 3, 4, 5, 7	5.04(2.92) 2, 3, 4, 6, 11	4.14(1.84) 2, 3, 4, 5, 7	2.79(0.91) 2, 2, 3, 3, 4	3.29(0.82) 2, 3, 3, 4, 5
-0.5	0.61		49.42(105.39) 7, 15, 27, 50, 152	38.13(93.67) 7, 12, 19, 34, 112	44.98(75.48) 5, 13, 24, 49, 146	41.01(80.62) 6, 11, 20, 41, 137	62.85(154.15) 7, 16, 29, 60, 202	43.46(97.44) 6, 12, 21, 42, 142	28.75(153.58) 4, 7, 12, 21, 65	17.12(34.2) 4, 7, 11, 18, 44
0.5	1.65		8.29(6.95) 2, 4, 6, 10, 21	10.69(8.65) 2, 5, 8, 13, 26	9.97(8.56) 2, 5, 8, 12, 25	10.64(8.90) 2, 5, 8, 13, 26	10.60(9.31) 2, 5, 8, 13, 27	14.60(12.63) 4, 7, 11, 18, 36	16.93(14.73) 3, 8, 13, 21, 44	19.34(16.66) 5, 9, 15, 24, 48

Continue

(continued)

Table 4 (continued)

Table 4 (Continue)

θ	δ	EC	cEC₁	cEC2	cEC3	EL	cEL	EWMA-CvM	EWMA-KS
1	2.72	2.80(1.52) 1, 2, 2, 4, 6	3.53(1.92) 1, 2, 3, 5, 7	2.98(1.61) 1, 2, 3, 4, 6	3.25(1.74) 1, 2, 3, 4, 6	3.55(1.85) 1, 2, 3, 4, 7	4.41(2.14) 2, 3, 4, 5, 8	6.58(3.88) 2, 4, 6, 8, 14	7.44(3.57) 3, 5, 7, 9, 14
1.5	4.48	1.80(0.82) 1, 1, 2, 2, 3	2.16(0.99) 1, 1, 2, 3, 4	1.87(0.85) 1, 1, 2, 2, 3	1.95(0.88) 1, 1, 2, 2, 4	2.37(1.0*) 1, 2, 2, 3, 4	2.71(1.05) 1, 2, 2, 3, 5	4.67(2.38) 1, 3, 4, 6, 9	5.26(2.04) 3, 4, 5, 6, 9
2	7.39	1.43(0.58) 1, 1, 1, 2, 2	1.64(0.68) 1, 1, 2, 2, 3	1.47(0.60) 1, 1, 1, 2, 2	1.50(0.61) 1, 1, 1, 2, 3	1.96(0.7?) 1, 2, 2, 2, 3	2.07(0.74) 1, 2, 2, 2, 3	3.98(1.81) 1, 3, 4, 5, 7	4.40(1.51) 2, 3, 4, 5, 7
Mixed location and scale shifts ($\theta, \delta = e^{\theta/2}$)									
-2	0.37	1.32(0.49) 1, 1, 1, 2, 2	1.88(0.71) 1, 1, 2, 2, 3	1.38(0.49) 1, 1, 1, 2, 2	1.97(0.24) 2, 2, 2, 2, 2	1.35(0.49) 1, 1, 1, 2, 2	2.01(0.16) 2, 2, 2, 2, 2	1.43(0.50) 1, 1, 1, 2, 2	2.05(0.22) 2, 2, 2, 2, 3
-1.5	0.47	2.23(0.96) 1, 2, 2, 3, 4	3.12(0.95) 2, 2, 3, 4, 5	2.12(0.73) 1, 2, 2, 2, 3	2.41(0.62) 2, 2, 2, 2, 3	2.22(0.89) 1, 2, 2, 3, 4	2.48(0.65) 2, 2, 2, 3, 4	2.03(0.52) 1, 2, 2, 2, 3	2.58(0.6) 2, 2, 3, 3, 3
-1	0.61	5.79(3.75) 2, 3, 5, 7, 13	5.97(2.25) 3, 4, 6, 7, 10	4.92(3.60) 2, 3, 4, 5, 11	4.81(2.45) 2, 3, 4, 6, 9	5.84(4.17) 2, 3, 5, 7, 13	5.02(2.70) 2, 3, 4, 6, 9	3.65(1.67) 2, 2, 3, 4, 7	4.3(1.57) 3, 3, 4, 5, 7
-0.5	0.78	91.45(251.67) 6, 15, 31, 75, 338	72.59(226.63) 8, 16, 27, 54, 239	67.85(174.93) 5, 11, 23, 56, 250	60.84(181.88) 5, 12, 21, 46, 216	96.64(273.70) 6, 15, 32, 78, 34	58.17(156.17) 6, 12, 23, 48, 206	31.32(118.16) 4, 8, 14, 26, 87	25.6(68.07) 5, 9, 14, 25, 70
0.5	1.28	17.91(28.18) 3, 6, 11, 20, 52	19.33(26.12) 3, 7, 13, 22, 55	25.93(39.38) 3, 9, 16, 29, 79	22.38(31.43) 4, 9, 15, 26, 64	21.96(31.99) 3, 8, 13, 25, 66	27.79(36.85) 5, 11, 18, 32, 80	22.54(28.03) 4, 8, 15, 26, 66	26.13(34.72) 5, 10, 17, 30, 73
1	1.65	4.30(2.94) 1, 2, 4, 6, 10	5.31(3.35) 2, 3, 5, 7, 12	5.44(3.77) 2, 3, 5, 7, 12	5.85(3.71) 2, 3, 5, 7, 13	5.10(3.31) 2, 3, 4, 6, 11	7.01(4.23) 2, 4, 6, 9, 15	6.62(4.28) 2, 4, 6, 8, 15	7.83(4.24) 3, 5, 7, 10, 16
1.5	2.12	2.51(1.35) 1, 2, 2, 3, 5	3.16(1.67) 1, 2, 3, 4, 6	2.83(1.53) 1, 2, 3, 4, 6	3.24(1.64) 1, 2, 3, 4, 6	2.97(1.52) 1, 2, 3, 4, 6	4.02(1.87) 2, 3, 4, 5, 8	4.19(2.32) 1, 3, 4, 5, 9	5.16(2.19) 3, 4, 5, 6, 9
2	2.72	1.88(0.90) 1, 1, 2, 2, 4	2.37(1.13) 1, 2, 2, 3, 4	2.02(0.95) 1, 1, 2, 2, 4	2.31(1.05) 1, 2, 2, 3, 4	2.28(1.05) 1, 2, 2, 3, 4	2.99(1.20) 2, 2, 3, 4, 5	3.44(1.79) 1, 2, 3, 4, 7	4.24(1.6) 2, 3, 4, 5, 7
Mixed location and scale shifts ($\theta, \delta = e^{2\theta}$)									
-1	0.14	4.24(1.54) 2, 3, 4, 5, 7	3.83(0.67) 3, 3, 4, 4, 5	3.96(2.33) 2, 3, 3, 4, 9	3.45(1.26) 2, 3, 3, 4, 6	4.49(2.59) 2, 3, 4, 5, 10	3.60(1.40) 2, 3, 3, 4, 6	2.17(0.40) 2, 2, 2, 2, 3	2.54(0.51) 2, 2, 3, 3, 3
-0.5	0.37	13.19(5.71) 7, 10, 12, 15, 23	13.13(7.96) 6, 8, 11, 16, 28	15.38(9.60) 5, 9, 13, 19, 33	15.47(9.97) 5, 9, 13, 20, 34	18.05(11.55) 6, 11, 15, 22, 38	18.32(15.48) 5, 9, 14, 23, 46	10.07(18.86) 3, 5, 7, 11, 23	7.85(5.23) 4, 5, 7, 9, 16
-0.25	0.61	433.24(904.67) 17, 46, 115, 342, 2153	348.60(745.39) 15, 39, 94, 284, 1591	144.59(370.74) 11, 24, 48, 115, 531	189.48(454.91) 11, 27, 59, 153, 757	318.26(763.46) 12, 33, 79, 226, 1613	163.23(427.05) 9, 22, 48, 126, 626	804.05(1490.64) 10, 32, 105, 588, 5000	261.19(636.71) 9, 23, 57, 181, 1243
0.25	1.65	10.69(9.94) 2, 5, 8, 13, 28	13.87(12.40) 3, 6, 11, 17, 36	12.89(12.12) 2, 6, 10, 16, 33	13.28(12.21) 3, 6, 10, 17, 35	14.72(14.21) 3, 6, 11, 18, 39	20.01(19.48) 4, 9, 15, 24, 53	29.26(29.99) 5, 12, 21, 36, 82	32.91(32.77) 7, 14, 23, 40, 91
0.5	2.72	3.35(1.92) 1, 2, 3, 4, 7	4.08(2.32) 1, 2, 4, 5, 8	3.63(2.05) 1, 2, 3, 5, 7	3.73(2.08) 1, 2, 3, 5, 8	4.46(2.46) 2, 3, 4, 6, 9	5.25(2.74) 2, 3, 5, 7, 10	9.72(6.16) 3, 5, 8, 13, 21	10.43(5.6) 4, 7, 9, 13, 21
1	7.39	1.51(0.63) 1, 1, 1, 2, 3	1.69(0.72) 1, 1, 2, 2, 3	1.58(0.67) 1, 1, 1, 2, 3	1.57(0.66) 1, 1, 1, 2, 3	2.09(0.76) 1, 2, 2, 2, 3	2.18(0.80) 1, 2, 2, 3, 4	4.54(2.03) 2, 3, 4, 6, 8	4.92(1.72) 3, 4, 5, 6, 8

(continued)

Table 5 Comparison of the eight schemes under Rayleigh(θ, δ) distribution for $m = 100$, $n = 5$, $\lambda = 0.05$, and $ARL_0 = 500$

θ	δ	EC	cEC1	cEC2	cEC3	EL	cEL	EWMA-CvM	EWMA-KS
0	1	499.88(885.67) 12, 55, 170, 502, 2254	498.31(811.87) 16, 70, 201, 542, 2081	496.94(809.23) 16, 70, 200, 540, 2062	500.21(800.44) 17, 75, 208, 551, 2033	503.13(847.39) 15, 70, 196, 526, 2107	503.11(766.72) 21, 86, 230, 571, 1937	503.77(774.70) 18, 82, 225, 563, 1995	510.86(712.20) 22, 97, 258, 627, 1881
Pure location shifts									
−2	1	1.01(0.10) 1, 1, 1, 1, 1	1.02(0.15) 1, 1, 1, 1, 1	1.04(0.19) 1, 1, 1, 1, 2	1.15(0.36) 1, 1, 1, 1, 2	1.05(0.21) 1, 1, 1, 1, 1	1.25(0.43) 1, 1, 1, 1, 2	1.10(0.36) 1, 1, 1, 1, 2	2.01(0.10) 2, 2, 2, 2, 2
−1.5	1	1.11(0.31) 1, 1, 1, 1, 2	1.19(0.42) 1, 1, 1, 1, 2	1.20(0.41) 1, 1, 1, 1, 2	1.46(0.53) 1, 1, 1, 2, 2	1.22(0.43) 1, 1, 1, 1, 2	1.62(0.54) 1, 1, 2, 2, 2	1.36(0.51) 1, 1, 1, 2, 2	2.17(0.39) 2, 2, 2, 2, 3
−1	1	1.62(0.76) 1, 1, 2, 2, 4	1.93(0.96) 1, 1, 2, 2, 3	1.78(0.77) 1, 1, 2, 2, 3	2.21(0.87) 1, 2, 2, 3, 4	1.85(0.85) 1, 1, 2, 2, 3	2.41(0.86) 1, 2, 2, 3, 4	2.12(0.91) 1, 2, 2, 3, 4	3.07(0.92) 2, 2, 3, 4, 5
−0.5	1	4.83(3.30) 1, 3, 4, 6, 11	6.55(4.38) 2, 3, 6, 9, 15	5.04(3.11) 1, 3, 4, 6, 11	6.27(3.73) 2, 4, 5, 8, 13	5.70(3.81) 1, 3, 5, 7, 13	6.96(4.05) 2, 4, 6, 9, 15	6.36(4.02) 2, 4, 5, 8, 14	8.34(4.66) 3, 5, 7, 10, 17
0.5	1	9.30(8.54) 2, 4, 7, 12, 23	7.91(6.87) 2, 4, 6, 10, 19	11.76(9.97) 4, 7, 10, 14, 26	8.44(6.36) 3, 5, 7, 10, 19	9.75(8.57) 2, 5, 8, 12, 24	8.81(6.45) 3, 5, 7, 11, 20	7.52(5.78) 2, 4, 6, 9, 17	9.17(6.53) 4, 5, 8, 11, 20
1	1	2.04(0.93) 1, 1, 2, 3, 4	2.02(0.78) 1, 2, 2, 2, 3	2.91(1.21) 1, 2, 3, 4, 5	2.41(0.67) 2, 2, 2, 3, 4	2.19(0.95) 1, 2, 2, 3, 4	2.49(0.68) 2, 2, 2, 3, 4	2.15(0.66) 1, 2, 2, 2, 3	3.01(0.72) 2, 3, 3, 3, 4
1.5	1	1.10(0.31) 1, 1, 1, 1, 2	1.14(0.35) 1, 1, 1, 1, 2	1.34(0.52) 1, 1, 1, 2, 2	1.69(0.46) 1, 1, 2, 2, 2	1.14(0.35) 1, 1, 1, 1, 2	1.89(0.32) 1, 2, 2, 2, 2	1.32(0.47) 1, 1, 1, 2, 2	2.06(0.24) 2, 2, 2, 2, 3
2	1	1.00(0.02) 1, 1, 1, 1, 1	1.00(0.02) 1, 1, 1, 1, 1	1.01(0.09) 1, 1, 1, 1, 1	1.11(0.32) 1, 1, 1, 1, 1	1.00(0.02) 1, 1, 1, 1, 1	1.40(0.49) 1, 1, 1, 2, 2	1.00(0.07) 1, 1, 1, 1, 1	2.00(0.00) 2, 2, 2, 2, 2
Pure scale shifts									
0	0.5	3.62(1.96) 1, 2, 3, 5, 7	4.69(1.63) 2, 4, 5, 6, 7	3.27(1.63) 1, 2, 3, 4, 6	3.56(1.39) 2, 3, 3, 4, 6	3.78(1.98) 1, 2, 3, 5, 7	3.69(1.44) 2, 3, 3, 4, 6	3.09(1.24) 2, 2, 3, 4, 5	3.98(1.24) 2, 3, 4, 5, 6
0	1.25	20.20(27.91) 3, 7, 13, 24, 60	19.99(23.94) 3, 8, 14, 24, 56	28.86(42.72) 4, 10, 18, 33, 87	26.00(33.11) 3, 8, 15, 29, 74	24.80(33.35) 3, 8, 15, 29, 74	31.46(40.31) 5, 12, 20, 36, 93	28.22(36.68) 4, 10, 18, 33, 84	35.96(49.29) 6, 13, 22, 41, 106
0	1.5	5.30(3.77) 1, 3, 4, 7, 12	5.66(3.63) 2, 3, 5, 7, 12	7.04(5.05) 2, 4, 6, 9, 16	6.98(4.36) 2, 4, 6, 9, 15	6.28(4.42) 2, 3, 5, 8, 15	7.88(4.87) 3, 5, 7, 10, 17	7.15(4.83) 2, 4, 6, 9, 16	9.20(5.44) 3, 6, 8, 11, 20
0	1.75	2.97(1.77) 1, 2, 3, 4, 7	3.28(1.77) 1, 2, 3, 4, 7	3.78(2.26) 1, 2, 3, 5, 8	4.02(2.05) 2, 3, 4, 5, 8	3.48(2.01) 1, 2, 3, 4, 7	4.50(2.21) 2, 3, 4, 6, 9	4.02(2.20) 1, 2, 4, 5, 8	5.38(2.38) 3, 4, 5, 7, 10
0	2	2.14(1.14) 1, 2, 3, 4, 6	2.42(1.19) 1, 2, 3, 4, 5	2.61(1.43) 1, 2, 3, 5, 8	2.91(1.32) 1, 2, 3, 4, 5	2.49(1.29) 1, 2, 3, 5	3.29(1.43) 2, 2, 3, 4, 6	2.92(1.43) 1, 2, 3, 4, 6	4.01(1.48) 2, 3, 4, 5, 7
Mixed location and scale shifts (θ, $\delta = e^{\theta}$)									
−2	0.14	1.00(0.00) 1, 1, 1, 1, 1	1.00(0.00) 1, 1, 1, 1, 1	1.00(0.00) 1, 1, 1, 1, 1	1.00(0.00) 1, 1, 1, 1, 1	1.00(0.00) 1, 1, 1, 1, 1	1.00(0.00) 1, 1, 1, 1, 1	1.00(0.01) 1, 1, 1, 1, 1	2.00(0.00) 2, 2, 2, 2, 2
−1.5	0.22	1.00(0.00) 1, 1, 1, 1, 1	1.00(0.00) 1, 1, 1, 1, 1	1.00(0.00) 1, 1, 1, 1, 1	1.00(0.00) 1, 1, 1, 1, 1	1.00(0.00) 1, 1, 1, 1, 1	1.00(0.00) 1, 1, 1, 1, 1	1.00(0.00) 1, 1, 1, 1, 1	2.00(0.00) 2, 2, 2, 2, 2
−1	0.37	1.00(0.00) 1, 1, 1, 1, 1	1.00(0.00) 1, 1, 1, 1, 1	1.00(0.00) 1, 1, 1, 1, 1	1.00(0.00) 1, 1, 1, 1, 1	1.00(0.00) 1, 1, 1, 1, 1	1.00(0.06) 1, 1, 1, 1, 1	1.00(0.06) 1, 1, 1, 1, 1	2.00(0.00) 2, 2, 2, 2, 2
−0.5	0.61	1.34(0.54) 1, 1, 2, 2, 3	1.66(0.75) 1, 1, 2, 2, 3	1.42(0.54) 1, 1, 1, 2, 2	1.86(0.54) 1, 1, 2, 2, 3	1.48(0.60) 1, 1, 1, 2, 2	2.01(0.47) 1, 2, 2, 2, 3	1.66(0.58) 1, 1, 2, 2, 3	2.50(0.59) 2, 2, 3, 3, 3
0.5	1.65	1.83(0.89) 1, 1, 2, 2, 3	1.93(0.83) 1, 1, 2, 2, 3	2.38(1.23) 1, 1, 2, 3, 5	2.43(0.87) 1, 2, 2, 3, 4	2.06(0.97) 1, 2, 2, 3, 4	2.55(0.85) 2, 2, 3, 3, 4	2.23(0.92) 1, 2, 2, 3, 4	3.22(0.96) 2, 3, 3, 4, 5

Table 5 (continued)

θ	δ	EC	cEC1	cEC2	cEC3	EL	cEL	EWMA-CvM	EWMA-KS
1	2.72	1.03(0.16) 1, 1, 1, 1, 1	1.05(0.21) 1, 1, 1, 1, 1	1.08(0.27) 1, 1, 1, 1, 2	1.27(0.45) 1, 1, 1, 2, 2	1.07(0.25) 1, 1, 1, 1, 2	1.44(0.50) 1, 1, 1, 2, 2	1.16(0.37) 1, 1, 1, 1, 2	2.34(0.19) 2, 2, 2, 2
1.5	4.48	1.00(0.01) 1, 1, 1, 1, 1	1.00(0.01) 1, 1, 1, 1, 1	1.00(0.59) 1, 1, 1, 1, 1	1.02(0.13) 1, 1, 1, 1, 1	1.00(0.01) 1, 1, 1, 1, 1	1.07(0.26) 1, 1, 1, 1, 2	1.00(0.06) 1, 1, 1, 1, 1	2.00(0.01) 2, 2, 2, 2
2	7.39	1.00(0.00) 1, 1, 1, 1, 1	1.00(0.00) 1, 1, 1, 1, 1	1.00(0.00) 1, 1, 1, 1, 1	1.00(0.01) 1, 1, 1, 1, 1	1.00(0.00) 1, 1, 1, 1, 1	1.00(0.06) 1, 1, 1, 1, 1	1.00(0.00) 1, 1, 1, 1, 1	2.00(0.00) 2, 2, 2, 2

Mixed location and scale shifts $(\theta,\ \delta = e^{\frac{\theta}{2}})$

θ	δ	EC	cEC1	cEC2	cEC3	EL	cEL	EWMA-CvM	EWMA-KS
-2	0.37	1.00(0.00) 1, 1, 1, 1, 1	1.00(0.00) 1, 1, 1, 1, 1	1.00(0.00) 1, 1, 1, 1, 1	1.00(0.00) 1, 1, 1, 1, 1	1.00(0.00) 1, 1, 1, 1, 1	1.00(0.00) 1, 1, 1, 1, 1	1.00(0.00) 1, 1, 1, 1, 1	2.00(0.00) 2, 2, 2, 2
-1.5	0.47	1.00(0.00) 1, 1, 1, 1, 1	1.00(0.00) 1, 1, 1, 1, 1	1.00(0.00) 1, 1, 1, 1, 1	1.00(0.01) 1, 1, 1, 1, 1	1.00(0.00) 1, 1, 1, 1, 1	1.00(0.04) 1, 1, 1, 1, 1	1.00(0.01) 1, 1, 1, 1, 1	2.00(0.00) 2, 2, 2, 2
-1	0.61	1.01(0.09) 1, 1, 1, 1, 1	1.03(0.17) 1, 1, 1, 1, 1	1.02(0.14) 1, 1, 1, 1, 1	1.16(0.37) 1, 1, 1, 1, 2	1.03(0.27) 1, 1, 1, 1, 1	1.31(0.46) 1, 1, 1, 2, 2	1.08(0.27) 1, 1, 1, 1, 2	2.01(0.1) 2, 2, 2, 2
-0.5	0.78	2.19(1.14) 1, 1, 2, 3, 4	2.97(1.59) 1, 2, 3, 4, 6	2.26(1.02) 1, 2, 2, 3, 4	2.80(1.09) 1, 2, 3, 3, 5	2.47(1.23) 1, 2, 2, 3, 5	2.97(1.10) 2, 2, 3, 4, 5	2.61(1.12) 1, 2, 2, 3, 5	2.69(1.21) 2, 3, 3, 6
0.5	1.28	3.18(1.86) 1, 2, 3, 4, 7	3.05(1.55) 1, 2, 3, 4, 6	4.59(2.41) 1, 3, 4, 6, 9	3.67(1.59) 2, 3, 3, 4, 7	3.50(1.95) 1, 2, 3, 4, 7	3.83(1.63) 2, 3, 4, 5, 7	3.39(1.61) 1, 2, 3, 4, 6	4.63(1..7) 3, 3, 4, 5, 8
1	1.65	1.19(0.41) 1, 1, 1, 1, 2	1.26(0.45) 1, 1, 1, 2, 2	1.42(0.59) 1, 1, 1, 2, 2	1.69(0.50) 1, 1, 2, 2, 2	1.29(0.48) 1, 1, 1, 2, 2	1.86(0.42) 1, 2, 2, 2, 2	1.47(0.52) 1, 1, 1, 2, 2	2.28(0.4) 2, 2, 2, 3,
1.5	2.12	1.00(0.05) 1, 1, 1, 1, 1	1.00(0.07) 1, 1, 1, 1, 1	1.02(0.13) 1, 1, 1, 1, 1	1.13(0.34) 1, 1, 1, 1, 2	1.01(0.07) 1, 1, 1, 1, 1	1.32(0.47) 1, 1, 1, 2, 2	1.03(0.18) 1, 1, 1, 1, 1	2.00(0.0) 2, 2, 2, 2
2	2.72	1.00(0.00) 1, 1, 1, 1, 1	1.00(0.00) 1, 1, 1, 1, 1	1.00(0.00) 1, 1, 1, 1, 1	1.00(0.05) 1, 1, 1, 1, 1	1.00(0.00) 1, 1, 1, 1, 1	1.03(0.18) 1, 1, 1, 1, 1	1.00(0.00) 1, 1, 1, 1, 1	2.00(0.00) 2, 2, 2, 2

Mixed location and scale shifts $(\delta = e^{\theta})$

θ	δ	EC	cEC1	cEC2	cEC3	EL	cEL	EWMA-CvM	EWMA-KS
-1	0.14	1.00(0.00) 1, 1, 1, 1, 1	1.00(0.00) 1, 1, 1, 1, 1	1.00(0.00) 1, 1, 1, 1, 1	1.00(0.00) 1, 1, 1, 1, 1	1.00(0.00) 1, 1, 1, 1, 1	1.00(0.00) 1, 1, 1, 1, 1	1.00(0.00) 1, 1, 1, 1, 1	2.00(0.00) 2, 2, 2, 2
-0.5	0.37	1.00(0.04) 1, 1, 1, 1, 1	1.01(0.10) 1, 1, 1, 1, 1	1.00(0.05) 1, 1, 1, 1, 1	1.10(0.29) 1, 1, 1, 1, 2	1.00(0.06) 1, 1, 1, 1, 1	1.27(0.45) 1, 1, 1, 2, 2	1.02(0.16) 1, 1, 1, 1, 1	2.06(0.03) 2, 2, 2, 2
-0.25	0.61	2.30(1.16) 1, 1, 2, 3, 4	3.29(1.55) 1, 2, 3, 4, 6	2.28(0.98) 1, 2, 2, 3, 4	2.74(0.95) 2, 2, 3, 3, 4	2.52(1.22) 1, 2, 2, 3, 5	2.85(0.95) 2, 2, 3, 3, 5	2.50(0.95) 1, 2, 2, 3, 4	3.49(1.03) 2, 3, 3, 4, 5
0.25	1.65	2.51(1.39) 1, 2, 2, 3, 5	2.61(1.29) 1, 2, 3, 4, 6	3.33(1.89) 1, 2, 3, 4, 7	3.27(1.45) 2, 2, 3, 4, 6	2.84(1.52) 1, 2, 3, 4, 6	3.49(1.47) 2, 2, 3, 4, 6	3.09(1.50) 1, 2, 3, 4, 6	4.27(1.51) 2, 3, 4, 5, 7
0.5	2.72	1.15(0.37) 1, 1, 1, 1, 2	1.25(0.44) 1, 1, 1, 2, 2	1.27(0.48) 1, 1, 1, 2, 2	1.55(0.55) 1, 1, 2, 2, 2	1.28(0.47) 1, 1, 1, 2, 2	1.72(0.54) 1, 1, 2, 2, 2	1.42(0.54) 1, 1, 1, 2, 2	2.24(0.65) 2, 2, 2, 3
1	7.39	1.00(0.02) 1, 1, 1, 1, 1	1.00(0.04) 1, 1, 1, 1, 1	1.00(0.04) 1, 1, 1, 1, 1	1.03(0.17) 1, 1, 1, 1, 1	1.01(0.07) 1, 1, 1, 1, 1	1.07(0.25) 1, 1, 1, 1, 2	1.02(0.13) 1, 1, 1, 1, 1	2.00(0.01) 2, 2, 2, 2

(continued)

Table 6 Comparison of the eight schemes under SE(θ, δ) distribution for $m = 100$, $n = 5$, $\lambda = 0.05$, and ARL$_0 = 500$

θ	δ	EC	cEC1	cEC2	cEC3	EL	cEL	EWMA-CvM	EWMA-KS
0	1	503.61(887.87) 12, 58, 173, 510, 2293	498.60(815.12) 17, 71, 200, 543, 2057	499.44(819.04) 17, 71, 198, 539, 2057	495.51(797.22) 18, 75, 205, 542, 2019	501.64(840.68) 15, 70, 197, 528, 2087	503.37(772.48) 21, 87, 231, 566, 1956	500.84(769.15) 18, 83, 225, 575, 1930	511.27(734.16) 22, 96, 257, 617, 1931
					Pure location shifts				
−2	1	1.10(0.31) 1, 1, 1, 1, 2	1.11(0.33) 1, 1, 1, 1, 2	1.28(0.47) 1, 1, 1, 2, 2	1.36(0.51) 1, 1, 1, 2, 2	1.21(0.43) 1, 1, 1, 1, 2	1.52(0.60) 1, 1, 1, 2, 2	1.45(0.61) 1, 1, 1, 2, 2	2.13(0.35) 2, 2, 2, 2, 3
−1.5	1	1.24(0.48) 1, 1, 1, 1, 2	1.28(0.51) 1, 1, 1, 1, 2	1.49(0.61) 1, 1, 1, 1, 2	1.59(0.63) 1, 1, 1, 2, 3	1.42(0.59) 1, 1, 1, 2, 2	1.87(0.77) 1, 1, 2, 2, 3	1.76(0.76) 1, 1, 2, 2, 3	2.40(0.61) 2, 2, 2, 3, 4
−1	1	1.63(0.79) 1, 1, 1, 2, 3	1.71(0.85) 1, 1, 2, 2, 3	1.99(0.94) 1, 1, 2, 2, 4	2.11(0.95) 1, 1, 2, 3, 4	1.93(0.96) 1, 1, 2, 2, 4	2.61(1.16) 1, 2, 2, 3, 5	2.49(1.21) 1, 2, 2, 3, 5	3.23(1.14) 2, 2, 3, 4, 5
−0.5	1	3.18(1.95) 1, 2, 3, 4, 7	3.45(2.05) 1, 2, 3, 4, 7	3.98(2.31) 1, 2, 4, 5, 8	4.16(2.36) 1, 2, 4, 5, 9	4.04(2.45) 1, 2, 4, 5, 9	5.57(3.05) 2, 3, 5, 7, 11	5.64(3.41) 2, 3, 5, 7, 12	7.05(3.48) 3, 5, 6, 9, 14
0.5	1	11.17(8.38) 4, 7, 10, 13, 23	12.96(10.70) 3, 6, 10, 16, 31	12.42(11.06) 5, 7, 10, 14, 28	11.94(9.80) 4, 6, 9, 14, 29	9.99(19.90) 4, 6, 9, 12, 20	11.08(8.60) 4, 6, 9, 13, 24	9.40(15.79) 3, 5, 7, 11, 22	8.66(11.63) 4.5, 7, 10, 17
1	1	3.47(1.55) 1, 2, 3, 4, 6	3.21(1.56) 2, 2, 3, 4, 6	4.06(0.91) 3, 4, 4, 4, 5	3.24(1.10) 2, 3, 3, 4, 5	3.60(1.64) 2, 2, 3, 4, 7	3.36(1.14) 2, 3, 3, 4, 5	2.58(0.73) 2, 2, 2, 3, 4	3.11(0.54) 2, 3, 3, 4
1.5	1	1.82(0.70) 1, 1, 2, 2, 3	1.82(0.57) 1, 1, 2, 2, 3	2.57(0.79) 1, 2, 3, 3, 4	2.13(0.37) 2, 2, 2, 2, 3	1.85(0.65) 1, 1, 2, 2, 3	2.17(0.40) 2, 2, 2, 2, 3	1.87(0.36) 1, 2, 2, 2, 2	2.19(0.39) 2, 2, 2, 2, 3
2	1	1.20(0.41) 1, 1, 1, 2, 3	1.26(0.44) 1, 1, 1, 2, 2	1.66(0.64) 1, 1, 2, 2, 3	1.92(0.28) 1, 2, 2, 2, 2	1.23(0.42) 1, 1, 1, 1, 2	1.99(0.13) 2, 2, 2, 2, 2	1.37(0.48) 1, 1, 1, 2, 2	2.00(0.04) 2, 2, 2, 2, 2
					Pure scale shifts				
0	0.5	22.61(41.28) 3, 8, 14, 25, 66	28.16(67.10) 6, 11, 16, 28, 75	19.39(44.63) 3, 7, 11, 20, 57	19.70(47.53) 4, 7, 12, 21, 53	22.11(42.39) 4, 8, 14, 24, 62	20.07(34.92) 4, 8, 13, 21, 55	15.20(24.97) 3, 6, 10, 17, 40	16.76(24.23) 4, 8, 12, 19, 42
0	1.25	108.59(248.65) 5, 18, 42, 103, 404	106.12(242.01) 7, 20, 44, 104, 390	148.36(300.85) 8, 26, 59, 146, 563	130.62(261.03) 8, 25, 56, 131, 474	135.08(292.46) 7, 23, 54, 132, 496	161.75(285.87) 11, 32, 73, 172, 594	142.98(260.05) 9, 27, 61, 152, 524	173.70(296.2) 12, 34, 79, 186, 647
0	1.5	26.00(41.99) 3, 8, 15, 29, 80	25.24(35.50) 4, 9, 16, 29, 74	37.74(62.17) 4, 11, 21, 42, 118	33.09(47.72) 5, 11, 20, 38, 99	32.33(50.48) 4, 10, 19, 36, 103	41.08(60.57) 6, 14, 25, 46, 125	35.49(47.67) 5, 11, 21, 41, 112	47.23(76.8) 7, 15, 27, 52, 146
0	1.75	11.66(11.45) 2, 5, 8, 14, 31	11.82(10.39) 2, 5, 9, 15, 30	16.11(16.35) 3, 7, 12, 20, 43	15.00(13.54) 3, 7, 11, 19, 38	13.97(13.81) 3, 6, 10, 17, 38	17.58(16.74) 4, 8, 13, 21, 45	15.98(15.66) 3, 7, 12, 20, 43	20.34(20.69) 5, 9, 15, 24, 54
0	2	7.22(5.72) 2, 3, 6, 9, 18	7.56(5.38) 2, 4, 6, 10, 18	9.85(7.83) 2, 5, 8, 13, 24	9.51(6.81) 3, 5, 8, 12, 22	8.65(6.84) 2, 4, 7, 11, 21	10.82(7.84) 3, 6, 9, 14, 25	9.85(7.82) 2, 5, 8, 12, 24	12.58(9.03) 4, 7, 10, 15, 29
					Mixed location and scale shifts (θ, $\delta = e^{\theta}$)				
−2	0.14	1.00(0.00) 1, 1, 1, 1, 1	1.00(0.00) 1, 1, 1, 1, 1	1.00(0.00) 1, 1, 1, 1, 1	1.00(0.00) 1, 1, 1, 1, 1	1.00(0.00) 1, 1, 1, 1, 1	1.00(0.00) 1, 1, 1, 1, 1	1.00(0.02) 1, 1, 1, 1, 1	2.00(0.00) 2, 2, 2, 2
−1.5	0.22	1.00(0.00) 1, 1, 1, 1, 1	1.00(0.00) 1, 1, 1, 1, 1	1.00(0.02) 1, 1, 1, 1, 1	1.00(0.04) 1, 1, 1, 1, 1	1.00(0.02) 1, 1, 1, 1, 1	1.00(0.06) 1, 1, 1, 1, 1	1.00(0.04) 1, 1, 1, 1, 1	2.00(0.00) 2, 2, 2, 2
−1	0.37	1.01(0.12) 1, 1, 1, 1, 1	1.02(0.15) 1, 1, 1, 1, 1	1.05(0.22) 1, 1, 1, 1, 1	1.17(0.38) 1, 1, 1, 1, 2	1.06(0.24) 1, 1, 1, 1, 2	1.24(0.43) 1, 1, 1, 1, 2	1.13(0.34) 1, 1, 1, 1, 2	2.01(0.1) 2, 2, 2, 2, 2
−0.5	0.61	1.77(0.90) 1, 1, 2, 2, 3	2.02(1.07) 1, 1, 2, 3, 4	2.01(0.93) 1, 1, 2, 2, 4	2.42(1.05) 1, 1, 2, 3, 4	2.08(1.04) 1, 1, 2, 3, 4	2.72(1.14) 1, 2, 3, 3, 5	2.43(1.12) 1, 2, 2, 3, 4	3.41(1.16) 2, 3, 3, 4, 6
0.5	1.65	4.50(2.65) 1, 3, 4, 6, 9	4.14(2.35) 2, 3, 4, 5, 9	5.89(2.48) 2, 4, 6, 7, 10	4.48(2.07) 2, 3, 4, 5, 8	4.81(2.64) 2, 3, 4, 6, 10	4.67(2.15) 2, 3, 4, 6, 9	4.00(1.91) 2, 3, 4, 5, 7	5.10(1.91) 3, 4, 5, 6, 9

Table 6 (continued)

θ	δ	EC	cEC1	cEC2	cEC3	EL	cEL	EWMA-C M	EWMA-KS
1	2.72	1.52(0.63) 1,1,1,2,3	1.58(0.58) 1,1,2,2,2	2.08(0.93) 1,1,2,3,4	2.01(0.45) 1,2,2,2,3	1.64(0.65) 1,1,2,2,3	2.10(0.40) 2,2,2,2,3	1.75(0.54) 1,1,2,2,3	2.56(0.54) 2,2,3,3,3
1.5	4.48	1.05(0.22) 1,1,1,1,2	1.07(0.26) 1,1,1,1,2	1.19(0.41) 1,1,1,1,2	1.52(0.50) 1,2,2,2,2	1.07(0.26) 1,1,1,1,2	1.77(0.42) 1,2,2,2,2	1.21(0.41) 1,1,1,1,2	2.03(0.16) 2,2,2,2,2
2	7.39	1.00(0.03) 1,1,1,1,1	1.00(0.04) 1,1,1,1,1	1.01(0.11) 1,1,1,1,1	1.12(0.33) 1,1,1,1,2	1.00(0.04) 1,1,1,1,1	1.39(0.49) 1,1,1,1,2	1.01(0.11) 1,1,1,1,1	2.00(0.00) 2,2,2,2,2
				Mixed location and scale shifts $(\theta, \delta = e^{\theta/2})$					
-2	0.37	1.00(0.01) 1,1,1,1,1	1.00(0.01) 1,1,1,1,1	1.00(0.05) 1,1,1,1,1	1.01(0.10) 1,1,1,1,1	1.00(0.06) 1,1,1,1,1	1.02(0.13) 1,1,1,1,1	1.01(0.10) 1,1,1,1,1	2.00(0.00) 2,2,2,2,2
-1.5	0.47	1.01(0.08) 1,1,1,1,1	1.01(0.10) 1,1,1,1,1	1.04(0.29) 1,1,1,1,1	1.12(0.32) 1,1,1,1,1	1.05(0.21) 1,1,1,1,1	1.17(0.37) 1,1,1,1,1	1.10(0.30) 1,1,1,1,1	2.00(0.07) 2,2,2,2,2
-1	0.61	1.16(0.38) 1,1,1,1,2	1.21(0.45) 1,1,1,1,2	1.32(0.56) 1,1,1,2,2	1.52(0.58) 1,1,1,2,2	1.31(0.50) 1,1,1,2,2	1.68(0.64) 1,1,2,2,3	1.52(0.60) 1,1,2,2,3	2.23(0.46) 2,2,2,2,3
-0.5	0.78	2.34(1.32) 1,1,2,3,5	2.64(1.51) 1,2,2,3,5	2.75(1.43) 1,2,2,3,5	3.17(1.61) 1,2,3,4,6	2.83(1.58) 1,2,3,4,6	3.80(1.84) 2,2,3,5,7	3.49(1.87) 1,2,3,4,7	4.65(1.92) 2,3,4,6,8
0.5	1.28	7.17(4.65) 2,4,6,9,15	6.96(5.22) 2,4,6,9,16	8.13(4.55) 4,6,7,9,15	6.76(4.38) 3,4,6,8,14	7.13(4.35) 2,4,6,9,14	6.94(4.39) 3,4,6,8,14	5.71(4.19) 2,4,5,7,12	6.39(3.07) 3,5,5,7,12
1	1.65	2.17(0.95) 1,2,2,3,4	2.10(0.77) 1,2,2,3	3.08(1.07) 1,2,3,4,4	2.42(0.64) 2,2,2,3,4	2.27(0.95) 2,2,3,4	2.50(0.66) 2,2,2,3,4	2.12(0.54) 1,2,2,2,3	2.85(0.53) 2,3,3,3,4
1.5	2.12	1.25(0.45) 1,1,1,2,2	1.31(0.47) 1,1,1,2,2	1.66(0.68) 1,1,2,2,3	1.88(0.35) 1,2,2,2,2	1.31(0.48) 1,1,1,2,2	1.38(0.22) 2,2,2,2,2	1.54(0.56) 1,2,2,2,2	2.09(0.28) 2,2,2,2,3
2	2.72	1.02(0.13) 1,1,1,1,1	1.03(0.17) 1,1,1,1,1	1.13(0.34) 1,1,1,1,2	1.50(0.50) 1,1,1,2,2	1.03(0.16) 1,1,1,1,1	1.80(0.40) 1,2,2,2,2	1.09(0.29) 1,1,1,1,2	2.00(0.02) 2,2,2,2,2
				Mixed location and scale shifts $(\theta, \delta = e^{2\theta})$					
-1	0.14	1.00(0.00) 1,1,1,1,1	1.00(0.00) 1,1,1,1,1	1.00(0.00) 1,1,1,1,1	1.00(0.03) 1,1,1,1,1	1.00(0.00) 1,1,1,1,1	1.00(0.04) 1,1,1,1,1	1.00(0.02) 1,1,1,1,1	2.00(0.00) 2,2,2,2,2
-0.5	0.37	1.19(0.41) 1,1,1,1,2	1.29(0.52) 1,1,1,2,2	1.29(0.48) 1,1,1,2,2	1.60(0.58) 1,1,2,2,2	1.34(0.52) 1,1,1,2,2	1.75(0.58) 1,1,2,2,2	1.51(0.57) 1,1,2,2,2	2.35(0.51) 2,2,3,3
-0.25	0.61	3.28(2.01) 1,2,3,4,7	4.17(2.60) 1,2,4,5,9	3.50(1.94) 1,2,3,4,7	4.38(2.31) 2,3,4,5,9	3.91(2.35) 1,2,3,5,8	4.84(2.46) 2,3,4,6,9	4.42(2.45) 2,3,4,6,9	6.0?(2.75) 3,4,?,7,11
0.25	1.65	8.57(7.31) 2,4,7,11,21	7.33(5.51) 2,4,6,9,17	12.20(9.97) 3,7,10,15,28	8.56(6.20) 3,5,7,10,19	9.33(7.70) 2,5,7,12,23	8.87(6.31) 3,5,7,11,20	7.98(6.03) 2,4,6,10,18	10.1?(6.96) 4,6,8,14,22
0.5	2.72	2.29(1.16) 1,1,2,3,4	2.28(1.00) 1,2,2,3,4	3.27(1.59) 1,2,3,4,6	2.75(0.96) 2,2,3,3,4	2.56(.24) 1,2,2,3,5	2.85(0.97) 2,2,3,3,5	2.52(1.00) 1,2,2,3,4	4.6?(2.22) 2,3,4,4,5
1	7.39	1.08(0.28) 1,1,1,1,2	1.12(0.32) 1,1,1,1,2	1.22(0.43) 1,1,1,1,2	1.50(0.50) 1,1,1,2,2	1.14(0.35) 1,1,1,1,2	1.69(0.47) 1,1,2,2,2	1.31(0.47) 1,1,2,2,2	2.13(0.34) 2,2,2,2,3

the shifted exponential distribution, when a small upward pure location shift ($\theta = 0.5$) occurs, EWMA-KS is the best while for $\theta = 0.5$ along with some small to moderate scale shift, EWMA-CvM is the best. For a pure downward scale shift ($\delta = 0.5$), EWMA-CvM also performs better than the other schemes.

Finally, we also see the IC robustness of various schemes from Tables 2, 3, 4, 5, and 6. In particular, the cEL and EWMA-KS have slightly higher IC values of the 5th percentile of corresponding run length distributions. That indicates, they could be more effective in preventing early false alarms.

4.3   OOC Performance of the Proposed Procedures at Macrolevel

In practice, the size of possible shifts is usually unknown and therefore, practitioners prefer a monitoring scheme that has overall good performance irrespective of the exact size of shift. To this end, Ryu et al. (2010) considered an expected value of weighted run length used to evaluate the overall performance. In the present case, we are monitoring two parameters simultaneously. Therefore, we use a simple uniform weighting scheme in the line of Ryu et al. (2010), Mukherjee and Marozzi (2017b), and Mukherjee and Sen (2018). This simplified index for measuring overall performance of a monitoring scheme is known as Expected Average Run Length (EARL) and is given by

$$\text{EARL} = \int\limits_{-\infty}^{\infty} \int\limits_{0}^{\infty} \omega(\theta, \delta) \text{ARL}(\theta, \delta | F) \mathrm{d}F_1(\theta) \mathrm{d}F_2(\delta), \tag{4}$$

where $\omega(\theta, \delta)$ is a suitable weight and F_1 and F_2 are the cdf of θ and δ, respectively. $\text{ARL}(\theta, \delta | F)$ is the OOC-ARL value for a shift (θ, δ) under process distribution F. Usually, in manufacturing and operation context, reduction of the scale parameter is considered as process improvement and does not act as an assignable cause. Therefore, we restrict ourselves to the upward shifts of the location and scale parameters. Note that the EWMA schemes are usually designed to detect small to moderate shifts in the process parameters. The Shewhart-type schemes are usually preferable over the EWMA schemes when larger shifts are of interest. Therefore, we consider $0 \leq \theta \leq 2$ and $1 \leq \delta \leq 2$ in the present case for all the symmetric distributions. For asymmetric distributions, we consider $0.25 \leq \theta \leq 2$ and $1 \leq \delta \leq 2$, as very small positive θ values, in these cases sometimes lead to ARL bias which in turn makes EARL practically unusable. For larger shifts in either parameter, the six schemes considered are almost equally efficient for the selected IC distribution. Thus, in a non-informative situation regarding $F_1(\theta)$ and $F_2(\delta)$ and taking $\omega(\theta, \delta) = 1$, we may approximate (4) using the functions "smooth.2d" and "sintegral", inbuilt in "fields" and "Bolstad" packages of R software, respectively. We provide the computational

Table 7 EARL values of the six schemes when $ARL_0 = 500$ and $\lambda = 0.05$

Distributions	m=100, n=5, λ=0.05						m=300, n=5, λ=0.05						m=500, n=5, λ=0.05					
	EC	cEC1	cEC2	cEC3	EL	cEL	EC	cEC1	cEC2	cEC3	EL	cEL	EC	cEC1	cEC2	cEC3	EL	cEL
Uniform	15.8	16.0	17.1	18.7	19.1	22.7	15.8	15.3	16.1	17.9	18.8	21.4	15.8	15.2	15.8	17.8	18.8	21.0
Normal	33.3	33.6	39.3	36.4	37.0	42.4	30.4	29.9	35.1	32.7	34.0	37.7	29.8	29.2	33.7	32.0	33.4	36.8
Laplace	43.6	44.1	51.5	46.8	49.0	55.0	39.2	38.9	46.4	41.6	44.2	48.4	38.3	38.0	44.8	40.6	43.1	47.1
Rayleigh	4.4	3.5	6.3	4.5	4.3	4.6	3.7	2.9	5.3	3.7	3.7	3.8	3.6	2.8	5.1	3.5	3.6	3.7
shifted exponential	8.0	6.9	10.3	7.8	7.7	7.3	6.7	6.0	9.1	6.5	6.0	5.9	6.6	5.9	8.8	6.3	5.9	5.8

Distributions	m=100, n=10, λ=0.05						m=300, n=10, λ=0.05						m=500, n=10, λ=0.05					
	EC	cEC1	cEC2	cEC3	EL	cEL	EC	cEC1	cEC2	cEC3	EL	cEL	EC	cEC1	cEC2	cEC3	EL	cEL
Uniform	14.1	14.4	15.8	16.3	16.4	18.7	13.2	12.7	14.2	14.4	15.1	16.5	13.0	12.3	13.8	14.0	14.8	16.0
Normal	30.0	30.7	34.7	33.0	32.5	36.4	24.9	24.5	28.3	26.5	27.1	29.9	23.9	23.3	26.7	25.2	26.1	28.5
Laplace	39.0	40.4	44.3	42.9	42.3	46.7	31.8	31.7	36.1	33.8	34.6	38.0	30.2	30.0	34.0	31.9	33.1	36.1
Rayleigh	2.2	2.1	2.6	2.2	2.2	2.2	1.9	1.8	2.2	1.9	1.9	1.9	1.9	1.7	2.1	1.8	1.9	1.9
shifted exponential	2.6	2.7	3.1	2.7	2.8	2.8	2.5	2.5	2.7	2.5	2.5	2.5	2.5	2.5	2.7	2.5	2.5	2.4

Distributions	m=100, n=15, λ=0.05						m=300, n=15, λ=0.05						m=500, n=15, λ=0.05					
	EC	cEC1	cEC2	cEC3	EL	cEL	EC	cEC1	cEC2	cEC3	EL	cEL	EC	cEC1	cEC2	cEC3	EL	cEL
Uniform	13.3	13.5	14.7	15.1	15.0	16.8	11.9	11.6	12.8	12.8	13.3	14.3	11.5	11.0	12.3	12.9	13.7	
Normal	28.5	29.8	31.9	31.4	30.5	33.4	22.4	22.1	24.7	23.6	24.0	25.9	20.8	20.4	23.1	21.9	22.5	24.3
Laplace	37.1	39.2	40.9	40.9	39.2	43.1	28.5	28.5	31.4	30.0	30.4	32.9	26.2	26.1	29.0	27.7	28.5	30.6
Rayleigh	1.6	1.6	1.8	1.6	1.7	1.7	1.5	1.4	1.6	1.5	1.5	1.5	1.5	1.4	1.6	1.4	1.5	1.4
shifted exponential	1.7	1.9	1.9	1.8	1.8	1.9	1.7	1.8	1.9	1.7	1.8	1.7	1.7	1.8	1.9	1.7	1.8	1.7

Table 8 EARL values of the six schemes when $ARL_0 = 500$ and $\lambda = 0.1$

Distributions	m=100, n=5, λ=0.1						m=300, n=5, λ=0.1						m=500, n=5, λ=0.1					
	EC	cEC1	cEC2	cEC3	EL	cEL	EC	cEC1	cEC2	cEC3	EL	cEL	EC	cEC1	cEC2	cEC3	EL	cEL
Uniform	16.2	16.4	17.7	19.2	19.8	23.5	16.2	15.8	16.6	18.5	19.3	22.1	16.2	15.8	16.5	18.4	19.2	21.7
Normal	34.9	34.3	42.0	37.8	39.0	44.4	32.1	32.2	37.1	34.3	35.8	39.5	31.6	30.9	36.4	33.6	35.1	38.9
Laplace	46.3	45.6	55.9	49.3	52.1	58.3	42.0	40.7	50.0	44.0	47.1	51.8	41.2	40.5	49.4	43.0	45.7	50.3
Rayleigh	4.9	3.7	8.0	4.9	4.7	4.7	4.1	3.0	6.6	3.9	3.9	3.8	3.9	2.9	6.5	3.7	3.8	3.6
shifted exponential	10.6	8.3	15.6	9.5	8.9	7.8	8.9	7.1	14.9	8.0	7.0	6.3	8.6	6.9	15.2	7.8	6.8	6.1

Distributions	m=100, n=10, λ=0.1						m=300, n=10, λ=0.1						m=500, n=10, λ=0.1					
	EC	cEC1	cEC2	cEC3	EL	cEL	EC	cEC1	cEC2	cEC3	EL	cEL	EC	cEC1	cEC2	cEC3	EL	cEL
Uniform	14.5	14.7	16.6	16.7	17.0	19.6	13.3	13.0	14.8	14.8	15.6	17.3	13.1	12.6	14.4	14.5	15.2	16.8
Normal	30.8	31.3	36.1	33.9	33.9	38.1	25.8	25.2	30.1	27.6	28.8	31.8	25.0	24.2	28.9	26.3	27.6	30.4
Laplace	40.5	41.5	46.7	44.1	44.1	49.3	33.1	32.7	38.5	35.2	37.0	40.9	31.9	31.3	37.0	33.4	35.2	38.8
Rayleigh	2.3	2.1	2.9	2.3	2.4	2.3	2.0	1.7	2.3	1.9	2.0	1.9	1.9	1.7	2.2	1.8	1.9	1.8
shifted exponential	2.8	2.8	3.3	2.8	3.0	2.9	2.5	2.5	2.8	2.4	2.5	2.4	2.5	2.5	2.7	2.4	2.4	2.3

Distributions	m=100, n=15, λ=0.1						m=300, n=15, λ=0.1						m=500, n=15, λ=0.1					
	EC	cEC1	cEC2	cEC3	EL	cEL	EC	cEC1	cEC2	cEC3	EL	cEL	EC	cEC1	cEC2	cEC3	EL	cEL
Uniform	13.7	13.8	15.5	15.4	15.7	17.7	12.1	11.8	13.5	13.1	13.7	14.9	11.7	11.3	13.0	12.7	13.2	14.2
Normal	29.7	30.3	33.4	31.7	31.9	35.1	23.0	22.9	26.4	24.4	25.2	27.6	21.6	21.2	24.8	22.8	23.5	25.9
Laplace	38.8	39.8	42.8	41.5	41.3	45.3	29.4	29.5	33.6	31.2	32.1	35.4	27.4	27.3	31.4	29.1	29.8	32.9
Rayleigh	1.7	1.7	1.9	1.7	1.7	1.7	1.5	1.4	1.6	1.4	1.5	1.4	1.5	1.4	1.5	1.4	1.5	1.4
shifted exponential	1.8	1.9	1.9	1.8	1.9	1.9	1.7	1.7	1.7	1.7	1.7	1.7	1.7	1.7	1.7	1.6	1.7	1.6

results of EARL for overall comparison. Our study considers the nominal ARL_0 of 500 and the five distributions listed in the previous subsection.

To investigate the effect of the reference and test sample sizes on the performance of the proposed monitoring schemes, we consider m as 100, 300, and 500 and $n = 5$, 10, and 15. The results for $\lambda = 0.05$ and 0.1 are shown in Tables 7 and 8, respectively. We highlight the schemes with the best OOC overall performance in the sense that the least EARL is observed with a dark gray shade. From Tables 7 and 8, we find that either EC or cEC1 scheme has the minimum of EARL in most of the situations. Further, we see that the reference sample size (m) and test sample size (n) exhibit significant effect on the performance of the six schemes. In general, for fixed n, increasing m produces a decreasing trend in EARL of each scheme under all environments. Similarly, for fixed m, the EARL decreases as the test sample size n increases in all cases. These findings indicate that the OOC performance of these schemes improves as the size of the reference sample or the test sample increases. We also observe that when $m = 100$, the EC scheme is, in general, a good choice for detecting upward shifts of the location and scale parameters. Interestingly, for asymmetric distributions, and symmetric distributions with $m \geq 300$, the cEC1 appears to be better than its competitors.

5 Illustration and Data Monitoring

Fuze 117 is an important product mass-produced at a manufacturing industry. Fuze 117 is a direct action and gaze fuze that is mainly used with howitzers and field guns (high explosive type) for low-angle impact. There are more than 30 individual components in a fuze 117 MK 20 like guide bush, striker head, striker pin, shutter, and arming sleeve. Here, we discuss the process monitoring of a guide bush component. The raw material for bush guide is in form of brass rod procured from some relevant supplier. Guide bush holds in place the striker pin which under impact force over the striker head compresses the striker spring. This leads to the pin head piercing through the detonator. This detonation impulse is then carried to the explosive pellet in the magazine and ultimately to the explosive charge in the shell. Critical monitoring of the guide bush inner and outer diameters is an important aspect of fuze 117 MK 20, as nominal increment in diameter can lead to insufficient force conveyed during impact and the shell will not detonate. Reduction in this diameter can play similar role only with friction to the striker pin during its movement leaving it with insufficient forces to pierce through the detonator. Owing to such complexities and critical dimensional nature of guide bush most of the manufacturers go for 100 percent inspection. Rod section is machined in a Computer Numerically Controlled (CNC) machine. The successive parts are loaded and unloaded from the CNC manually. Regular monitoring of the machined component in the form of online inspection is done by the quality control department. Gauging of all the major dimensions is done in this stage. Any sort of anomaly detected in inspection is conveyed immediately to the CNC department and corrected as early as possible. The data collected is for the outer diameter of the bush guide. The target diameter for bush guide is 27.03 mm and the tolerance limit is ± 0.05 mm. Individual units are examined using gauges and vernier. Any unit exceeding the tolerance limit on the higher side is considered to further machining; however, units falling below the lower limit are rejected as scrape. Since Fuze 117 MK 20 is an assembly comprised of several units, strict quality control on each unit becomes extremely important to ensure uncompromised functioning.

Table 9 Outer diameter of the bush guide data set

SERIAL NO	Phase-I samples
1	27.02, 27.05, 27.06, 27.09, 27.02, 27.05, 27.01, 26.82, 26.85, 27.02
2	27.05, 27.01, 27.04, 27.03, 27.02, 27.02, 27.05, 27.02, 27.05, 27.06
3	27.03, 27.06, 27.05, 27.04, 27.02, 27.03, 27.02, 27.05, 27.06, 27.02
4	27.07, 26.99, 27.02, 27.06, 27.05, 27.05, 27.03, 27.06, 27.02, 27.04
5	27.05, 27.02, 27.04, 27.06, 27.05, 27.04, 27.08, 27.03, 26.98, 27.02
6	26.97, 27.01, 27.05, 27.08, 27.11, 27.03, 27.04, 27.06, 27.04, 27.01
7	26.99, 27.11, 27.13, 27.18, 27.02, 27.06, 27.05, 27.04, 27.01, 27.05
8	27.03, 26.99, 26.99, 27.01, 27.06, 27.08, 27.07, 27.01, 27.06, 27.05
9	27.07, 27.09, 27.11, 27.00, 27.03, 27.02, 26.97, 27.04, 27.06, 27.07
10	27.09, 27.01, 27.00, 27.09, 27.04, 27.02, 27.01, 27.00, 27.05, 27.06
11	27.04, 27.07, 27.09, 27.02, 27.11, 27.04, 27.00, 27.00, 27.02, 27.04
12	26.98, 27.00, 27.03, 27.00, 27.03, 27.01, 27.06, 26.99, 26.98, 27.00
13	27.03, 27.01, 27.00, 27.03, 27.02, 27.01, 27.04, 27.02, 27.02, 27.05
14	27.01, 27.03, 27.01, 26.98, 26.97, 27.05, 27.08, 27.03, 27.02, 27.06
15	27.04, 27.04, 27.07, 27.02, 27.03, 27.06, 27.05, 27.08, 27.00, 27.06
16	27.07, 27.05, 27.05, 27.07, 27.08, 27.02, 27.06, 27.05, 27.00, 27.03
17	27.06, 27.02, 27.06, 27.07, 27.05, 27.08, 27.00, 27.02, 27.05, 27.05
18	27.09, 27.02, 27.06, 26.99, 27.03, 27.05, 26.98, 26.96, 27.05, 27.10
19	26.98, 26.92, 26.99, 26.96, 27.00, 27.01, 26.97, 27.03, 27.06, 27.01
20	27.05, 27.04, 27.00, 27.02, 27.03, 26.98, 27.03, 27.04, 27.01, 27.04
21	27.02, 26.97, 26.98, 27.03, 27.04, 27.01, 27.04, 27.02, 26.97, 27.17

SERIAL NO	Phase-II samples
1	26.98, 26.98, 26.99, 26.98, 27.02, 27.03, 27.00, 27.03, 27.01, 27.05
2	27.04, 27.02, 26.98, 27.03, 27.01, 27.04, 27.01, 27.02, 26.99, 26.99
3	27.03, 27.05, 27.02, 27.01, 27.04, 27.03, 27.03, 27.02, 27.03, 27.05
4	27.03, 27.06, 27.06, 27.09, 27.08, 27.11, 27.09, 27.08, 27.06, 27.00
5	27.01, 27.02, 26.99, 26.98, 26.99, 26.99, 27.03, 27.03, 27.08, 27.05
6	27.02, 27.05, 27.06, 27.04, 26.93, 27.01, 27.03, 27.06, 27.08, 26.98
7	26.98, 26.96, 27.03, 27.02, 27.00, 27.05, 27.03, 27.05, 27.01, 27.06
8	27.05, 27.03, 27.08, 26.99, 27.05, 27.06, 27.05, 27.03, 27.08, 27.04
9	27.03, 27.00, 26.98, 27.03, 27.06, 27.02, 27.05, 27.01, 26.96, 26.95
10	27.05, 27.01, 27.03, 27.05, 27.06, 26.98, 27.03, 27.05, 27.06, 26.98
11	27.03, 27.05, 27.06, 26.98, 27.03, 27.05, 27.01, 27.06, 27.05, 27.03
12	27.01, 27.02, 26.99, 26.95, 26.98, 27.04, 27.01, 27.06, 27.02, 27.03
13	27.02, 27.05, 27.06, 27.01, 27.05, 27.02, 26.96, 26.99, 26.98, 26.95
14	27.03, 27.01, 27.06, 27.02, 27.00, 27.05, 27.04, 27.01, 26.96, 27.00
15	26.98, 27.04, 27.03, 27.01, 26.95, 27.01, 26.95, 26.98, 27.03, 27.04
16	27.01, 27.06, 27.02, 27.00, 27.05, 27.03, 26.98, 27.04, 27.01, 27.06
17	27.02, 27.02, 27.05, 27.08, 26.98, 27.03, 27.01, 27.05, 27.03, 27.05
18	26.98, 27.05, 27.00, 27.04, 27.01, 27.04, 27.02, 26.98, 27.03, 27.01
19	26.95, 27.05, 27.02, 27.00, 26.98, 27.05, 27.08, 27.03, 27.06, 27.04
20	27.08, 27.04, 26.98, 26.95, 27.03, 27.06, 27.01, 27.03, 27.02, 27.00
21	27.05, 27.03, 27.00, 27.05, 27.06, 27.09, 27.01, 27.05, 27.03, 27.02
22	27.03, 26.99, 27.01, 27.00, 27.03, 27.06, 27.01, 27.05, 27.13, 27.02
23	27.06, 26.95, 26.98, 26.93, 27.07, 27.03, 27.01, 27.00, 27.03, 27.00
24	27.03, 27.05, 27.03, 27.06, 27.07, 27.05, 27.06, 27.07, 27.07, 27.09
25	27.08, 27.09, 27.07, 27.07, 27.09, 27.07, 27.08, 27.06, 27.08, 27.10
26	27.11, 27.09, 27.05, 27.01, 26.98, 27.08, 27.08, 27.07, 27.10, 27.05
27	27.13, 27.12, 27.05, 27.07, 27.11, 27.05, 27.08, 27.03, 27.06, 27.08
28	27.01, 26.99, 26.98, 27.00, 27.01, 27.01, 27.03, 27.00, 27.03, 27.01
29	27.05, 27.01, 27.02, 26.98, 26.95, 27.01, 27.05, 27.03, 27.05, 27.08
30	27.09, 27.03, 27.05, 27.01, 26.98, 26.98, 27.03, 27.00, 27.05, 27.08

In this section, we use the proposed schemes to monitor the outer diameter data of the bush guide. The data set consists of 510 observations of the outer diameters, see Table 9. In practice, the Phase-I collection and analysis of data form the basis for the statistical process monitoring and are crucial to the success of Phase-II monitoring. In this context, we consider the first 210 observations for Phase-I analysis. First, we use one of the most popular existing distribution-free monitoring schemes for Phase-I analysis of location and scale, namely, the RS/P method, proposed by Capizzi and Masarotto (2013). The RS/P method is based on recursive segmentation and

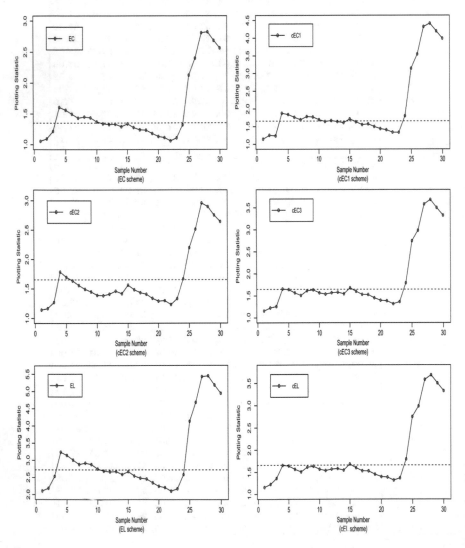

Fig. 1 Six monitoring schemes for the outer diameter data of the bush guide

permutation. This can be implemented via "dfphase1" R package. The results show that the 210 observations all fall within the IC region. Recently, Li et al. (2019) proposed a distribution-free Phase-I monitoring scheme based on the multi-sample Lepage statistic. We further employ this Phase-I Lepage scheme to check the process stability. The results provide no indication of an OOC signal. Therefore, we may consider the first 210 outer diameters as the reference sample. Consequently, $m = 210$. Examining the data set, we find that the underlying data distribution appears to be non-normal. The p-value of the Shapiro test for normality is very low and is almost 0. In this case, we strongly recommend the proposed distribution-free monitoring schemes. The following 300 observations of the outer diameters may be regarded as the Phase-II data that consists of 30 subgroups each of size $n = 10$.

To implement six monitoring schemes discussed in Sects. 2 and 3, we first compute the UCLs of these schemes for $m = 210, n = 10, \lambda = 0.05$, and a target $ARL_0 = 500$. The values of UCL for each scheme are computed by Monte Carlo simulation and given in Table 10. In the same Table, we also present 30 plotting statistics for each scheme as well as the OOC signals indicated with dark gray shades. We further display the observed values of 30 plotting statistics for the EC, cEC1, cEC2,

Table 10 Plotting statistics for the outer diameter data of the bush guide

SERIAL NO	EC $UCL = 1.351$	cEC1 $UCL = 1.655$	cEC2 $UCL = 1.655$	cEC3 $UCL = 1.648$	EL $UCL = 2.723$	cEL $UCL = 1.659$
1	1.054	1.146	1.144	1.158	2.108	1.158
2	1.094	1.254	1.168	1.227	2.187	1.227
3	1.217	1.238	1.271	1.260	2.535	1.362
4	1.604	1.877	1.787	1.656	3.237	1.656
5	1.560	1.842	1.698	1.642	3.145	1.642
6	1.494	1.763	1.637	1.567	3.007	1.567
7	1.430	1.695	1.559	1.508	2.879	1.508
8	1.450	1.783	1.494	1.614	2.917	1.614
9	1.435	1.771	1.450	1.635	2.877	1.635
10	1.370	1.695	1.390	1.567	2.747	1.567
11	1.338	1.642	1.385	1.538	2.687	1.538
12	1.326	1.668	1.410	1.569	2.666	1.569
13	1.332	1.640	1.462	1.582	2.675	1.582
14	1.292	1.609	1.419	1.545	2.589	1.545
15	1.337	1.714	1.564	1.682	2.673	1.682
16	1.278	1.632	1.486	1.599	2.546	1.599
17	1.240	1.554	1.436	1.531	2.480	1.531
18	1.235	1.573	1.409	1.528	2.456	1.528
19	1.181	1.499	1.339	1.453	2.345	1.453
20	1.132	1.438	1.292	1.398	2.246	1.398
21	1.115	1.407	1.300	1.387	2.210	1.387
22	1.061	1.337	1.235	1.318	2.105	1.318
23	1.109	1.335	1.335	1.367	2.175	1.367
24	1.320	1.801	1.669	1.794	2.586	1.794
25	2.120	3.142	2.200	2.749	4.131	2.749
26	2.398	3.546	2.516	2.986	4.676	2.986
27	2.807	4.320	2.960	3.583	5.419	3.583
28	2.822	4.414	2.900	3.686	5.443	3.686
29	2.683	4.194	2.756	3.504	5.177	3.504
30	2.562	3.988	2.645	3.329	4.938	3.329

cEC3, EL, and cEL schemes in Fig. 1. From Table 10 and Fig. 1, we observe that all the monitoring schemes considered in this context detect no change in the process distribution for the first three test samples. These NSPM schemes further indicate that the process is OOC from sample 25 onward until sample 30. Apart from that, the EL and EC schemes produce OOC signals from sample number 4 until sample number 10. The cEC1 scheme signals from the 4th test sample until the 10th test sample and the 12th, 15th, and 24th test samples. We find OOC signals at the 4th, 5th, and 24th test samples from the cEC2 scheme. The cEC3 scheme signals at the 4th, 15th, and 24th test samples. The cEL scheme produces only two additional signals at sample numbers 15 and 24. Almost all charts are working efficiently but the cEC1 scheme produces more signals compared to the others for this data set.

6 Concluding Remarks

In this chapter, we showed that the Cucconi statistic can be decomposed as a quadratic combination of a location statistic and a scale statistic as the well-known Lepage statistic. Moreover, such decomposition is not unique and we presented three different decompositions. Motivated by this, we introduced six distribution-free EWMA monitoring schemes for joint monitoring of location and scale parameters. Two of the six schemes are the traditional EL and EC schemes. The other new schemes are based on a maximum of EWMA of two individual components of the Lepage statistic and the Cucconi statistic, referred to as the cEL, cEC1, cEC2, and cEC3 schemes, respectively. We discussed the implementation procedures, IC and OOC performance of the proposed schemes. The comparative study reveals that no individual monitoring scheme is the best in all cases. Nevertheless, the overall performance of the three cEWMA schemes based on the decompositions of the Cucconi statistic, cEC1, cEC2, and cEC3, is very good for a broad class of location–scale models. Especially for the upward shifts of process parameters, we recommend cEC1 scheme when there is no information related to process distribution.

Mukherjee and Marozzi (2017a) introduced a new graphical device, the circular-grid scheme, for simultaneous monitoring of process location and process scale, based on Lepage-type statistics. Now, we have succeeded to identify the orthogonal location and scale components of the Cucconi statistic and therefore, the development of the circular-grid Cucconi schemes will be feasible in near future. Further, the majority of the existing distribution-free schemes on joint monitoring are intended for two-sided shifts in location–scale. However, more often, only one-sided shifts are important. Chong et al. (2018) proposed a class of Shewhart–Lepage-type schemes for monitoring one-sided shifts. As a future research problem, we may also consider one-sided Cucconi schemes based on the three different decompositions of the Cucconi statistic discussed in this chapter.

Acknowledgements The first author was supported by the Scientific Research Fund of Liaoning Provincial Education Department of China [grant number LSNQN201912] for carrying out this research. The data were collected by Mr. Divyangshu Singh, a former student of BIT Mesra, India as part of his project during an internship under Second Author. Authors are also grateful to an anonymous reviewer for constructing comments and suggestions.

References

Bakir, S. T. (2004). A distribution-free Shewhart quality control chart based on signed-ranks. *Quality Engineering, 16*(4), 613–623.

Bakir, S. T. (2006). Distribution-free quality control charts based on signed-rank-like statistics. *Communications in Statistics-Theory and Methods, 35*(4), 743–757.

Bhattacharya, P., & Frierson, D., Jr. (1981). A nonparametric control chart for detecting small disorders. *The Annals of Statistics, 9*(3), 544–554.

Bonnini, S., Corain, L., Marozzi, M., & Salmaso, L. (2014). *Nonparametric hypothesis testing: rank and permutation methods with applications in R*. Wiley.

Capizzi, G., & Masarotto, G. (2013). Phase I distribution-free analysis of univariate data. *Journal of Quality Technology, 45*(3), 273–284.

Capizzi, G., & Masarotto, G. (2018). Phase I distribution-free analysis with the R package dfphase1. In *Frontiers in statistical quality control* (Vol. 12, pp. 3–19). Springer.

Celano, G., Castagliola, P., & Chakraborti, S. (2016). Joint Shewhart control charts for location and scale monitoring in finite horizon processes. *Computers & Industrial Engineering, 101*, 427–439.

Chakraborti, S., & Graham, M. (2019a). Nonparametric (distribution-free) control charts: An updated overview and some results. *Quality Engineering, 31*(4), 523–544.

Chakraborti, S., & Graham, M. A. (2019b). *Nonparametric statistical process control*. Wiley.

Chakraborti, S., Van der Laan, P., & Van de Wiel, M. (2004). A class of distribution-free control charts. *Journal of the Royal Statistical Society: Series C (Applied Statistics), 53*(3), 443–462.

Chatterjee, S., Qiu, P., et al. (2009). Distribution-free cumulative sum control charts using bootstrap-based control limits. *The Annals of applied statistics, 3*(1), 349–369.

Chong, Z. L., Mukherjee, A., & Khoo, M. B. (2017). Distribution-free Shewhart-Lepage type premier control schemes for simultaneous monitoring of location and scale. *Computers & Industrial Engineering, 104*, 201–215.

Chong, Z. L., Mukherjee, A., & Khoo, M. B. (2018). Some distribution-free Lepage-type schemes for simultaneous monitoring of one-sided shifts in location and scale. *Computers & Industrial Engineering, 115*, 653–669.

Chowdhury, S., Mukherjee, A., & Chakraborti, S. (2014). A new distribution-free control chart for joint monitoring of unknown location and scale parameters of continuous distributions. *Quality and Reliability Engineering International, 30*(2), 191–204.

Chowdhury, S., Mukherjee, A., & Chakraborti, S. (2015). Distribution-free phase II CUSUM control chart for joint monitoring of location and scale. *Quality and Reliability Engineering International, 31*(1), 135–151.

Cucconi, O. (1968). Un nuovo test non parametrico per il confronto fra due gruppi di valori campionari. *Giornale degli Economisti e Annali di Economia, 27*, 225–248.

Graham, M. A., Mukherjee, A., & Chakraborti, S. (2012). Distribution-free exponentially weighted moving average control charts for monitoring unknown location. *Computational Statistics & Data Analysis, 56*(8), 2539–2561.

Hájek, J., Šidák, Z., & Sen, P. K. (1999). *Theory of rank tests*. Academic Press.

Jones-Farmer, L., Woodall, W. H., Steiner, S., & Champ, C. (2014). An overview of phase I analysis for process improvement and monitoring. *Journal of Quality Technology, 46*(3), 265–280.

Lepage, Y. (1971). A combination of Wilcoxon's and Ansari-Bradley's statistics. *Biometrika, 58*(1), 213–217.

Li, C., Mukherjee, A., & Su, Q. (2019). A distribution-free phase I monitoring scheme for subgroup location and scale based on the multi-sample Lepage statistic. *Computers & Industrial Engineering, 129*, 259–273.

Li, S.-Y., Tang, L.-C., & Ng, S.-H. (2010). Nonparametric CUSUM and EWMA control charts for detecting mean shifts. *Journal of Quality Technology, 42*(2), 209–226.

Li, Z., Xie, M., & Zhou, M. (2018). Rank-based EWMA procedure for sequentially detecting changes of process location and variability. *Quality Technology & Quantitative Management, 15*(3), 354–373.

Mahmood, T., Nazir, H. Z., Abbas, N., Riaz, M., & Ali, A. (2017). Performance evaluation of joint monitoring control charts. *Scientia Iranica, 24*(4), 2152–2163.

Marozzi, M. (2009). Some notes on the location-scale Cucconi test. *Journal of Nonparametric Statistics, 21*(5), 629–647.

Marozzi, M. (2013). Nonparametric simultaneous tests for location and scale testing: A comparison of several methods. *Communications in Statistics-Simulation and Computation, 42*(6), 1298–1317.

Marozzi, M. (2014). The multisample Cucconi test. *Statistical Methods & Applications, 23*(2), 209–227.

Montgomery, D. C. (2009). *Statistical quality control*. New York: Wiley.

Mukherjee, A. (2017a). Distribution-free phase-II exponentially weighted moving average schemes for joint monitoring of location and scale based on subgroup samples. *The International Journal of Advanced Manufacturing Technology, 92*(1–4), 101–116.

Mukherjee, A. (2017b). Recent developments in phase-II monitoring of location and scale—An overview and some new results. 61st ISI World Statistics Congress, Marrakesh, Morocco.

Mukherjee, A., & Chakraborti, S. (2012). A distribution-free control chart for the joint monitoring of location and scale. *Quality and Reliability Engineering International, 28*(3), 335–352.

Mukherjee, A., Graham, M. A., & Chakraborti, S. (2013). Distribution-free exceedance CUSUM control charts for location. *Communications in Statistics-Simulation and Computation, 42*(5), 1153–1187.

Mukherjee, A., & Marozzi, M. (2017a). Distribution-free Lepage type circular-grid charts for joint monitoring of location and scale parameters of a process. *Quality and Reliability Engineering International, 33*(2), 241–274.

Mukherjee, A., & Marozzi, M. (2017b). A distribution-free phase-II CUSUM procedure for monitoring service quality. *Total Quality Management & Business Excellence, 28*(11–12), 1227–1263.

Mukherjee, A., & Sen, R. (2018). Optimal design of Shewhart-Lepage type schemes and its application in monitoring service quality. *European Journal of Operational Research, 266*(1), 147–167.

Park, C., Park, C., Reynolds, M. R, Jr., & Reynolds, M. R, Jr. (1987). Nonparametric procedures for monitoring a location parameter based on linear placement statistics. *Sequential Analysis, 6*(4), 303–323.

Qiu, P. (2014). *Introduction to statistical process control*. Chapman and Hall/CRC.

Qiu, P. (2018). Some perspectives on nonparametric statistical process control. *Journal of Quality Technology, 50*, 49–65.

Qiu, P., & Hawkins, D. (2001). A rank-based multivariate CUSUM procedure. *Technometrics, 43*(2), 120–132.

Qiu, P., & Hawkins, D. (2003). A nonparametric multivariate cumulative sum procedure for detecting shifts in all directions. *Journal of the Royal Statistical Society: Series D (The Statistician), 52*(2), 151–164.

Qiu, P., & Li, Z. (2011a). Distribution-free monitoring of univariate processes. *Statistics & Probability Letters, 81*(12), 1833–1840.

Qiu, P., & Li, Z. (2011b). On nonparametric statistical process control of univariate processes. *Technometrics, 53*(4), 390–405.

Roberts, S. (1959). Control chart tests based on geometric moving averages. *Technometrics, 1*(3), 239–250.

Ryu, J.-H., Wan, G., & Kim, S. (2010). Optimal design of a CUSUM chart for a mean shift of unknown size. *Journal of Quality Technology, 42*(3), 311–326.

Song, Z., Mukherjee, A., Liu, Y., & Zhang, J. (2019). Optimizing joint location-scale monitoring— An adaptive distribution-free approach with minimal loss of information. *European Journal of Operational Research, 274*, 1019–1036.

Stromberg, A. J. (2005). Nonparametric control chart for the range. US Patent 6,980,875 B1.

Woodall, W. H., & Montgomery, D. C. (1999). Research issues and ideas in statistical process control. *Journal of Quality Technology, 31*(4), 376–386.

Zafar, R. F., Mahmood, T., Abbas, N., Riaz, M., & Hussain, Z. (2018). A progressive approach to joint monitoring of process parameters. *Computers & Industrial Engineering, 115*, 253–268.

Distribution-Free Phase II Control Charts Based on Order Statistics with Runs-Rules

Ioannis S. Triantafyllou and Nikolaos I. Panayiotou

Abstract In this article, we introduce two nonparametric Shewhart-type control charts based on order statistics with signaling runs-type rules. The proposed monitoring schemes enhance the control charts established by Triantafyllou (2018). Exact formulae for the alarm rate, the variance of the run length distribution and the average run length (*ARL*) for both charts are all derived. Tables are provided for the implementation of the proposed schemes for some typical *ARL*-values. In addition, several numerical comparisons against competitive nonparametric control charts reveal that the new monitoring schemes, under different out-of-control situations, are quite efficient in detecting the shift of the underlying distribution.

Keywords Average run length · Distribution-free control charts · Lehmann alternatives · Nonparametric methods · Runs-type rules · Statistical process control

1 Introduction

Statistical process control is widely used to monitor the quality of a production process, where regardless of how carefully it is maintained, a natural variability always occurs. Control charts help the practitioners to identify assignable causes so that the state of statistical control can be achieved. Tentatively, in the event of observing an undesirable shift in the process, a control chart should detect it as quickly as possible and produce an out-of-control signal.

In control charting practice, two distinct phases have been used in the literature: Phase I and Phase II; see, for example, Woodall (2000). In Phase I, the basic aim is to test historical data for identifying whether they were sampled from an in-control process or not, while Phase II focuses on testing future data for identifying

I. S. Triantafyllou (✉) · N. I. Panayiotou
Department of Computer Science & Biomedical Informatics, University of Thessaly, Volos, Greece
e-mail: itriantafyllou@uth.gr

N. I. Panayiotou
e-mail: nipanagiotou@uth.gr

© Springer Nature Switzerland AG 2020
M. V. Koutras and I. S. Triantafyllou (eds.), *Distribution-Free Methods for Statistical Process Monitoring and Control*,
https://doi.org/10.1007/978-3-030-25081-2_7

whether the process is in-control or has shifted to an out-of-control state. It is often that the practitioner may be interested in determining *both* whether the past data came from a process that was in-control and whether future (test) samples from this process indicate statistical control. In the literature, several types of control charts have been introduced, such as the *Shewhart*-type, the *Cumulative* (*CUSUM*), the *Exponentially Weighted Moving Average* (*EWMA*) or control charts based on change-point detection. Shewhart-type control charts were introduced in the early work of Shewhart (1926) and since then several modifications have been established and studied in detail. The main difference between the *Shewhart*-type and the *CUSUM* or *EWMA* chart is that the former one takes advantage only on the data observed at the current time point for detecting possible shift of the underlying distribution, while the latter charts utilize all available data being observed either currently or at earlier time points. Compared to the *Shewhart*-type, *CUSUM*, or *EWMA* control charts, monitoring schemes based on change-point detection allow the estimation of the occurrence time of a special cause deviation directly when a signal of the special cause deviation is delivered. For a thorough study on Statistical Process Control, the interested reader is referred to the classical textbooks of Montgomery (2009) or Qiu (2014).

Most of the monitoring schemes are distribution-based procedures, even though this presumption is not always realized in practice. To overcome this obstacle and yet keep on the primary formation of the traditional control charts, several nonparametric (or distribution-free) monitoring schemes have been proposed in the literature. The plotting statistics which are often utilized for constructing such kind of control charts are related to well-known nonparametric testing procedures. Among others, a variety of distribution-free control charts appeared already in the literature are based on order statistics; see, e.g., Chakraborti et al. (2004), Balakrishnan et al. (2010), Triantafyllou (2018, 2019a, b). For an up-to-date account on nonparametric Statistical Process Control, the reader is referred to the recent monograph of Chakraborti and Graham (2019) or Qiu (2018, 2019).

In the present chapter, we introduce two distribution-free *Shewhart*-type monitoring schemes based on order statistics. More specifically, we apply the framework established by Triantafyllou (2018) and the resulting chart is empowered by adding two well-known runs-rules. In Sects. 2 and 3, the setup of the proposed monitoring schemes is presented in detail, while explicit formulae for calculating the false alarm rate, the average and the variance of the corresponding run length are also derived. In Sect. 4, several numerical results reveal the efficacy of the proposed charts in comparison to the competitive nonparametric control scheme introduced by Triantafyllou (2018).

2   The 3-*of*-3 *DR* Control Chart Based on Order Statistics

In this section, we establish a new monitoring scheme based on order statistics which utilizes the well-known runs-type rule introduced by Derman and Ross (1997). Three different monitoring statistics are plotted in separate control charts, while the control limits are based on reference data drawn from the in-control process. The proposed scheme is constructed by following the general framework introduced by Triantafyllou (2018) and enhancing its performance with the aid of runs-rules.

Let us denote by $X_1, X_2, \ldots, X_m$ a random sample of size m from the process with distribution F when it is supposed to be in-control, while two specific order statistics, say $X_{a:m}, X_{b:m}$, are appropriately determined and utilized as control limits ($1 \leq a < b \leq m$). Suppose next that test samples are picked out independently of each other (and also of the reference sample) and that we are interested in checking whether the process is still in-control or not. In statistical terms, if $Y_1, Y_2, \ldots, Y_n$ denote the test sample of size n with cumulative distribution function G, our aim is to detect a possible shift in the underlying distribution from F to G. After the test sample is collected, the j-th and the k-th order statistic $Y_{j:n}, Y_{k:n}$ are chosen and made use of along with the statistic

$$R = R(Y_1, Y_2, \ldots, Y_n; X_{a:m}, X_{b:m}) = |\{i \in \{1, 2, \ldots, n\} : X_{a:m} \leq Y_i \leq X_{b:m}\}|,$$

where R is simply the number of test observations between the control limits LCL, UCL. According to the monitoring scheme introduced by Triantafyllou (2018), the process is declared to be in-control, if the following conditions hold true

$$LCL \leq Y_{j:n} \leq Y_{k:n} \leq UCL \quad \text{and} \quad R \geq r,$$

where r is a positive integer. For more details about the general setup of the abovementioned monitoring scheme, the interested reader is referred to Triantafyllou (2018) and Sect. 2 therein.

In order to improve the performance of the aforementioned nonparametric control chart, we activate the 3-*of*-3 runs-rule proposed by Derman and Ross (1997). Under this scenario, an out-of-control signal is produced from the new monitoring scheme (3-of-3 *DR* chart, hereafter) whenever three consecutive plotting points: (*a*) fall all of them on or above the UCL, (*b*) fall all of them on or below the LCL, (*c*) one falls on or above the UCL and the other two fall on or below the LCL, (*d*) two of them fall on or above the UCL and the other one falls on or below the LCL.

The signaling indicator for the h-th test sample is defined as

$$Z_h = \begin{cases} 0, \text{ if } Y_{j:n}^h, Y_{k:n}^h \in (LCL, UCL) \\ 1, \text{ otherwise} \end{cases}, h = 1, 2, 3, \ldots \tag{1}$$

If we denote by T_3 the run length of the proposed monitoring scheme, namely the waiting time until the first signal, the following ensues

$$T_3 = \min\{t : Z_{t-2} = 1, Z_{t-1} = 1, Z_t = 1\} \text{ or } \min\left\{t : \sum_{i=t-2}^{t} Z_i = 3\right\}. \quad (2)$$

Employing analogous arguments than those implemented by Chakraborti et al. (2009), we next deduce the unconditional distribution of the positive random variable T_3. More specifically, we examine the distribution of the random variable T_3 by conditioning on the total number of successes $S_n = \sum_{i=1}^{n} Z_i$ in the sequence of exchangeable binary random variables $Z_1, Z_2, \ldots$.

Proposition 1 *The probability mass function of the unconditional distribution of T_3 is given by*

$$P(T_3 = x) = \begin{cases} 0, & \text{if } 0 \le x < 3 \\ \lambda_3, & \text{if } x = 3 \end{cases}$$

and for $x \ge 4$

$$P(T_3 = x) = \sum_{y=1}^{x-3} \sum_{j=0}^{\min\left[y, \left[\frac{x-y-3}{3}\right]\right]} \sum_{i=0}^{y} (-1)^{i+j} \binom{y}{j}\binom{y}{i}\binom{x-3(j+1)-1}{y-1}\lambda_{x-y+i}, \quad (3)$$

where

$$\lambda_w = P(Z_1 = 1, Z_2 = 1, \ldots, Z_w = 1), w = 1, 2, \ldots, n.$$

Proof For $0 \le x < 3$, the proposed monitoring scheme could not produce by definition a signal, therefore we have that

$$P(T_3 = x) = 0, \text{ for } x \in [0, 3).$$

In addition, when $x = 3$ it is straightforward that the following holds true

$$P(T_3 = 3) = P(Z_1 = 1, Z_2 = 1, Z_3 = 1) = \lambda_3.$$

In case of $x \ge 4$, the unconditional distribution of the random variable T_3 can be written as

$$P(T_3 = x) = \sum_{y=1}^{x-3} P(T_3 = x | S_x = x - y) P(S_x = x - y). \quad (4)$$

Since (see, e.g., George and Bowman (1995))

$$P(S_n = d) = \binom{n}{d} \sum_{i=0}^{n-d} (-1)^i \binom{n-d}{i} \lambda_{d+i}, \quad d = 1, 2, \ldots, n$$

we may write

$$P(S_x = x - y) = \binom{x}{y} \sum_{i=0}^{y} (-1)^i \binom{y}{i} \lambda_{x-y+i}. \tag{5}$$

In addition, it is well known that, given the number of successes, the conditional distribution of T_3 for the exchangeable case is the same as the one for a sequence of independent and identically distributed binary variables (see, e.g. Kingman (1978)). Hence, the conditional probability $P(T_3 = x | S_x = x - y)$ can be expressed as (see, e.g., Balakrishnan and Koutras (2002))

$$P(T_3 = x | S_x = x - y) = \binom{x}{y}^{-1} \sum_{j=0}^{\left[\frac{x-y}{3}\right]} (-1)^j \binom{y}{j} \binom{x - 3(j+1) - 1}{y - 1}$$

$$= \binom{x}{y}^{-1} N(x - y - 3, y, 2). \tag{6}$$

Substituting (6) and (5) in (4) the proof is complete.   □

The following proposition offers explicit expressions for two important characteristics of the run length of the proposed 3-of-3 DR chart.

Proposition 2

(i) *The unconditional* Average Run Length *of the proposed 3-of-3* DR *chart is given by*

$$ARL_{DR} = \int_0^1 \int_0^t \frac{1 - p^3}{(1 - p) \cdot p^3} f_{a,b}(s, t) ds dt. \tag{7}$$

(ii) *The unconditional* Variance *of the* Run Length *of the proposed 3-of-3 DR chart is given by*

$$VAR_{DR} = \int_0^1 \int_0^t \frac{1 - 7(1 - p)p^3 - p^7}{(1 - p)^2 p^6} f_{a,b}(s, t) ds dt, \tag{8}$$

where the probability $p = 1 - q(GF^{-1}(s), GF^{-1}(t); r)$ *is expressed as*

$$q(v, w; r) = \sum_{c_1=0}^{n-2} \sum_{c_3=(0,r-c_1-c_2)}^{n-c_1-c_2-2} \frac{n!}{(j - c_1 - 1)!(n - k - c_3)!(c_1 + c_2 + c_3 + 2)!}$$
$$\times\ v^{j-c_1-1}(u_1 - v)^{c_1}(u_2 - u_1)^{c_2}(w - u_2)^{c_3}(1 - w)^{n-k-c_3}$$

$$(9)$$

where $0 \leq v < w < 1$, $c_2 = k - j - 1$.

Proof Given $X_{a:m} = x_1$ and $X_{b:m} = x_2$, the random variable T_3, defined earlier as the run length of the proposed 3-of-3 *DR* chart, has a geometric distribution of order 3. Therefore, the conditional expected value and variance of the run length T_3 are given in terms of the probability of producing a signal (p) as (see, e.g. Balakrishnan and Koutras (2002))

$$E(T_3|X_{a:m} = x_1, X_{b:m} = x_2) = \frac{1 - p^3}{(1 - p)p^3}$$

and

$$VAR(T_3|X_{a:m} = x_1, X_{b:m} = x_2) = \frac{1 - 7(1 - p)p^3 - p^7}{(1 - p)^2 p^6}$$

respectively. Since the $p = 1 - q(GF^{-1}(s), GF^{-1}(t); r)$ can be expressed via formula (9) (for more details, see Triantafyllou (2018) and Proposition 1 therein), the desired result is effortlessly derived by averaging over the distribution of $X_{a:m}$ and $X_{b:m}$. □

It is straightforward that the unconditional *Average* and *Variance* of the in-control *Run Length* can be obtained by substituting $F = G$ in the expressions proved in the previous proposition. In Table 1, we present the in-control *ARL* of the proposed control chart for several designs corresponding to different values of a, b, m, n, j, k, r. The calculations were carried out with the aid of Proposition 2. Table 1 can be used to design a distribution-free control chart that attains a pre-specified in-control level of performance (ARL_0). The use of seven design parameters in this chart offers the flexibility to fix some of them and then look for the optimal choice of the others, or alternatively search for an acceptable combination of them that meets our special needs.

For example, if we draw a reference sample of size $m = 50$, an in-control *Average Run Length* equal to 370 (approximately) can be achieved by

- utilizing the 3rd and the 44th ordered observation from the reference sample ($a = 3, b = 44$), working with test samples of size $n = 5$ and determining the remaining parameters as $j = 1, k = 3, r = 2$ (with $ARL_{in} = 369.19$) or
- utilizing the 13th and the 46th ordered observation from the reference sample ($a = 13, b = 46$), working with test samples of size $n = 11$ and determining the remaining parameters as $j = 4, k = 8, r = 3$ (with $ARL_{in} = 369.40$) or

Table 1 In-control *Average Run Length* of the proposed 3-of-3 *DR* chart for a given design. Reference sample size m

ARL$_o$	n	50		100		200		500	
		(a, b, j, k, r)	ARL$_{in}$	(a, b, j, k, r)	ARL$_{in}$	(a, b, j, k, r)	ARL$_{in}$	(a, b, j, k, r)	ARL$_{in}$
370	5	(3, 44, 1, 3, 2)	369.19	(5, 95, 1, 3, 2)	366.10	(8, 182, 1, 3, 3)	369.37	(17, 463, 1, 3, 2)	370.45
	11	(13, 46, 4, 8, 3)	369.40	(8, 71, 2, 5, 3)	372.87	(15, 179, 2, 7, 5)	370.76	(34, 497, 2, 9, 6)	369.80
	25	(2, 39, 2, 20, 11)	373.75	(6, 95, 3, 23, 12)	372.46	(7, 193, 4, 24, 15)	370.55	(16, 385, 2, 14, 11)	370.65
500	5	(3, 45, 1, 3, 2)	495.27	(4, 96, 1, 5, 2)	499.90	(29, 195, 2, 4, 3)	500.37	(15, 455, 1, 3, 3)	501.00
	11	(3, 44, 1, 3, 3)	501.14	(9, 93, 2, 6, 4)	501.88	(26, 184, 3, 8, 5)	501.33	(32, 465, 2, 7, 5)	499.72
	25	(14, 44, 14, 23, 7)	494.45	(7, 94, 3, 22, 15)	499.67	(8, 175, 2, 17, 13)	501.93	(15, 453, 2, 19, 16)	501.82

- utilizing the 2nd and the 39th ordered observation from the reference sample ($a = 2, b = 39$), working with test samples of size $n = 25$ and determining the remaining parameters as $j = 2, k = 20, r = 11$ (with $ARL_{in} = 373.75$).

The out-of-control performance could be evaluated via the corresponding ARL that the control chart attains. If the process shifts out-of-control, Proposition 2 offers an explicit expression for computing the out-of-control ARL of the proposed 3-of-3 DR chart. It is evident that this quantity depends on both the in-control and out-of-control distributions F and G. Consequently, since the final result depends on the form of the function $G \circ F^{-1}$, it will clearly be unreasonable to expect an explicit expression for it in the general case. However, there is a wide class of alternatives, the so-called *Lehmann alternatives*, for which the expression provided by Proposition 2 can be shown to simplify to a neat exact formula. The Lehmann alternatives (see Lehmann (1953)) have been extensively used to assess the power of nonparametric control charts, see, e.g., Triantafyllou (2019a, b) or Koutras and Triantafyllou (2018). Under the Lehmann-type alternative, the out-of-control distribution function takes on the form $G = F^{\gamma}$ for some fixed, positive number $\gamma > 0$. If γ is a positive integer, the Lehmann alternative states that the Y random variables are distributed as the largest of γ of the X variables. Table 2 displays the out-of-control ARL-*values* achieved by the new control chart under the Lehmann-type alternatives for $\gamma = 0.5$ and $\gamma = 0.9$. The designs which are implemented for producing the out-of-control ARL-values in Table 2 are the same with the ones presented already in Table 1.

One may draw interesting conclusions based on the numerical results displayed in Table 2. For example, let us consider the same case study mentioned earlier, namely let us assume that the practitioner works with a reference sample of size $m = 50$ in order to reach an in-control ARL equal to 370. Then, under the Lehmann alternatives with parameter γ equal to 0.9 (0.5) the proposed 3-of-3 DR chart achieves an out-of-control ARL equal to 156.51 (7.06), 93.69 (4.14) and 107.76 (4.46) when test sample of size $n = 5, 11, 25$ are drawn respectively.

Let us next denote by G_j the cumulative distribution function of the j-th order statistic in a random sample of size n from a continuous distribution with cumulative distribution function G. It is evident that

$$G_j(x) = I_{G(x)}(j, n - j + 1),$$

where $I_a(b, c)$ denotes the incomplete beta function. The following proposition provides an explicit formula for the computation of the *False Alarm Rate* of the proposed 3-of-3 DR chart.

Table 2 Out-of-control *Average Run Length* of the proposed 3-of-3 *DR* chart for a given design

Reference sample size *m*

ARL_0	n	50		100		200		500	
		(a, b, j, k, r)	ARL_{out}	(a, b, j, k, r)	ARL_{out}	(a, b, j, k, r)	ARL_{out}	(a, b, j, k, r)	ARL_{out}
370	5	(3, 44, 1, 3, 2)	156.51 / 7.06	(5, 95, 1, 3, 2)	134.22 / 7.04	(8, 188, 1, 3, 3)	144.35 / 7.71	(17, 463, 1, 3, 2)	151.21 / 8.21
	11	(13, 46, 4, 8, 3)	93.69 / 4.14	(8, 71, 2, 5, 3)	131.40 / 4.51	(15, 179, 2, 7, 5)	103.97 / 4.45	(34, 497, 2, 9, 6)	104.35 / 4.59
	25	(2, 39, 2, 20, 11)	107.76 / 4.46	(6, 95, 3, 23, 12)	118.89 / 3.49	(7, 193, 4, 24, 15)	336.33 / 7.73	(16, 385, 2, 14, 11)	79.45 / 3.38
500	5	(3, 45, 1, 3, 2)	187.49 / 7.08	(4, 96, 1, 5, 2)	209.24 / 8.60	(29, 190, 2, 4, 3)	183.02 / 8.65	(15, 455, 1, 3, 3)	200.96 / 9.09
	11	(3, 44, 1, 3, 3)	56.06 / 3.64	(9, 93, 2, 6, 4)	103.74 / 4.15	(26, 184, 3, 8, 5)	140.38 / 4.76	(32, 465, 2, 7, 5)	133.58 / 4.79
	25	(14, 44, 14, 23, 7)	242.50 / 16.27	(7, 94, 3, 22, 15)	113.58 / 3.31	(8, 175, 2, 17, 13)	83.07 / 3.28	(15, 453, 2, 19, 16)	102.59 / 3.44

Each ARL_{out} cell contains the values attained for $\gamma = 0.9$ and 0.5 respectively

Proposition 3 *The* False Alarm Rate *of the proposed 3-of-3* DR *chart is given by*

FAR_{DR}

$$= \int_0^1 [I_x(j, n-j+1)]^3 f_a(x)dx + 3\int_0^1 [I_x(j, n-j+1)]^2 I_x(k, n-k+1) f_a(x)dx$$

$$+ 3\int_0^1 I_x(j, n-j+1)[I_x(k, n-k+1)]^2 f_a(x)dx + \int_0^1 [I_x(k, n-k+1)]^3 f_a(x)dx$$

$$+ \int_0^1 [1 - I_y(k, n-k+1)]^3 f_b(y)dy + 3\int_0^1 [1 - I_y(k, n-k+1)]^2 I_y(j, n-j+1) f_b(y)dy$$

$$+ 3\int_0^1 (1 - I_y(k, n-k+1))[1 - I_y(j, n-j+1)]^2 f_b(y)dy + \int_0^1 [1 - I_y(j, n-j+1)]^3 f_b(y)dy$$

$$+ 3\int_0^1 \int_0^y [I_x(j, n-j+1)]^2 (2 - I_y(k, n-k+1) - I_y(j, n-j+1)) f_{a,b}(x, y)dxdy$$

$$+ 6\int_0^1 \int_0^y I_x(j, n-j+1)I_x(k, n-k+1)(2 - I_y(k, n-k+1) - I_y(j, n-j+1)) f_{a,b}(x, y)dxdy$$

$$+ 3\int_0^1 \int_0^y [I_x(k, n-k+1)]^2 (2 - I_y(k, n-k+1) - I_y(j, n-j+1)) f_{a,b}(x, y)dxdy$$

$$+ 3\int_0^1 \int_0^y [1 - I_y(j, n-j+1)]^2 (I_x(k, n-k+1) + I_x(j, n-j+1)) f_{a,b}(x, y)dxdy$$

$$+ 6\int_0^1 \int_0^y (1 - I_y(j, n-j+1))(1 - I_y(k, n-k+1))(I_x(k, n-k+1) + I_x(j, n-j+1)) f_{a,b}(x, y)dxdy$$

$$+ 3\int_0^1 \int_0^y [1 - I_y(k, n-k+1)]^2 (I_x(k, n-k+1) + I_x(j, n-j+1)) f_{a,b}(x, y)dxdy$$

where f_a, f_b denote the probability density functions of the beta $(a, m - a + 1)$ and $(b, m - b + 1)$ distribution respectively, while $f_{a,b}$ corresponds to the joint probability density function of the a-th and the b-th order statistics of a random sample of size m from the Uniform $(0, 1)$ distribution.

Proof The proposed monitoring scheme produces an out-of-control signal, whenever three consecutive plotting points violate either the lower or the upper control limit. Therefore, the *False Alarm Rate* of the 3-of-3 *DR* chart, namely the probability for the chart to produce a signal under the assumption that the process is still in-control can be simply expressed as

$$FAR_{DR} = P(3 \text{ plotting points below } X_{a:m}|F = G) + P(3 \text{ plotting points over } X_{b:m}|F = G)$$
$$+ P(2 \text{ plotting points below } X_{a:m} \text{ and } 1 \text{ plotting point over } X_{b:m}|F = G)$$
$$+ P(1 \text{ plotting point below } X_{a:m} \text{ and } 2 \text{ plotting points over } X_{b:m}|F = G) \quad (10)$$

The first term of the above summation can be explicated as follows:

$P(3 \text{ plotting points below } X_{a:m}|F = G)$

$$= P\left(Y_{j:n}^{h-2}, Y_{j:n}^{h-1}, Y_{j:n}^h \le X_{a:m}|F = G\right) + P\left(Y_{j:n}^{h-2}, Y_{j:n}^{h-1}, Y_{k:n}^h \le X_{a:m}|F = G\right)$$

$$+ P\left(Y_{j:n}^{h-2}, Y_{k:n}^{h-1}, Y_{k:n}^h \le X_{a:m}|F = G\right) + P\left(Y_{k:n}^{h-2}, Y_{k:n}^{h-1}, Y_{k:n}^h \le X_{a:m}|F = G\right)$$

$$+ P\left(Y_{j:n}^{h-2}, Y_{k:n}^{h-1}, Y_{j:n}^{h} \leq X_{a:m}|F = G\right) + P\left(Y_{k:n}^{h-2}, Y_{j:n}^{h-1}, Y_{j:n}^{h} \leq X_{a:m}|F = G\right)$$

$$+ P\left(Y_{k:n}^{h-2}, Y_{j:n}^{h-1}, Y_{k:n}^{h} \leq X_{a:m}|F = G\right) + P\left(Y_{k:n}^{h-2}, Y_{k:n}^{h-1}, Y_{j:n}^{h} \leq X_{a:m}|F = G\right)$$

$$(11)$$

We next average on $X_{a:m} = x_1$ and consequently each one of the above conditional probabilities is expressed via the integral $I_a(b, c)$ defined earlier. Indeed, we conclude that

$$P\left(Y_{j:n}^{h-2}, Y_{j:n}^{h-1}, Y_{j:n}^{h} \leq X_{a:m}|F = G\right) = \int_0^1 [I_x(j, n - j + 1)]^3 f_a(x) dx$$

$$P\left(Y_{j:n}^{h-2}, Y_{j:n}^{h-1}, Y_{k:n}^{h} \leq X_{a:m}|F = G\right) = \int_0^1 [I_x(j, n - j + 1)]^2 I_x(k, n - k + 1) f_a(x) dx$$

$$P\left(Y_{j:n}^{h-2}, Y_{k:n}^{h-1}, Y_{k:n}^{h} \leq X_{a:m}|F = G\right) = \int_0^1 I_x(j, n - j + 1)[I_x(k, n - k + 1)]^2 f_a(x) dx$$

$$P\left(Y_{k:n}^{h-2}, Y_{k:n}^{h-1}, Y_{k:n}^{h} \leq X_{a:m}|F = G\right) = \int_0^1 [I_x(k, n - k + 1)]^3 f_a(x) dx.$$

Substituting the above expressions in (11), the probability of observing three consecutive plotting points below the lower control limit $X_{a:m}$ becomes quite convenient. We next apply analogous arguments for the remaining terms of the summation in (10) and the desired result for the *False Alarm Rate* of the proposed scheme is deduced after some straightforward manipulations. □

3 The 3-*of*-3 *KL* Control Chart Based on Order Statistics

In this section, we establish a new monitoring scheme based on order statistics which utilizes the well-known runs-type rule introduced by Klein (2000). Three different monitoring statistics are plotted in separate control charts, while the control limits are based on reference data drawn from the in-control process. The proposed scheme is constructed by following the general framework introduced by Triantafyllou (2018) and enhancing its performance with the aid of runs-rules.

Let us denote once again by $X_1, X_2, \ldots, X_m$ a random sample of size m from the in-control distribution F, while test samples $Y_1, Y_2, \ldots, Y_n$ with cumulative distribution function G are drawn independently. After each test sample is picked out, the j-th and the k-th order statistic $Y_{j:n}$, $Y_{k:n}$ are chosen and made use of along with the statistic R defined earlier. According to the monitoring scheme introduced by Triantafyllou (2018), the process is declared to be in-control, if the following conditions hold true

$$LCL \leq Y_{j:n} \leq Y_{k:n} \leq UCL \quad \text{and} \quad R \geq r,$$

where r is a positive integer (see also Triantafyllou (2018) and Sect. 2 therein).

In order to improve the performance of the aforementioned nonparametric control chart, we activate the 3-*of*-3 runs-rule proposed by Klein (2000). Under this scenario, an out-of-control signal is produced from the new monitoring scheme (3-of-3 *KL* chart, hereafter) whenever three consecutive plotting points fall all of them either on or above the *UCL* or all of them fall on or below the *LCL*.

The signaling indicator for the h-th test sample is defined as

$$Z'_h = \begin{cases} 0, & \text{if } Y^h_{j:n}, Y^h_{k:n} \in (LCL, UCL) \\ 1, & \text{if at least one of } Y^h_{j:n}, Y^h_{k:n} \geq UCL, h = 1, 2, 3, \dots \\ 2, & \text{if at least one of } Y^h_{j:n}, Y^h_{k:n} \leq LCL \end{cases} \quad (11)$$

If we denote by T^*_3 the run length of the proposed monitoring scheme, namely the waiting time for three consecutive 1's or three consecutive 2's in the sequence of independent and identically distributed trials $Z'_1, Z'_2, \dots$. Thus is called a compound pattern $\Lambda = \Lambda_1 \cup \Lambda_2$, where $\Lambda_1 = \{1\,1\,1\}$ and $\Lambda_2 = \{2\,2\,2\}$. We next denote by $\mathbf{I}_\nu$ the $\nu \times \nu$ identity matrix while the quantities p_L, p_U are defined as

$$p_L = \left(\text{at least one of } Y_{j:n}, Y_{k:n} \leq X_{a:m} | X_{a:m} = x_1\right)$$

and

$$p_U = \left(\text{at least one of } Y_{j:n}, Y_{k:n} \geq X_{b:m} | X_{b:m} = x_2\right)$$

respectively. In other words, p_L corresponds to the probability of a single plotting point falling below or on the *LCL*, while p_U expresses the probability of a single plotting point falling above or on the *UCL*.

The following proposition offers an explicit formula for determining the probability mass function of the random variable T^*_3.

Proposition 4 *The probability mass function of the unconditional distribution of T^*_3 is given by*

$$P(T^*_3 = x) = \int_{-\infty}^{+\infty} \int_{-\infty}^{x_2} \xi \cdot \mathbf{N}^{x-1} \cdot (\mathbf{I}_6 - \mathbf{N}) \cdot \mathbf{1}' f_{a,b}(x_1, x_2) dx_1 dx_2, \ \mathrm{x} \geq 3 \quad (12)$$

where

$$\mathbf{N} = \begin{bmatrix} 0 & 1 - p_L - p_U & p_U & p_L & 0 & 0 \\ 0 & 1 - p_L - p_U & p_U & p_L & 0 & 0 \\ 0 & 1 - p_L - p_U & 0 & p_L & p_U & 0 \\ 0 & 1 - p_L - p_U & p_U & 0 & 0 & p_L \\ 0 & 1 - p_L - p_U & 0 & p_L & 0 & 0 \\ 0 & 1 - p_L - p_U & p_U & 0 & 0 & 0 \end{bmatrix},$$

$$\xi = [1\,0\,0\,0\,0\,0\,0\,0], \mathbf{1} = [1\,1\,1\,1\,1\,1\,1\,1],$$

while $f_{a,b}$ denotes the joint probability density function of the a-th and the b-th order statistics of a random sample of size m from the Uniform (0, 1) distribution.

Proof In order to derive the probability mass function of the random variable T_3^*, we next implement the general approach established by Fu and Lou (2003) for the distribution of the waiting time till the first occurrence of a compound pattern in a sequence of independent and identically distributed or homogeneous Markov dependent k-state trials. In our case, the parameter k equals to 8 and the imbedded Markov chain is defined on the state space

$$\Omega = \{\emptyset, \{0\}, \{1\}, \{2\}, (1\,1), (2\,2), a_1, a_2\}$$

where $a_1 = (1\ 1\ 1)$ and $a_2 = (2\ 2\ 2)$ correspond to the absorbing states. The transition probability matrix of the Markov chain is denoted as $\mathbf{M}_{8\times8} = \begin{bmatrix} \mathbf{N} & \mathbf{C} \\ \mathbf{O} & \mathbf{I}_2 \end{bmatrix}$, where $\mathbf{N}$ is the essential transition probability sub-matrix defined above, while

$$\mathbf{C} = \begin{bmatrix} 0 & 0 \\ 0 & 0 \\ 0 & 0 \\ 0 & 0 \\ p_U & 0 \\ 0 & p_L \end{bmatrix}, \mathbf{O} = \begin{bmatrix} 0\,0\,0\,0\,0\,0 \\ 0\,0\,0\,0\,0\,0 \end{bmatrix}.$$

Appling the Theorem 5.2 of Fu and Lou (2003), the conditional distribution of T_3^* is given by

$$P\big(T_3^* = x | X_{a:m} = x_1, X_{b:m} = x_2\big) = \xi \cdot \mathbf{N}^{x-1} \cdot (\mathbf{I}_6 - \mathbf{N}) \cdot \mathbf{1}', x \geq 3. \tag{13}$$

Consequently, the unconditional distribution of T_3^* is effortlessly derived by averaging the conditional distribution of T_3^* over the joint probability density function of the order statistics $X_{a:m}$ and $X_{b:m}$

$$P(T_3^* = x) = E_{X_{a:m}, X_{b:m}} \big(P\big(T_3^* = x | X_{a:m} = x_1, X_{b:m} = x_2\big) \big)$$
$$= \int_{-\infty}^{+\infty} \int_{-\infty}^{x_2} P\big(T_3^* = x | X_{a:m} = x_1, X_{b:m} = x_2\big) f_{a,b}(x_1, x_2) dx_1 dx_2, \ x \geq 3.$$

and the proof is complete by substituting the expression (13) in the last formula. $\square$

The following proposition offers explicit expressions for two important characteristics of the run length of the proposed 3-of-3 *KL* chart.

Proposition 5

(i) *The unconditional* Average Run Length *of the proposed 3-of-3* KL *chart is given by*

$$ARL_{KL} = \int_0^1 \int_0^t \frac{(p_L^2 + p_L + 1)(p_U^2 + p_U + 1)}{p_L^3 p_U^2 + p_L^3 p_U + p_L^3 + p_L^2 p_U^3 + p_L p_U^3 + p_U^3} f_{a,b}(s,t) ds dt. \quad (14)$$

(ii) *The unconditional* Variance *of the* Run Length *of the proposed 3-of-3* KL *chart is given by*

VAR_{KL}

$$= \int_0^1 \int_0^t -\left(\frac{p_L^5(p_U^4 + 2p_U^3 + 3p_U^2 + 2p_U + 1) + p_L^4(p_U^5 + 5p_U^4 + 9p_U^3 + 6p_U^2 + 4p_U + 2)}{(p_L^3(p_U^2 + p_U + 1) + p_U^3(p_L^2 + p_L + 1))^2} \right.$$

$$- \frac{p_L^3(2p_U^5 + 9p_U^4 + 16p_U^3 + 9p_U^2 + 6p_U + 3) + p_L^2(3p_U^5 + 6p_U^4 + 9p_U^3 - 9p_U^2 - 6p_U - 3)}{(p_L^3(p_U^2 + p_U + 1) + p_U^3(p_L^2 + p_L + 1))^2}$$

$$\left. - \frac{-2p_L(p_U^5 + 4p_U^4 + 6p_U^3 - 6p_U^2 - 4p_U - 2) + (p_U^5 + 2p_U^4 + 3p_U^3 - 3p_U^2 - 2p_U - 1)}{(p_L^3(p_U^2 + p_U + 1) + p_U^3(p_L^2 + p_L + 1))^2} \right)$$

$$\times f_{a,b}(s,t) ds dt. \quad (15)$$

Proof

(i) Given $X_{a:m} = x_1$ and $X_{b:m} = x_2$, the conditional expected value of the waiting time T_3^* is derived by applying Theorem 7.4 of Fu and Lou (2003). More specifically, the conditional average value of the random variable T_3^* can be expressed as

$$E(T_3^* | X_{a:m}, X_{b:m}) = \xi \cdot (\mathbf{I} - \mathbf{N})^{-1} \cdot \mathbf{1}',$$

where $\xi, \mathbf{N}$ are given in Proposition 4. As it has been already mentioned, the above expected value coincides to the conditional *Average Run Length* of the proposed 3-of-3 *KL* chart. Consequently, the unconditional *Average Run Length* of the proposed 3-of-3 *KL* chart is readily obtained by averaging the above expression over the joint distribution of the order statistics $X_{a:m}, X_{b:m}$.

(ii) Given $X_{a:m} = x_1$ and $X_{b:m} = x_2$, the conditional variance of the waiting time T_3^* is derived by applying Theorem 7.4 of Fu and Lou (2003) as

$$Var(T_3^* | X_{a:m}, X_{b:m}) = \xi(\mathbf{I} + \mathbf{N})(I - \mathbf{N})^{-2}\mathbf{1}' - (E(T_3^* | X_{a:m}, X_{b:m}))^2.$$

Since the matrices $\xi, \mathbf{N}$ are given in Proposition 4, the proof is complete by substituting the expression (14) in the above formula. □

In Table 3, we present the in-control *ARL* of the proposed control chart for several designs corresponding to different values of a, b, m, n, j, k, r. The calculations were

Table 3 In-control *Average Run Length* of the proposed 3-of-3 KL chart for a given design

Reference sample size m

ARL_o	n	50		100		200		500	
		(a, b, j, k, r)	ARL_{in}	(a, b, j, k, r)	ARL_{in}	(a, b, j, k, r)	ARL_{in}	(a, b, j, k, r)	ARL_{in}
370	5	(2, 43, 1, 4, 2)	369.21	(7, 99, 1, 2, 2)	369.07	(12, 192, 1, 3, 3)	369.62	(25, 434, 1, 3, 2)	369.48
	11	(15, 47, 4, 8, 3)	372.75	(11, 95, 2, 8, 5)	367.93	(16, 193, 2, 10, 5)	369.58	(78, 459, 3, 8, 4)	369.39
	25	(6, 39, 4, 19, 12)	365.31	(19, 94, 6, 22, 14)	370.23	(26, 181, 5, 21, 15)	369.60	(60, 437, 5, 20, 14)	369.39
500	5	(4, 46, 1, 3, 2)	497.79	(6, 78, 1, 2, 2)	503.40	(10, 192, 1, 4, 3)	499.05	(23, 482, 1, 4, 3)	500.57
	11	(14, 44, 5, 9, 3)	503.72	(10, 99, 2, 10, 5)	498.98	(30, 188, 3, 9, 4)	502.90	(41, 486, 2, 9, 5)	498.73
	25	(17, 48, 10, 24, 14)	494.00	(22, 94, 7, 22, 12)	500.10	(37, 192, 6, 21, 13)	500.92	(85, 468, 6, 20, 15)	500.30

carried out with the aid of Proposition 5. Table 3 can be used to design a distribution-free control chart that attains a pre-specified in-control level of performance (ARL_0).

For example, if we draw a reference sample of size $m = 100$, an in-control *Average Run Length* equal to 500 (approximately) can be achieved by

- utilizing the 6th and the 78th ordered observation from the reference sample ($a = 6, b = 78$), working with test samples of size $n = 5$ and determining the remaining parameters as $j = 1, k = 2, r = 2$ (with $ARL_{in} = 503.40$) or
- utilizing the 10th and the 99th ordered observation from the reference sample ($a = 10, b = 99$), working with test samples of size $n = 11$ and determining the remaining parameters as $j = 2, k = 10, r = 5$ (with $ARL_{in} = 498.98$) or
- utilizing the 22nd and the 94th ordered observation from the reference sample ($a = 22, b = 94$), working with test samples of size $n = 25$ and determining the remaining parameters as $j = 7, k = 22, r = 12$ (with $ARL_{in} = 500.10$).

The out-of-control performance could be evaluated via the corresponding ARL that the control chart attains. If the process shifts out-of-control, Proposition 5 offers an explicit expression for computing the out-of-control ARL of the proposed 3-of-3 KL chart. Table 4 displays the out-of-control ARL-*values* achieved by the proposed 3-of-3 KL chart under the Lehmann-type alternatives for $\gamma = 0.3$ and $\gamma = 0.8$. The designs which are implemented for producing the out-of-control ARL-values in Table 4 are the same with the ones presented already in Table 3.

One may draw interesting conclusions based on the numerical results displayed in Table 4. For example, let us consider the same case study mentioned earlier, namely let us assume that the practitioner works with a reference sample of size $m = 100$ in order to reach an in-control ARL equal to 500. Then, under the Lehmann alternatives with parameter γ equal to 0.8 (0.3) the proposed 3-of-3 KL chart achieves an out-of-control ARL equal to 105.33 (8.40), 57.50 (7.15) and 35.37 (7.00) when test sample of size $n = 5, 11, 25$ are drawn respectively.

The following proposition provides an explicit formula for the computation of the *False Alarm Rate* of the proposed 3-of-3 KL chart.

Proposition 6 *The* False Alarm Rate *of the proposed 3-of-3* KL *chart is given by*

$$
\begin{aligned}
FAR_{KL} \\
&= \int_0^1 [I_x(j, n - j + 1)]^3 f_a(x)dx \\
&+ 3 \int_0^1 [I_x(j, n - j + 1)]^2 I_x(k, n - k + 1) f_a(x)dx \\
&+ 3 \int_0^1 I_x(j, n - j + 1)[I_x(k, n - k + 1)]^2 f_a(x)dx \\
&+ \int_0^1 [I_x(k, n - k + 1)]^3 f_a(x)dx + \int_0^1 [1 - I_y(k, n - k + 1)]^3 f_b(y)dy \\
&+ 3 \int_0^1 [1 - I_y(k, n - k + 1)]^2 I_y(j, n - j + 1) f_b(y)dy
\end{aligned}
$$

Table 4 Out-of-control *Average Run Length* of the proposed 3-of-3 KL chart for a given design

Reference sample size m

ARL_O	n	50		100		200		500	
		(a, b, j, k, r)	ARL_{out}	(a, b, j, k, r)	ARL_{out}	(a, b, j, k, r)	ARL_{out}	(a, b, j, k, r)	ARL_{out}
370	5	(3, 44, 1, 3, 2)	194.02 / 9.98	(5, 95, 1, 3, 2)	76.25 / 8.13	(8, 183, 1, 3, 3)	82.01 / 8.27	(17, 463, 1, 3, 2)	93.58 / 8.49
	11	(13, 46, 4, 8, 3)	39.95 / 7.14	(8, 71, 2, 5, 3)	44.56 / 7.11	(15, 179, 2, 7, 5)	67.95 / 7.21	(34, 497, 2, 9, 6)	54.36 / 7.18
	25	(2, 39, 2, 20, 11)	44.58 / 7.01	(6, 95, 3, 23, 12)	29.62 / 7.00	(7, 193, 4, 24, 15)	41.07 / 7.00	(16, 385, 2, 14, 11)	44.90 / 7.00
500	5	(3, 45, 1, 3, 2)	89.60 / 8.11	(4, 96, 1, 5, 2)	105.33 / 8.40	(29, 199, 2, 4, 3)	111.54 / 8.59	(15, 455, 1, 3, 3)	112.05 / 8.65
	11	(3, 44, 1, 3, 3)	134.32 / 7.74	(9, 93, 2, 6, 4)	57.50 / 7.15	(26, 184, 3, 8, 5)	70.53 / 7.23	(32, 465, 2, 7, 5)	63.77 / 7.18
	25	(14, 44, 14, 23, 7)	27.41 / 7.01	(7, 94, 3, 22, 15)	35.37 / 7.00	(8, 175, 2, 17, 13)	27.35 / 7.00	(15, 453, 2, 19, 16)	31.55 / 7.00

Each ARL_{out} cell contains the values attained for $\gamma = 0.8$ and 0.3 respectively

$$+ 3 \int_0^1 (1 - I_y(k, n - k + 1))[1 - I_y(j, n - j + 1)]^2 f_b(y) dy$$

$$+ \int_0^1 [1 - I_y(j, n - j + 1)]^3 f_b(y) dy$$

where f_a, f_b denote the probability density functions of the beta $(a, m - a + 1)$ and $(b, m - b + 1)$ distribution respectively.

Proof The proposed monitoring scheme produces an out-of-control signal, whenever three consecutive plotting points violate either all of them the lower or all of them the upper control limit. Therefore, the *False Alarm Rate* of the 3-of-3 *KL* chart, namely the probability for the chart to produce a signal under the assumption that the process is still in-control can be simply expressed as

$$FAR_{KL} = P(3 \text{ plotting points above } X_{b:m} | F = G)$$
$$+ P(3 \text{ plotting points below } X_{a:m} | F = G) \qquad (16)$$

The first term of the above summation can be explicated as follows:

$P(3 \text{ plotting points above } X_{b:m} | F = G)$

$$= P\left(Y_{j:n}^{h-2}, Y_{j:n}^{h-1}, Y_{j:n}^{h} \geq X_{b:m} | F = G\right) + P\left(Y_{j:n}^{h-2}, Y_{j:n}^{h-1}, Y_{k:n}^{h} \geq X_{b:m} | F = G\right)$$

$$+ P\left(Y_{j:n}^{h-2}, Y_{k:n}^{h-1}, Y_{k:n}^{h} \geq X_{b:m} | F = G\right) + P\left(Y_{k:n}^{h-2}, Y_{k:n}^{h-1}, Y_{k:n}^{h} \geq X_{b:m} | F = G\right)$$

$$+ P\left(Y_{j:n}^{h-2}, Y_{k:n}^{h-1}, Y_{j:n}^{h} \geq X_{b:m} | F = G\right) + P\left(Y_{k:n}^{h-2}, Y_{j:n}^{h-1}, Y_{j:n}^{h} \geq X_{b:m} | F = G\right)$$

$$+ P\left(Y_{k:n}^{h-2}, Y_{j:n}^{h-1}, Y_{k:n}^{h} \geq X_{b:m} | F = G\right) + P\left(Y_{k:n}^{h-2}, Y_{k:n}^{h-1}, Y_{j:n}^{h} \geq X_{b:m} | F = G\right)$$

$$\qquad (17)$$

We next average on $X_{b:m} = x_2$ and consequently each one of the above conditional probabilities is expressed via the integral $I_a(b, c)$ defined earlier. Indeed, we conclude that

$$P\left(Y_{j:n}^{h-2}, Y_{j:n}^{h-1}, Y_{j:n}^{h} \geq X_{b:m} | F = G\right) = \int_0^1 [1 - I_x(j, n - j + 1)]^3 f_b(x) dx$$

$$P\left(Y_{j:n}^{h-2}, Y_{j:n}^{h-1}, Y_{k:n}^{h} \geq X_{b:m} | F = G\right) = \int_0^1 [1 - I_x(j, n - j + 1)]^2 (1 - I_x(k, n - k + 1)) f_b(x) dx$$

$$P\left(Y_{j:n}^{h-2}, Y_{k:n}^{h-1}, Y_{k:n}^{h} \geq X_{b:m} | F = G\right) = \int_0^1 (1 - I_x(j, n - j + 1))[1 - I_x(k, n - k + 1)]^2 f_b(x) dx$$

$$P\left(Y_{k:n}^{h-2}, Y_{k:n}^{h-1}, Y_{k:n}^{h} \geq X_{b:m} | F = G\right) = \int_0^1 [1 - I_x(k, n - k + 1)]^3 f_b(x) dx.$$

Substituting the above expressions in (17), the probability of observing three consecutive plotting points over the upper control limit $X_{b:m}$ becomes quite convenient. We next apply analogous arguments for the other term of the summation in (16) and the desired result for the *False Alarm Rate* of the proposed scheme is deduced after some straightforward manipulations.   □

4 Numerical Comparisons

In this section, we carry out an extensive numerical experimentation to illustrate the efficacy of the new control charts and their robustness features under both in-control and out-of-control situations. The computations are accomplished with the aid of theoretical results presented in previous sections. It is important to mention that the proposed distribution-free chart is quite capable of identifying not only a location shift but also shift through the variance and this is due to the incorporation into the control chart of the additional condition based on the number of observations from the test sample falling between the control limits.

A typical way of comparing two different control charts is to use a common in-control average length (ARL_{in}) and then to examine their out-control average lengths (ARL_{out}'s). In what follows, we compare the performance of the proposed 3-of-3 DR chart to the one established by Triantafyllou (2018).

Table 5 offers some numerical comparisons between the proposed control scheme and the nonparametric chart introduced by Triantafyllou (2018). We consider the case of a process with underlying in-control Normal distribution with parameters 0 and 1, while as out-of-control distribution is assumed to be Normal distribution with mean and standard deviation equal to θ and δ respectively. In Table 5, two different ARL_{in} levels have been considered. For each choice, Proposition 2 was implemented to determine the design parameters of the new control chart, viz., $LCL = X_{a:m}$, $UCL = X_{b:m}$, j, k and r, so that the exact ARL_{in} of the resulting chart is as close to the desired ARL_{in} as possible. Then, the ARL_{out}-values for the new chart are evaluated numerically and the corresponding results are summarized in Table 5 under the label *New chart*. Under the label *Triantafyllou chart* (2018), we have presented the respective results for his chart; since we have considered the same parameter choices for m, n, ARL_0 as those used there, this part of the table is simply reproduced from their Table 5.

Table 5 clearly reveals that, under the same ARL_{in}, the new control chart performs better than the chart introduced by Triantafyllou (2018), in terms of ARL_{out} values, in some cases considered. For example, when a reference sample of size $m = 100$ is drawn and the practitioner decides to work with test samples of size $n = 5$ and pre-fixed $ARL_{in} = 500$, the chart by Triantafyllou (2018) achieves ARL_{out} equal to 15.00 (9.80) for out-of-control Normal distribution with $\theta = 0.5, \delta = 1.25$ ($\theta = 1, \delta = 1.25$), while the corresponding ARL_{out}-values for the 3-of-3 DR chart are 12.37 (9.24) respectively.

Table 5 *ARL-values* of different control charts under the $N(\theta, \delta)$ distribution ($m = 100, n = 5$)

θ	δ	3-of-3 DR chart	Triantafyllou chart (2018)
0	1	456.52	446.6
0.25	1	360.99	163.9
0.5	1	145.81	51.64
1	1	14.78	7.4
1.5	1	4.66	2.1
2	1	3.27	1.2
0	1.25	45.19	61.4
0.25	1.25	33.58	35.7
0.5	1.25	9.24	17.9
1	1.25	7.03	5.0
1.5	1.25	4.02	2.1
2	1.25	3.23	1.3
0	1.5	15.15	20.2
0.25	1.5	12.37	15.0
0.5	1.5	9.24	9.8
1	1.5	5.28	4.1
1.5	1.5	3.75	2.1
2	1.5	3.22	1.4
0	1.75	8.55	10.0
0.25	1.75	7.50	8.5
0.5	1.75	6.32	6.5
1	1.75	4.52	3.5
1.5	1.75	3.60	2.1
2	1.75	3.20	1.5
0	2	6.16	6.2
0.25	2	5.65	5.7
0.5	2	5.07	4.8
1	2	4.10	3.1
1.5	2	3.50	2.0
2	2	3.20	1.5

Generally speaking, the nonparametric control charts are robust by definition; that is, their in-control behavior remains the same for all continuous distributions. However, it is useful to examine their out-of-control performance for different underlying distributions. We next compare the performance of 3-of-3 *DR* chart with the one established by Triantafyllou (2018). Table 6 depicts the *ARL-values* not only of the proposed control scheme but also of the abovementioned nonparametric chart under Laplace distribution (θ, δ). The proposed distribution-free control scheme performs better than the competitive chart established by Triantafyllou (2018) for almost all

Table 6 *ARL-values* of different control charts under the *Laplace* (θ, δ) distribution $(m = 100, n = 5)$

θ	δ	3-of-3 DR chart	Triantafyllou chart (2018)
0	1	456.52	446.6
0.25	1	266.31	276.9
0.5	1	120.70	159.2
1	1	15.98	45.7
1.5	1	4.46	12.2
2	1	3.29	3.6
0	1.25	81.74	107.5
0.25	1.25	61.27	75.9
0.5	1.25	38.52	50.4
1	1.25	11.15	19.6
1.5	1.25	4.66	7.2
2	1.25	3.46	2.9
0	1.5	30.17	43.1
0.25	1.5	25.76	33.2
0.5	1.5	19.47	24.3
1	1.5	8.80	11.7
1.5	1.5	4.72	5.3
2	1.5	3.60	2.6
0	1.75	16.22	22.8
0.25	1.75	14.82	18.7
0.5	1.75	12.44	14.7
1	1.75	7.35	8.2
1.5	1.75	4.66	4.4
2	1.75	3.69	2.4
0	2	10.77	14.3
0.25	2	10.21	12.2
0.5	2	9.12	10.0
1	2	6.36	6.3
1.5	2	4.54	3.8
2	2	3.73	2.3

the cases considered. More specifically, the in-control reference sample in each case is drawn from the corresponding standard distribution with $\theta = 0$ and $\delta = 1$, while several combinations of parameters θ, δ have been examined. When the underlying distribution of the process is assumed to be Laplace, the new control scheme is superior compared to the other chart for shifts of the location parameter θ no greater than 1.5, while for the remaining cases both charts seem to be almost equivalent.

5 Conclusions

In the present chapter, two nonparametric Shewhart-type control charts are introduced and studied in some detail. The plotting statistics which are monitored, are related to specified order statistics from successive test samples which are drawn from the underlying process. For the enhancement of the proposed charts, two different well-known runs-rules are also activated. The main performance characteristics of the new distribution-free monitoring schemes are studied, while several numerical outcomes reveal the ability of the proposed charts for detecting possible shift of the underlying distribution process. It is of some future research interest, to investigate the performance of the aforementioned control chart under the presence of alternative runs-rules.

References

Balakrishnan, N., & Koutras, M. V. (2002). *Runs and scans with applications*. New York: Wiley.
Balakrishnan, N., Triantafyllou, I. S., & Koutras, M. V. (2010). A distribution-free control chart based on order statistics. *Communication in Statistics: Theory & Methods, 39*, 3652–3677.
Chakraborti, S., Eryilmaz, S., & Human, S. W. (2009). A phase II nonparametric control chart based on precedence statistics with runs-type signaling rules. *Computational Statistics & Data Analysis, 53*, 1054–1065.
Chakraborti, S., & Graham, M. (2019). *Nonparametric statistical process control*. USA: Wiley.
Chakraborti, S., van der Laan, P., & van der Wiel, M. A. (2004). A class of distribution-free control charts. *Journal of the Royal Statistical Society, Series C-Applied Statistics, 53*, 443–462.
Derman, C., & Ross, S. M. (1997). *Statistical aspects of quality control*. San Diego: Academic Press.
Fu, J. C., & Lou, W. Y. W. (2003). *Distribution theory of runs and patterns and its applications: A finite Markov chain imbedding approach*. Singapore: World Scientific Publishing.
George, E. O., & Bowman, D. (1995). A full likelihood procedure for analyzing exchangeable binary data. *Biometrics, 51*, 512–523.
Klein, M. (2000). Two alternatives to the Shewhart $\bar{X}$ control chart. *Journal of Quality Technology, 32*, 427–431.
Kingman, J. F. C. (1978). Uses of exchangeability. *Annals of Probability, 6*, 183–197.
Koutras, M. V., & Triantafyllou, I. S. (2018). A general class of nonparametric control charts. *Quality and Reliability Engineering International, 34*, 427–435.
Lehmann, E. L. (1953). The power of rank tests. *Annals of Mathematical Statistics, 24*, 23–43.
Montgomery, D. C. (2009). *Introduction to statistical quality control* (6th ed.). New York: Wiley.
Qiu, P. (2014). *Introduction to statistical process control*. New York: CRC Press, Taylor & Francis Group.
Qiu, P. (2018). Some perspectives on nonparametric statistical process control. *Journal of Quality Technology, 50*, 49–65.
Qiu, P. (2019). Some recent studies in statistical process control. In *Statistical quality technologies*, pp. 3–19.
Shewhart, W. A. (1926). Quality control charts. *Bell System Technical Journal, 2*, 593–603.
Triantafyllou, I. S. (2018). Nonparametric control charts based on order statistics. *Communication in Statistics: Simulation and Computation, 47*, 2684–2702.
Triantafyllou, I. S. (2019a). A new distribution-free control scheme based on order statistics, *Journal of Nonparametric Statistics, 31*, 1–30.

Triantafyllou, I. S. (2019b). Wilcoxon-type rank-sum control charts based on progressively censored reference data, *Communication in Statistics: Theory and Methods.* https://doi.org/10.1080/03610926.2019.1634816.

Woodall, W. H. (2000). Controversies and contradictions in statistical process control. *Journal of Quality Technology, 32,* 341–350.

A Nonparametric Control Chart for Dynamic Disease Risk Monitoring

Lu You and Peihua Qiu

Abstract Some deadly diseases can be treated or even prevented if they or some of their symptoms are detected early. Disease early detection and prevention is thus important for our health improvement. In this paper, we suggest a novel and effective new method for disease early detection. By this method, a patient's risk to the disease is first quantified at each time point by survival data analysis of a training dataset that contains patients' survival information and longitudinally observed disease predictors (e.g., disease risk factors and other covariates). To improve the effectiveness of the proposed method, variable selection is used in the survival analysis to keep only important disease predictors in disease risk quantification. Then, the longitudinal pattern of the quantified risk is monitored sequentially over time by a nonparametric control chart. A signal will be given by the chart once the cumulative difference between the risk pattern of the patient under monitoring and the risk pattern of a typical person without the disease in concern exceeds a control limit.

Keywords Disease screening · Disease early detection · Dynamic process · Longitudinal data · Statistical process control · Survival data

1 Introduction

One of the primary objectives of a disease screening program is to give early signals to patients who have the disease in concern or who are at high risk of having the disease, so that these patients can receive timely intervention and treatment (Qiu and Xiang 2014). This paper aims to develop a novel and effective method for disease screening.

Medical research has identified major predictors of many diseases. For instance, the major predictors for cardiovascular diseases include high blood pressure, high cholesterol level, obesity, tobacco use, lack of physical activity, diabetes, unhealthy

L. You · P. Qiu (✉)
Department of Biostatistics, University of Florida, Gainesville, FL 32610, USA
e-mail: pqiu@ufl.edu

© Springer Nature Switzerland AG 2020
M. V. Koutras and I. S. Triantafyllou (eds.), *Distribution-Free
Methods for Statistical Process Monitoring and Control*,
https://doi.org/10.1007/978-3-030-25081-2_8

diet, age, gender, family history, and some others (e.g., Mendis et al. 2011). For disease screening, patients often take scheduled disease screening examinations over time to have their medical conditions evaluated. To identify high-risk patients through the data collected during the screening examinations, an effective statistical tool is needed. This type of research problem is called dynamic screening (DS) problem in Qiu and Xiang (2014), because medical data are collected sequentially over time from patients, data distribution would change over time, and decisions about the disease status need to be made sequentially as well during the process of data collection. Qiu and Xiang (2014) proposed a dynamic screening system (DySS) to monitor a single disease predictor over time for handling the DS problem. In their method, they first model the regular longitudinal pattern of the disease predictor by a nonparametric longitudinal model estimated from an in-control (IC) dataset that contains observed data of the disease predictor of patients without the disease in concern. Then, to monitor the disease predictor of a new individual, they constructed a statistical process control (SPC) chart to detect undesirable deviations and/or changes in the longitudinal pattern of the disease predictor of the individual under monitoring from the estimated regular longitudinal pattern. By employing a cumulative sum (CUSUM) control chart, this method makes use of the observed data at the current time point and all history data efficiently, and it has been demonstrated to good performance in many applications. In subsequent research, Qiu and Xiang (2015) further extended the DySS method to multivariate cases where multiple disease predictors are considered. A multivariate control chart was proposed to jointly monitor all disease predictors. Some other extensions of the DySS method include those discussed in Li and Qiu (2016, 2017) and You and Qiu (2018) where serially correlated data are considered, and the one discussed in Qiu et al. (2018) where unequally spaced observation times were accommodated in the construction of the control chart. Qiu et al. (2019) proposed a new metric for evaluating the numerical performance of DS methods.

In practice, there could be many different disease predictors involved. Some of them might be more important than the others in predicting the occurrence of the disease in question. But, in the multivariate DySS methods mentioned above, all disease predictors are treated equally in constructing the related multivariate control charts, which would make the charts less effective in predicting the disease. To overcome this limitation, You and Qiu (2019) recently proposed a new method consisting of the following two steps: (i) estimation of a survival model from a training dataset and the estimated survival model is then used for quantifying the disease risk of a person, where the quantified disease risk is a linear combination of all disease predictors, and (ii) sequential monitoring of the quantified risks over time using a control chart. In the estimated survival model, more important covariates will receive more weights in the linear combination of the disease predictors. Thus, the effectiveness of the control chart is improved. Nonetheless, the aforementioned method still uses all disease predictors when defining disease risk. Intuitively, if certain disease predictors actually contain little useful information about the disease in concern, then they should be removed from disease screening. Based on this intuition, we propose a new method in this paper, in which variable selection by LASSO is incorporated in survival data modeling, so that the redundant disease predictors are deleted during survival model estimation. It will be shown that this new method is more effective than the original one by You and Qiu (2019) in various different cases.

The remaining part of the article is organized as follows. In Sect. 2, the proposed model and its estimation for disease risk quantification will be introduced. In Sect. 3, some simulation studies will be presented to evaluate the performance of the proposed method. The proposed method will be demonstrated in a real data example in Sect. 4. Finally, Sect. 5 will conclude the article with some discussions about certain future research topics.

2 Proposed Method

In this section, we describe the proposed disease screening method in detail. Our proposed method consists of two main steps. In the first step, a survival model is fitted from a training dataset that contains observations of the survival times and disease predictors of certain individuals. The fitted model can then be used to quantify people's disease risk at a given time. In this step, we will also discuss how to select important disease predictors by a LASSO variable selection method. In the second step, the quantified disease risk of a specific individual is monitored sequentially over time by a control chart. These two steps are discussed in detail in the following two parts.

2.1 Risk Estimation and Variable Selection

Suppose that a training dataset containing observations of the longitudinal disease predictors and survival times of n individuals. The survival outcomes of the ith individual are described by (δ_i, T_i), where T_i is the last follow-up time and δ_i is the survival indicator with $\delta_i = 1$ indicating the occurrence of an disease at the last follow-up time T_i, and $\delta_i = 0$ otherwise. Following the notations of survival models in the literature, we use D_i to denote the true disease time and C_i to denote the censoring time. Then, the survival outcomes can be expressed as $T_i = \min\{D_i, C_i\}$ and $\delta_i = I(D_i \leq C_i)$. For simplicity of presentation, we will also use $R(t) = \{i : T_i \geq t\}$ to denote the set of all individuals who are at risk of disease at a given time t (i.e., they are still under monitoring in the study at time t). The q-dimensional longitudinal disease predictor of the ith individual is denoted as $\mathbf{x}_i(t)$, and it is repeatedly and sequentially observed at times $t_{i1}, \ldots, t_{im_i}$, where these observation times can be unequally spaced and $t_{im_i} = T_i$. Let $\lambda_i(t) = \lim_{dt \to 0} P\{D_i \in [t, t + dt] | D_i \geq t\}/dt$ be the hazard function of the disease in question for the ith individual. Then, the following Cox proportional hazard model is assumed (cf., Klein and Moeschberger 1997):

$$\lambda_i(t) = \lambda_0(t) \exp(\boldsymbol{\beta}' \mathbf{x}_i(t)), \qquad (1)$$

where β is a q-dimensional vector of coefficients and $\lambda_0(t)$ is the baseline hazard function. By using model (1), the linear combination $\beta'\mathbf{x}_i(t)$ can measure the disease risk of the ith individual at time t, and it is denoted as $r_i(t)$. Namely, we define

$$r_i(t) = \beta'\mathbf{x}_i(t).$$

To estimate model (1), You and Qiu (2019) suggested using the following kernel-smoothed likelihood:

$$L(\beta) = \prod_{i:\delta_i=1} \frac{\exp(\beta'\mathbf{x}_i(T_i))}{\sum_{l \in R(T_i)} \sum_{j=1}^{m_l} K_h(T_i - t_{lj}) \exp(\beta'\mathbf{x}_l(t_{lj}))},$$

where $K_h(s) = K(s/h)/h$, $K(s)$ is a density kernel function, and $h > 0$ is a bandwidth. The use of kernel smoothing in Cox proportional hazards model is motivated by some existing works on estimating time-varying coefficients model in the literature (e.g., Cai and Sun 2003; Tian et al. 2005). The corresponding log-likelihood function is given by

$$l(\beta) = \sum_{i:\delta_i=1} \left[\beta'\mathbf{x}_i(T_i) - \log \left\{ \sum_{l \in R(T_i)} \sum_{j=1}^{m_l} K_h(T_i - t_{lj}) \exp(\beta'\mathbf{x}_l(t_{lj})) \right\} \right]. \quad (2)$$

Then, β can be estimated by the maximizer of (2), denoted as $\tilde{\beta}$, which can be obtained by using the Newton–Raphson algorithm.

So far, we assume that all disease predictors in $\mathbf{x}_i(t)$ have substantial impact on the disease risk. In reality, because we do not know which disease predictors are important and which are not, we often include many potential disease predictors in $\mathbf{x}_i(t)$, to avoid important disease predictors being overlooked. Thus, some disease predictors in $\mathbf{x}_i(t)$ may not have much prediction power for the specific disease. This will be reflected in the regression coefficients in β, which some of them could be 0 or small. However, the estimate $\tilde{\beta}$ given by the partial log-likelihood function in (2) usually would not contain elements that are exactly 0. Thus, it cannot serve the purpose of variable selection. To properly select important disease predictors and exclude unimportant ones, we need to identify zero elements in the regression coefficient β, which can be achieved by using the LASSO method (Tibshirani 1996). The main idea of LASSO is to add a penalty term on the regression coefficients to shrink the coefficients of unimportant disease predictors toward zero. In this paper, we choose to use the following L_1 adaptive LASSO penalty (cf., Zou 2006):

$$p_\gamma(\beta) = \gamma \sum_{k=1}^{q} w_k |\beta_k|,$$

where γ is a nonnegative regularization parameter, and $\mathbf{w} = (w_1, \dots, w_q)'$ is a vector of adaptive weights. The adaptive weights $\{w_k\}$ can be simply chosen to be $1/|\tilde{\beta}_k|$, where $\tilde{\boldsymbol{\beta}} = (\tilde{\beta}_1, \ldots, \tilde{\beta}_k)$ is the estimate of $\boldsymbol{\beta}$ obtained from (2), as discussed above. The LASSO penalized estimate of $\boldsymbol{\beta}$ is then defined to be the minimizer of the following penalized log-likelihood function:

$$-l(\boldsymbol{\beta}) + p_\gamma(\boldsymbol{\beta}). \tag{3}$$

The above penalized log-likelihood function is not differentiable with respect to $\boldsymbol{\beta}$ at $\mathbf{0}$, and thus $\widehat{\boldsymbol{\beta}}$ cannot be obtained by the Newton-type optimization algorithms. Here, we can use the coordinate optimization algorithm discussed in Friedman et al. (2007) and Simon et al. (2011) to find the estimate, which is denoted as $\widehat{\boldsymbol{\beta}}$.

Note that there are some components in $\widehat{\boldsymbol{\beta}}$ that are exactly 0. These components and the corresponding disease predictors can then be deleted from the subsequent analysis. Without loss of generality, after these components and the corresponding disease predictors are deleted, the estimated regression coefficient vector and the disease predictor vector are still denoted as $\widehat{\boldsymbol{\beta}}$ and $\mathbf{x}_i(t)$. Then, the estimated disease risk of the ith individual at time t is defined to be

$$\widehat{r}_i(t) = \widehat{\boldsymbol{\beta}}' \mathbf{x}_i(t).$$

Similar to the univariate DySS method discussed in Qiu and Xiang (2014), we can characterize the regular pattern of the disease risk by a nonparametric longitudinal model, with the mean to be $\mu(t) = E[r_i(t)|T_i \geq t]$ and the variance to be $\sigma^2(t) = \text{Var}[r_i(t)|T_i \geq t]$. Here, we only assume that $\mu(t)$ and $\sigma^2(t)$ are two smooth functions of time, and they can be estimated by the local linear kernel smoothing procedure, as discussed in Qiu and Xiang (2014) and Xiang et al. (2013). The corresponding local linear kernel estimates of $\mu(t)$ and $\sigma^2(t)$ are given by

$$\widehat{\mu}(t) = \frac{R_0(t) W_{2,h_\mu}(t) - R_1(t) W_{1,h_\mu}(t)}{W_{0,h_\mu}(t) W_{2,h_\mu}(t) - W_{1,h_\mu}(t)^2}, \tag{4}$$

$$\widehat{\sigma}^2(t) = \frac{Q_0(t) W_{2,h_\sigma}(t) - Q_1(t) W_{1,h_\sigma}(t)}{W_{0,h_\sigma}(t) W_{2,h_\sigma}(t) - W_{1,h_\sigma}(t)^2}, \tag{5}$$

where h_μ and h_σ are two bandwidths that could be different from h used in (2), $\widehat{\epsilon}_i(t_{ij}) = \widehat{r}_i(t_{ij}) - \widehat{\mu}(t_{ij})$, and

$$W_{l,h}(t) = \frac{1}{n} \sum_{i \in R(t)} \sum_{j=1}^{m_i} K_h(t_{ij} - t)\left(\frac{t_{ij} - t}{h}\right)^l,$$

$$R_l(t) = \frac{1}{n} \sum_{i \in R(t)} \sum_{j=1}^{m_i} K_{h_\mu}(t_{ij} - t)\left(\frac{t_{ij} - t}{h_\mu}\right)^l \widehat{r}_i(t_{ij}),$$

$$Q_l(t) = \frac{1}{n} \sum_{i \in R(t)} \sum_{j=1}^{m_i} K_{h_\sigma}(t_{ij} - t)\left(\frac{t_{ij} - t}{h_\sigma}\right)^l \widehat{\epsilon}_i^2(t_{ij}).$$

As a side note, in the above model, the regular disease risk pattern is characterized by the first and second moments $\mu(t)$ and $\sigma^2(t)$. Alternatively, we can characterize the regular disease risk pattern by the entire distribution of $r_i(t)$. To this end, let $F(y; t) = P(r_i(t) \leq y | T_i \geq t)$ be the conditional distribution function of the disease risk at time t. For given values of y and t, we can use the local linear kernel smoothing method to estimate this conditional distribution, as discussed in Fan et al. (1996) and Yu and Jones (1998). By following their ideas, we can consider minimizing the following objective function:

$$\sum_{i \in R(t)} \sum_{j=1}^{m_i} \left[\Psi_{h_\psi}\left(\widehat{r}_i(t_{ij}) - y\right) - \alpha_0 - \alpha_1(t_{ij} - t) \right]^2 K_{h_F}(t_{ij} - t),$$

where $\Psi(y)$ is a suitable kernel cumulative distribution function, h_ψ and h_F are two bandwidths, and $\Psi_h(y) = \Psi(y/h)$. Then, the estimate $\widehat{F}(y; t)$ of $F(y; t)$ can be defined by the minimizer with respect to α_0 in the above minimization problem, which has the expression

$$\widehat{F}(y; t) = \frac{S_0(y; t) W_{2,h_F}(t) - S_1(y; t) W_{1,h_F}(t)}{W_{0,h_F}(t) W_{2,h_F}(t) - W_{1,h_F}(t)^2}, \tag{6}$$

where for $l = 0, 1$,

$$S_l(y; t) = \frac{1}{n} \sum_{i \in R(t)} \sum_{j=1}^{m_i} K_{h_F}(t_{ij} - t) \left(\frac{t_{ij} - t}{h_F}\right)^l \Psi_{h_\psi}\left(\widehat{r}_i(t_{ij}) - y\right).$$

This idea will not be further explored in this paper and will be studied in our future research.

In (3)–(6), there are several bandwidths to use. To determine the bandwidth h used in (3) for estimating β, we can use the leave-one-out cross-validation (CV) criterion that is based on the martingale residuals (cf., Tian et al. 2005, You and Qiu 2019a, b). The selected bandwidth can be calculated by minimizing the following function of h:

$$CV_\beta(h) = \sum_{i=1}^{n} PE_i(h),$$

where

$$PE_i(h) = \left[\delta_i - \sum_{\substack{k \neq i, \delta_k=1 \\ T_k \leq T_i}} \frac{\sum_{j=1}^{m_i} K_h(T_k - t_{ij}) \exp(\tilde{\beta}'_{-i} \mathbf{x}_i(t_{ij}))}{\sum_{d \neq k, d \in R(T_k)} \sum_{j=1}^{m_d} K_h(T_k - t_{dj}) \exp(\tilde{\beta}'_{-i} \mathbf{x}_d(t_{dj}))} \right]^2,$$

β_{-i} is the estimate of β when the ith individual is excluded from model estimation, and PE_i is the square of some estimate of the integrated martingale residual. To choose the regularization parameter γ in the LASSO penalty in (3), we propose to use the Akaike information criterion (AIC) (cf., Akaike 1992; Tibshirani 1997). Let $c(\beta)$ be the number of nonzero elements of the vector β. Then, the AIC of the modified Cox partial likelihood is defined as

$$AIC(\gamma) = -2l(\widehat{\beta}) + 2c(\widehat{\beta})$$

where $\widehat{\beta}$ is the estimate of β with the regularization parameter being γ. The regularization parameter γ is then chosen to be the minimizer of $AIC(\gamma)$. The bandwidths h_μ and h_σ in (4) and (5) can be chosen using the leave-one-out CV procedure discussed in Qiu and Xiang (2014). In this chapter, all kernel functions are chosen to be the Epanechnikov kernel function (Epanechnikov 1969).

2.2 Online Disease Risk Monitoring

To monitor the quantified disease risk of a new individual, assume that the disease predictors are sequentially observed at times $t_1^*, t_2^*, \ldots$, and the corresponding observations are $\mathbf{x}(t_1^*), \mathbf{x}(t_2^*), \ldots$. For simplicity of presentation, we further assume that $t_1^*, t_2^*, \ldots$ are multiplications of a basic time ω. Thus, we can write $t_j^* = n_j^* \omega$, for $j \geq 1$. When the disease risk pattern is characterized by the estimated mean and variance function $\widehat{\mu}(t)$ and $\widehat{\sigma}^2(t)$, we can define the standardized value of the estimated disease risk as

$$\widehat{e}(t_j^*) = \frac{\widehat{r}(t_j^*) - \widehat{\mu}(t_j^*)}{\widehat{\sigma}(t_j^*)}, \qquad \text{for } j \geq 1.$$

To monitor the quantified disease risks of the new individual and detect an undesirable upward shift in the longitudinal pattern of the disease risk when observations are sequentially obtained, SPC charts can be used. To this end, we consider using the following upward EWMA chart, based on the exponential smoothing idea that was discussed in Wright (1986) and Qiu et al. (2018) to account for irregularly spaced observation times:

$$E_1 = V_1 \widehat{e}(t_1^*), \tag{7}$$

$$E_j = (1 - V_j)E_{j-1} + V_j \widehat{e}(t_j^*), \tag{8}$$

where $V_1 = 1 - (1 - \lambda)^{\bar{\Delta}}$, $V_j = V_{j-1}/[(1 - \lambda)^{n_j^* - n_{j-1}^*} + V_{j-1}]$, λ is a weighting parameter in $[0, 1)$, and $\bar{\Delta}$ is an estimate of the mean of $n_j^* - n_{j-1}^*$ obtained from the IC dataset. The chart gives a signal at time t_j^* if

$$E_j > \rho$$

where $\rho > 0$ is a control limit. It should be pointed out that the upward chart is considered here because we are mainly concerned about upward shifts in disease risk in the current disease screening problem. In other problems, downward or two-sided charts might be more appropriate. Also, the observation times $t_1^*, t_2^*, \ldots$ are often unequally spaced in disease screening applications, and the above EWMA chart can accommodate the unequally spaced observation times well. With other types of control charts (e.g., CUSUM), we still do not know how to accommodate this properly in their chart construction.

The performance of control charts is traditionally evaluated by the average run length (ARL), which is the average number of collected observations before triggering a signal. When observation times are unequally spaced, a more sensible measure is the average time to signal (ATS) (cf., Qiu and Xiang 2014). In disease monitoring problems, there is also interest in the receiver operating characteristics (ROC) of monitoring schemes in terms of sensitivity and specificity, which are for evaluating whether the monitoring scheme can correctly identify patients who may or may not have the disease in the future. Recently, Qiu et al. (2019) has proposed a new measure, called process monitoring ROC curve, which attempts to combine the ATS measure and the ROC measures.

The control limit ρ is usually chosen such that the nominal IC ATS value is fixed at a given level when the monitoring schemes are applied to an IC dataset. To accommodate the within-subject data correlation, the block bootstrap procedure discussed in Qiu and Xiang (2014) can be used for searching for the control limit. When there are enough training data, we can split them into two parts. The first part can be used for estimating model (1) to describe the regular pattern of the disease risk, and the IC individuals in the second part can be used for determining the control limit ρ. To expedite the searching algorithm, we can use the bisection method as discussed in Qiu (2014, Sect. 4.2).

3 Simulation Study

Simulations were conducted to evaluate the numerical performance of the proposed method. We used a simulated training dataset of $n = 500$ individuals to estimate the regular disease risk pattern. The whole design interval [0, 1] is discretized in to 1000 basic time units. We considered three different cases with the dimension of covariates being $q = 10, 20$, and 30. The processes of $x_{ik}(t)$ are generated from the following random process model:

$$x_{ik}(t) = -\sin(\pi t + \pi u_{ik1}) + 0.5\cos(10\pi t + 10\pi u_{ik2}) + \epsilon_{ik}(t), \quad \text{for } i = 1, \ldots, n, k = 1, \ldots, q, \tag{9}$$

where $\{u_{ik1}, u_{ik2}\}$ are independent realizations from the uniform [0, 1] distribution, $\epsilon_{ik}(t)$ are generated from the Ornstein–Uhlenbeck processes with $d\epsilon_{ik}(t) = -\theta(m_{ik} - \epsilon_{ik}(t))dt + \sigma dW_{ik}(t)$, $W_{ik}(t)$ are independent realizations from the Wiener process, $\theta = 50$, $\sigma = 20$, $\epsilon_{ik}(0) \sim N(0, 0.2^2)$, and the random mean vectors $\mathbf{m}_i = (m_{i1}, \ldots, m_{iq})'$ are realizations from the multivariate normal distribution with mean $\mathbf{0}$ and variance–covariance matrix being

$$\begin{pmatrix} 0.5 & 0.1 & \cdots & 0.1 \\ 0.1 & 0.5 & \cdots & 0.1 \\ \vdots & \vdots & \ddots & \vdots \\ 0.1 & 0.1 & \cdots & 0.5 \end{pmatrix}.$$

The baseline hazard function in model (1) is chosen to be $\lambda_0(t) = 0.25$. The true regression coefficients β are sparse with the first three dimensions being 0.2 and all the remaining dimensions being 0, i.e., $\beta = (0.2, 0.2, 0.2, 0, \ldots, 0)'$. In each simulation, we estimate the regular disease risk pattern using the simulated dataset of $n = 500$ individuals, and then determine the control limit ρ from another simulated dataset of 500 individuals. The control limit ρ here is chosen such that the nominal IC ATS is 370. Then, the proposed monitoring scheme is applied to simulated data of 10,000 new individuals to evaluate its performance. All the results presented in this section are based on 1000 replicated simulations.

We first present some results about the variable selection method using LASSO. In Table 1, we displayed the mean squared errors of estimated regression coefficients for different dimensions $q = 10, 20, 30$. From the table, we can see that when there is a substantial percentage of zeros in the true regression coefficients, the LASSO penalized estimate $\widehat{\beta}$ has a smaller mean squared error (MSE), compared to the MSE of the ordinary estimate $\tilde{\beta}$ in this example. The relative efficiency of $\widehat{\beta}$ with respect to $\tilde{\beta}$, defined as the ratio of their MSE values, decreases as the dimension q increases. The implication is that when many covariates are unrelated to the disease risk, applying the LASSO penalty can indeed improve the efficiency of parameter estimates.

Next, we present some results about the proposed online monitoring scheme. To examine whether the proposed method can effectively detect distributional shifts that lead to an increased disease risks, we considered a shift of $\mathbf{x}(t)$ in the direction of β, namely,

$$\mathbf{x}^*(t) = \mathbf{x}(t) + \delta\beta,$$

Table 1 MSE of LASSO penalized estimate $\widehat{\beta}$ and unpenalized estimate $\tilde{\beta}$ of β and their corresponding standard errors (in parentheses). The relative efficiency of $\widehat{\beta}$ with respect to $\tilde{\beta}$ is the ratio of the their MSE values, and the standard errors of relative efficiency are obtained by the delta method

	$q = 10$	$q = 20$	$q = 30$
MSE of $\tilde{\beta}$	0.0140 (0.0002)	0.0283 (0.0003)	0.0424 (0.0004)
MSE of $\widehat{\beta}$	0.0084 (0.0002)	0.0115 (0.0003)	0.0139 (0.0003)
Relative efficiency of $\widehat{\beta}$	0.6000 (0.0072)	0.4077 (0.0061)	0.3265 (0.0051)

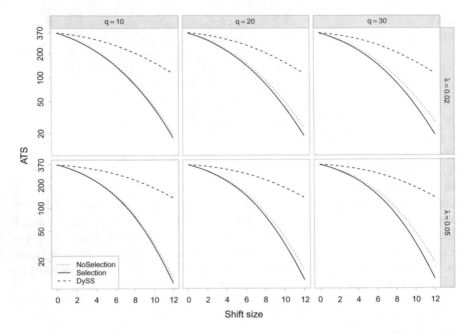

Fig. 1 OC ATS values of different monitoring methods when $\lambda = 0.02, 0.05$ and $q = 10, 20, 30$. NoSelection is the method by You and Qiu (2019) where β is estimated from (2) without using the LASSO penalty. Selection is the proposed method in this paper. DySS is the multivariate DySS by Qiu and Xiang (2015)

where $\mathbf{x}(t)$ is simulated from model (9). The out-of-control ATS values of three different methods are presented in Fig. 1, where DySS denotes the multivariate DySS by Qiu and Xiang (2015), NoSelection denotes the original risk monitoring method by You and Qiu (2019) without using the LASSO variable selection, and Selection is the proposed method in this paper. From the figure, we can see that (i) the proposed method Selection has the best performance among all three methods, (ii) DySS has the worst performance among all three methods, and (iii) the improvement from NoSelection to Selection is more pronounced as the dimensionality of $\mathbf{x}(t)$ increases.

We then compare the three different methods when they are applied to individuals in a simulated training dataset, using the metrics of true positive rate (TPR) and false positive rates (FPR). Here, TPR is defined to be the percentage of individuals receiving signals among all diseased people, and FPR is defined to be the percentage of individuals receiving signals among all non-diseased people. The results are presented in Table 2, from which we can see that i) DySS tends to have a high false positive rate in all cases considered, and ii) Selection and NoSelection have similar TPR values in most scenarios considered here, but the FPR values of NoSelection are lower than those of Selection in all cases considered here.

Table 2 TPR and FPR values of different monitoring methods when $\lambda = 0.02, 0.05$ and $q = 10, 20, 30$. Numbers in parentheses are the corresponding standard errors. NoSelection is the risk monitoring method by You and Qiu (2019) where β is estimated from (1) without using the LASSO penalty. Selection is the proposed risk monitoring method where β is estimated from (3). DySS is the multivariate DySS by Qiu and Xiang (2015)

	$q = 10$		$q = 20$		$q = 30$	
	TPR	FPR	TPR	FPR	TPR	FPR
$\lambda = 0.02$						
NoSelection	0.427	0.457	0.432	0.473	0.430	0.468
	(0.002)	(0.003)	(0.002)	(0.003)	(0.002)	(0.002)
Selection	0.427	0.453	0.432	0.465	0.433	0.453
	(0.002)	(0.003)	(0.002)	(0.003)	(0.002)	(0.003)
DySS	0.483	0.640	0.483	0.640	0.483	0.639
	(0.002)	(0.002)	(0.002)	(0.002)	(0.002)	(0.002)
$\lambda = 0.05$						
NoSelection	0.426	0.467	0.431	0.482	0.428	0.478
	(0.003)	(0.003)	(0.003)	(0.003)	(0.003)	(0.003)
Selection	0.426	0.463	0.430	0.471	0.430	0.462
	(0.002)	(0.003)	(0.003)	(0.003)	(0.003)	(0.003)
DySS	0.392	0.509	0.392	0.508	0.392	0.509
	(0.003)	(0.003)	(0.003)	(0.003)	(0.003)	(0.003)

4 Real Data Example

In this section, we apply the proposed method to a real data example from the Framingham heart study. In this study, participants are regularly examined for risk factors of cardiovascular diseases. The dataset contains observations of the cholesterol levels, systolic blood pressures, and diastolic blood pressures of 1,055 participants. Each participant was followed for 7 times at their different ages. Among the 1,055 participants, 27 of them had strokes at least once during the study. This dataset is displayed in Fig. 2. To implement and evaluate the proposed method, we randomly partition the original dataset into training and test datasets. The training dataset contains approximately two-thirds of all the participants, among which 18 of them had strokes during the study and 686 did not have any strokes. This training dataset is then used for estimating the regular disease risk pattern using (2)–(5). The test dataset contains approximately one-third of all the participants, among which 9 had strokes and 342 did not have any strokes. The test dataset is then used for evaluating the numerical performance of the proposed method. Its weighting parameter λ is chosen to be 0.2, and the nominal IC ATS is set to be 10 years.

The estimate of β by (2) without the LASSO penalty is $\tilde{\beta} = (-0.0013, 0.0178, 0.0099)'$, and the LASSO estimate of β is $\widehat{\beta} = (0.0000, 0.0169, 0.0092)'$ where the parameter γ is chosen to be 0.05. We can see that the first dimension of the LASSO estimate has been shrunk to 0. We then compare the performance of the

three monitoring methods: NoSelection, Selection, and DySS. A summary of the
results is presented in Table 3. From the table, we can see that (i) all the three methods
considered here correctly gave signals to 8 out of 9 stroke patients in the test dataset,
(ii) NoSelection gives 132 signals to 342 non-stroke patients, and Selection gives only

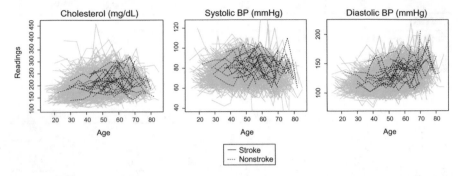

Fig. 2 Cholesterol levels, systolic blood pressure readings, and diastolic blood pressure readings
of 1,055 participants of the Framingham heart study. Gray solid lines are longitudinal observations
of 1,028 non-stroke participants, while black dashed lines are longitudinal observations of 27 stroke
participants

Table 3 Number of signals when different methods are used to monitor patients in the test dataset

	DySS		NoSelection		Selection	
	Signal	No signal	Signal	No signal	Signal	No signal
Stroke patients	8	1	8	1	8	1
Non-stroke patients	167	175	132	210	123	219

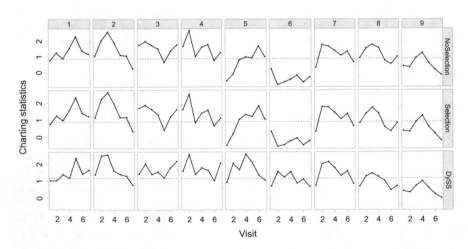

Fig. 3 Charting statistics of the three types of charts for monitoring the 9 stroke patients in the test
dataset. Horizontal dotted lines are the control limits

9 less signals to the non-stroke patients, and (iii) DySS has the worst performance because it gives more signals to the non-stroke patients than each of the other two methods. The three types of charts for monitoring the 9 stroke patients in the test dataset are shown in Fig. 3.

5 Concluding Remarks

In this chapter, we presented an improved version of the disease risk monitoring method suggested by You and Qiu (2019). The major contribution of the improved version is that variable selection is used when quantifying disease risks, in order to reduce variability of the quantified disease risks. Variable selection is achieved by using the LASSO penalty to reduce the dimensionality of the disease predictors in the related survival model. Through numerical simulations and a real data example, we have shown that when many disease predictors are included in the survival model, implementing variable selection before monitoring the quantified disease risks can often improve the performance of disease risk monitoring.

Our proposed method still has its own limitations, and there are many issues to be addressed in the future research. For instance, in real-life disease screening practice, it is quite common that patients may miss some medical tests or have some incomplete medical examinations during a clinic visit. Future research is needed to extend existing methods to allow for missing data of different types. Also, the effect of disease predictors can be time-varying. Though You and Qiu (2019) has provided a method for estimating time-varying regression coefficients, the variable selection problems in a time-varying-effect model will be much more challenging than the problem considered here, which has not properly discussed yet. Finally, the proposed variable selection method is based on the L_1 adaptive LASSO penalty. In the literature, there are a series of alternative penalized regression methods for variable selection. For example, one may consider the alternative penalty functions like the elastic net (Zou and Hastie 2005) and the smoothly clipped absolute deviation penalty (Fan and Li 2001) to reduce the bias of the LASSO estimates. When disease predictors come from many different groups, one may consider using the group LASSO (Yuan and Lin 2006) to select some groups of disease predictors for quantifying disease risks. It is of interest to study all these variable selection methods in the dynamic disease screening and monitoring problems in the future research.

Acknowledgements The authors thank the editors for the invitation of this contribution to the edited book. One referee provided some comments about the paper for improvements, which is greatly appreciated. This research is supported in part by an NSF grant in USA.

References

Akaike, H. (1992). Information theory and an extension of the maximum likelihood principle. *Second International Symposium on Information Theory Proceeding*, *1*, 610–624.

Cai, Z., & Sun, Y. (2003). Local linear estimation for time-dependent coefficients in cox's regression models. *Scandinavian Journal of Statistics*, *30*, 93–111.

Epanechnikov, V. A. (1969). Non-parametric estimation of a multivariate probability density. *Theory of Probability & Its Applications*, *14*, 153–158.

Fan, J., & Li, R. (2001). Variable selection via nonconcave penalized likelihood and its oracle properties. *Journal of the American Statistical Association*, *96*, 1348–1360.

Fan, J., Yao, Q., & Tong, H. (1996). Estimation of conditional densities and sensitivity measures in nonlinear dynamical systems. *Biometrika*, *83*, 189–206.

Friedman, J., Hastie, T., Höfling, H., & Tibshirani, R. (2007). Pathwise coordinate optimization. *The Annals of Applied Statistics*, *1*, 302–332.

Klein, J. P., & Moeschberger, M. L. (1997). *Survival Analysis: Techniques for Censored and Truncated Data*. New York: Springer.

Li, J., & Qiu, P. (2016). Nonparametric dynamic screening system for monitoring correlated longitudinal data. *IIE Transactions*, *48*, 772–786.

Li, J., & Qiu, P. (2017). Construction of an efficient multivariate dynamic screening system. *Quality and Reliability Engineering International*, *33*, 1969–1981.

Mendis, S., Puska, P., & Norrving, B. (2011). Global atlas on cardiovascular disease prevention and control. *World Health Organization*.,. https://apps.who.int/iris/handle/10665/44701.

Qiu, P. (2014). *Introduction to Statistical Process Control*. Boca Raton, FL: Chapman Hall/CRC.

Qiu, P., Xia, Z., & You, L. (2019). Process monitoring ROC curve for evaluating dynamic screening methods. *Technometrics*,. https://doi.org/10.1080/00401706.2019.1604434.

Qiu, P., & Xiang, D. (2014). Univariate dynamic screening system: an approach for identifying individuals with irregular longitudinal behavior. *Technometrics*, *56*, 248–260.

Qiu, P., & Xiang, D. (2015). Surveillance of cardiovascular diseases using a multivariate dynamic screening system. *Statistics in Medicine*, *34*, 2204–2221.

Qiu, P., Zi, X., & Zou, C. (2018). Nonparametric dynamic curve monitoring. *Technometrics*, *60*, 386–397.

Simon, N., Friedman, J., Hastie, T., & Tibshirani, R. (2011). Regularization paths for Cox's proportional hazards model via coordinate descent. *Journal of Statistical Software*, *39*, 1–13.

Tian, L., Zucker, D., & Wei, L. (2005). On the Cox model with time-varying regression coefficients. *Journal of the American Statistical Association*, *100*, 172–183.

Tibshirani, R. (1996). Regression shrinkage and selection via the lasso. *Journal of the Royal Statistical Society: Series B (Methodological)*, *58*, 267–288.

Tibshirani, R. (1997). The LASSO method for variable selection in the Cox model. *Statistics in Medicine*, *16*, 385–395.

Wright, D. J. (1986). Forecasting data published at irregular time intervals using an extension of Holt's method. *Management Science*, *32*, 499–510.

Xiang, D., Qiu, P., & Pu, X. (2013). Nonparametric regression analysis of multivariate longitudinal data. *Statistica Sinica*, *23*, 769–789.

You, L., & Qiu, P. (2019a). Fast computing for dynamic screening systems when analyzing correlated data. *Journal of Statistical Computation and Simulation*, *89*, 379–394.

You, L., & Qiu, P. (2019b). An effective method for online disease risk monitoring. *Technometrics*,. https://doi.org/10.1080/00401706.2019.1625813.

Yu, K., & Jones, M. (1998). Local linear quantile regression. *Journal of the American Statistical Association*, *93*, 228–237.

Yuan, M., & Lin, Y. (2006). Model selection and estimation in regression with grouped variables. *Journal of the Royal Statistical Society: Series B (Statistical Methodology)*, *68*, 49–67.

Zou, H. (2006). The adaptive LASSO and its oracle properties. *Journal of the American Statistical Association*, *101*, 1418–1429.

Zou, H., & Hastie, T. (2005). Regularization and variable selection via the elastic net. *Journal of The Royal Statistical Society: Series B (Statistical Methodology)*, *67*, 301–320.

Printed in the United States
by Baker & Taylor Publisher Services